Matter and Energy for Growth and Activity

Teacher Edition

Claire Reinburg, Director
Rachel Ledbetter, Managing Editor
Andrea Silen, Associate Editor
Jennifer Thompson, Associate Editor
Donna Yudkin, Book Acquisitions Manager

Art and Design
Will Thomas Jr., Director

Printing and Production
Catherine Lorrain, Director

National Science Teaching Association
David L. Evans, Executive Director

1840 Wilson Blvd., Arlington, VA 22201
www.nsta.org/store
For customer service inquiries, please call 800-277-5300.

23 22 21 20 4 3 2 1

Library of Congress Cataloging-in-Publication Data
Names: National Science Teaching Association.
Title: Matter and energy for growth and activity.
Description: Teacher edition. | Arlington, VA : National Science Teaching Association, [2019] | Includes bibliographical references.
Identifiers: LCCN 2019004323 (print) | LCCN 2019005895 (ebook) | ISBN 9781681406879 (e-book) | ISBN 9781681406855 (print)
Subjects: LCSH: Biological systems--Study and teaching.
Classification: LCC QH313 (ebook) | LCC QH313 .M384 2019 (print) | DDC 570.76--dc23
LC record available at *https://lccn.loc.gov/2019004323*

Contents

Introduction to the Teacher Edition v

Getting Started: Matter and Energy Changes in Our Bodies 1

Chapter 1: Making Sense of Matter Changes Involved in Human Growth 9

Lesson 1.1 – Matter Changes in the Human Body 14
Lesson 1.2 – Comparing Our Bodies to Our Food 24
Lesson 1.3 – Matter Changes During Chemical Reactions 50
Lesson 1.4 – Chemical Reactions and Systems 80
Lesson 1.5 – Converting Food to Body Structures 130

Chapter 2: Making Sense of Energy Changes Involved in Human Growth and Activity 183

Lesson 2.1 – Observing Energy Changes All Around Us 190
Lesson 2.2 – Representing Energy Changes and Transfers 218
Lesson 2.3 – Energy Changes During Chemical Reactions 250
Lesson 2.4 – Energy Changes During Other Chemical Reactions 292
Lesson 2.5 – Making Sense of Energy Changes During Chemical Reactions 326
Lesson 2.6 – Coupling Energy-Releasing to Energy-Requiring Chemical Reactions . . 366
Lesson 2.7 – Coupling Chemical Reactions to Motion and Growth in Living Organisms 392
Lesson 2.8 – Growing and Moving Without Oxygen 440
Lesson 2.9 – How Body Systems Respond to Exercise 470

AAAS Project 2061 Team

The AAAS Project 2061 team was responsible for the development of the Student and Teacher Editions and the testing of the unit's promise to promote student learning.

Jo Ellen Roseman, Ph.D., Director of Project 2061 and Principal Investigator
Cari Herrmann-Abell, Ph.D., Senior Research Associate (currently Research Scientist at BSCS Science Learning)
Joseph Hardcastle, Ph.D., Research Associate
Ana Cordova, Research Assistant (currently Research Analyst at AccessLex Institute)
Mary Koppal, Communications Director
Barbara Goldstein, Program Associate

GSLC Team

The Genetic Science Learning Center (GSLC) at the University of Utah was responsible for the development of accompanying visualizations, including print manipulatives and videos that are available as online resources for teachers.

Louisa A. Stark, Ph.D., GSLC Director and Co-Principal Investigator
Kevin Pompei, Associate Director of GSLC
Harmony Ann Starr, Multimedia Content Producer

Advisory Board

Melanie Cooper, Ph.D., Michigan State University, East Lansing, MI
Marlene Hilkowitz, Drexel University, Philadelphia, PA
Edward Redish, Ph.D., University of Maryland, College Park, MD
Mark Roseman, Ph.D., Uniformed Services University of the Health Sciences, Bethesda, MD
Elizabeth Tipton, Ph.D., Northwestern University, Evanston, IL
Mary Weller, Howard County Public School System, Ellicott City, MD
Catherine Wolfe, Brookline High School, Brookline, MA

Acknowledgments

We would like to thank the many excellent teachers and their students in the District of Columbia, Maryland, and New Jersey who participated in pilot- and field-testing the unit. Their insights and feedback have been invaluable in improving the unit for all students and teachers.

Development of *Matter and Energy for Growth and Activity* was funded by the U.S. Department of Education Institute of Education Sciences, Grant #IES-R305A150310. Any opinions, findings, conclusions, or recommendations expressed in this publication are those of the authors and do not necessarily reflect the views of the funding agency.

INTRODUCTION TO THE TEACHER EDITION

The Teacher Edition is designed to provide easy access to essential background information and support needed for using the *Matter and Energy for Growth and Activity* (*MEGA*) unit effectively in the classroom. Additional teacher support materials (e.g., videos and handouts) are provided online. The Teacher Edition and the online materials aim to provide both a "big picture" sense of the unit and its goals as well as the specific information and guidance needed to teach each lesson and carry out each activity.

Features of the Teacher Edition

At the unit level, the Teacher Edition lays out the overarching **Student Learning Goals for the Unit** and the central question students should be able to answer after completing all the lessons. It then outlines the disciplinary core ideas, crosscutting concepts, science practices, and performance expectations from *A Framework for K–12 Science Education* (the *Framework*) (National Research Council 2012) and the *Next Generation Science Standards* (*NGSS*) (NGSS Lead States 2013) that are targeted in the *MEGA* unit. In the **Design of the Unit** section, the Teacher Edition describes how research-based design principles influenced the development of the unit. The Content Storyline map provides a visual representation of the coherent story of science ideas that are drawn from *NGSS* core ideas and crosscutting concepts and take account of students' entering knowledge. A phenomena table summarizes the life and physical science phenomena that students explain using evidence from observations and data, models, and science ideas. This section also describes how science practices, including modeling and constructing explanations, are used to help students make sense of phenomena. **Unique Aspects of the Unit** describes the rationale for the unit's approach to the science content, including its choice of vocabulary used in the Student Edition. **Structure of the Unit** presents the overall design of the unit and describes the purpose and key features of each section of a lesson.

At the chapter level, each **Chapter Overview** describes the concepts developed in that chapter and provides a short synopsis of what students will do and think about in each lesson. At the lesson level, a **Lesson Guide** provides an overview of each lesson, including the key question, targeted science ideas and practices, materials, and advance preparation needed and a summary chart describing key phenomena—both firsthand observations and data—and representations used in each activity, the pedagogical purpose of each, and the intended observations students should make. The remaining pages of the lesson consist of facing **Student Edition and Answer Key** pages and **Teacher Talk and Actions** pages. Each Student Edition and Answer Key page reproduces the Student Edition page but also includes ideal student responses (in script text) written to reflect what students are expected to understand at that point in the lesson. The Teacher Talk and Actions pages provide a variety of information to help teachers facilitate the lessons, such as Safety Notes, background for teachers on the relevant science concepts, options for students who are more advanced or need more support, suggestions for managing classroom activities, and questions that can be used to elicit or clarify students' thinking. Each lesson concludes with a *Closure and Link* note on the Teacher Talk and Actions page that outlines a classroom discussion aimed at helping students gain perspective about what they now know and what questions remain.

A set of online resources available on the book's Extras page (*www.nsta.org/growthandactivity*) supplements the Teacher Edition with videos, photos, interactive media, handouts, and a list of supplies needed to teach the unit.

Student Learning Goals for the Unit

The overarching goal of the *MEGA* unit is for students to use important ideas about matter and energy, initially encountered in simple physical systems, to explain observable phenomena in the bodies of living organisms:

> To build body structures for growth and repair, living organisms produce polymers (and water molecules) during energy-requiring chemical reactions between monomers. Animals obtain monomers from their food, whereas plants make monomers from inorganic substances during energy-requiring chemical reactions that use energy from the Sun. The energy that both plants and animals use to produce polymers comes indirectly from energy-releasing chemical reactions such as the oxidation of glucose, fatty acid, and amino acid monomers and directly from the breakdown of ATP to form ADP and inorganic phosphate. The same energy-releasing chemical reactions provide energy for motion. While much of the energy released can be used for growth and motion, a lot of the energy is transferred as heat to the surroundings. Feedback mechanisms maintain a living system's internal conditions within certain limits and mediate behaviors, allowing it to remain alive and functional even as conditions change within some range.
>
> During both energy-releasing and energy-requiring chemical reactions, atoms are rearranged and conserved; therefore, total mass is conserved. The net amount of energy released or required reflects the difference between the total amount of energy that must be added to break bonds between atoms of reactant molecules and the total amount of energy that is released when bonds form between atoms of product molecules. As with atoms, energy is neither created nor destroyed during chemical reactions.

By the end of the unit, students will be able to answer the following unit question in terms of (a) the rearrangement and conservation of atoms during chemical reactions and (b) the use of energy-releasing chemical reactions to drive energy-requiring chemical reactions or processes:

> ***How do living things use food as a source of matter for building and repairing their body structures and as a source of energy for carrying out a wide range of activities?***

Alignment With National Science Education Framework and Standards

As noted earlier, the *MEGA* unit addresses important goals for science learning that are recommended in the *Framework* and *NGSS*. The science ideas displayed in text boxes in the content storyline that is shown in Figure 1 were drawn from the following disciplinary core ideas, crosscutting concepts, and science practices:

Disciplinary Core Ideas

Matter and Its Interactions (PS1)

Structure and Properties of Matter (PS1.A, Grades 8 and 12)

Chemical Reactions (PS1.B, Grades 8 and 12)

Energy (PS3)

Definitions of Energy (PS3.A, Grade 12)

Conservation of Energy and Energy Transfer (PS3.B, Grade 12)

Energy in Chemical Processes and Everyday Life (PS3.D, Grades 8 and 12)

From Molecules to Organisms: Structures and Processes (LS1)

Structure and Function (LS1.A, Grades 8 and 12)

Organization for Matter and Energy Flow in Organisms (LS1.C, Grades 8 and 12)

Ecosystems: Interactions, Energy, and Dynamics (LS2)

Cycles of Matter and Energy Transfer in Ecosystems (LS2.B, Grade 12)

Crosscutting Concepts

Systems and System Models (Grades 9-12)

Energy and Matter: Flows, Cycles, and Conservation (Grades 9-12)

Science Practices

Practice 2, Developing and Using Models (Grades 9-12)

Practice 4, Analyzing and Interpreting Data (Grades 9-12)

Practice 6, Constructing Explanations and Designing Solutions (Grades 9-12)

These *NGSS* elements are addressed throughout the *MEGA* unit, and all three dimensions of science learning—core disciplinary ideas, crosscutting concepts, and science practices—are carefully integrated to help students make sense of phenomena. Students model and construct evidence-based explanations about phenomena. The evidence comes from direct observations students make during classroom activities and from patterns in data reported in the scientific research literature that are simplified in the student text. Students use a variety of models to help them link the evidence to science ideas about atom rearrangement and conservation and about energy changes during chemical reactions within systems and energy transfers between systems.

NGSS Performance Expectations

The unit contributes to the following high school performance expectations:

- Develop and use a model to illustrate the hierarchical organization of interacting systems that provide specific functions within multicellular organisms. (HS-LS1-2)
- Construct and revise an explanation based on evidence for how carbon, hydrogen, and oxygen from sugar molecules may combine with other elements to form amino acids and/or other large carbon-based molecules. (HS-LS1-6)
- Develop a model to illustrate that the release or absorption of energy from a chemical reaction system depends upon the changes in total bond energy. (HS-PS1-4)
- Use a model to illustrate how photosynthesis transforms light energy into stored chemical energy. (HS-LS1-5)
- Use a model to illustrate that cellular respiration is a chemical process whereby the bonds of food molecules and oxygen molecules are broken and the bonds in new compounds are formed resulting in a net transfer of energy. (HS-LS1-7)
- Construct and revise an explanation based on evidence for the cycling of matter and flow of energy [within organisms] in aerobic and anaerobic conditions. (HS-LS2-3)

Figure 1

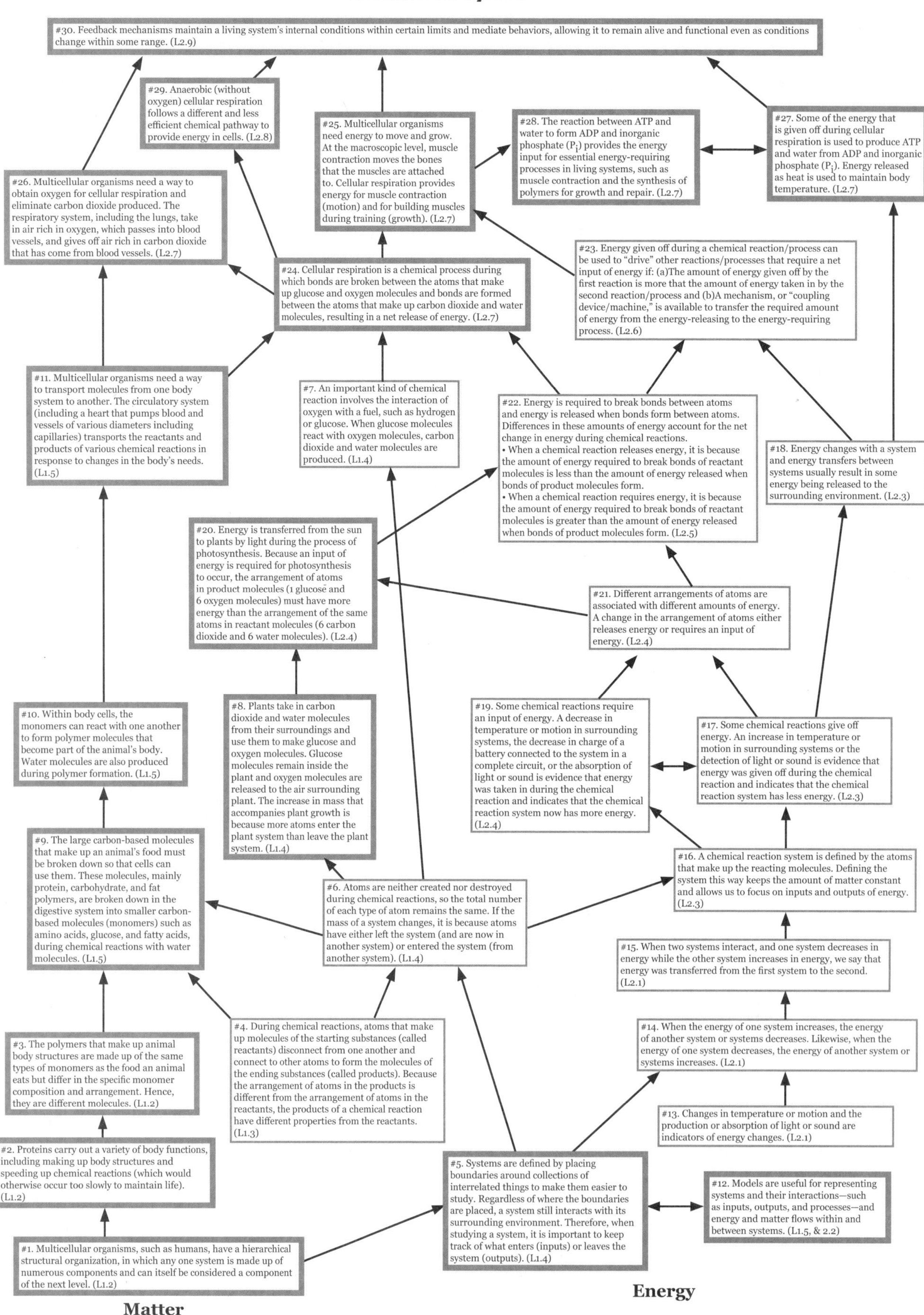

Matter and Energy for Growth and Activity
Content Storyline
#30. Feedback mechanisms maintain a living system's internal conditions within certain limits and mediate behaviors, allowing it to remain alive and functional even as conditions change within some range. (L2.9)
#29. Anaerobic (without oxygen) cellular respiration follows a different and less efficient chemical pathway to provide energy in cells. (L2.8)
#25. Multicellular organisms need energy to move and grow. At the macroscopic level, muscle contraction moves the bones that the muscles are attached to. Cellular respiration provides energy for muscle contraction (motion) and for building muscles during training (growth). (L2.7)
#28. The reaction between ATP and water to form ADP and inorganic phosphate (P_i) provides the energy input for essential energy-requiring processes in living systems, such as muscle contraction and the synthesis of polymers for growth and repair. (L2.7)
#27. Some of the energy that is given off during cellular respiration is used to produce ATP and water from ADP and inorganic phosphate (P_i). Energy released as heat is used to maintain body temperature. (L2.7)
#26. Multicellular organisms need a way to obtain oxygen for cellular respiration and eliminate carbon dioxide produced. The respiratory system, including the lungs, take in air rich in oxygen, which passes into blood vessels, and gives off air rich in carbon dioxide that has come from blood vessels. (L2.7)
#24. Cellular respiration is a chemical process during which bonds are broken between the atoms that make up glucose and oxygen molecules and bonds are formed between the atoms that make up carbon dioxide and water molecules, resulting in a net release of energy. (L2.7)
#23. Energy given off during a chemical reaction/process can be used to "drive" other reactions/processes that require a net input of energy if: (a)The amount of energy given off by the first reaction is more that the amount of energy taken in by the second reaction/process and (b)A mechanism, or "coupling device/machine," is available to transfer the required amount of energy from the energy-releasing to the energy-requiring process. (L2.6)
#11. Multicellular organisms need a way to transport molecules from one body system to another. The circulatory system (including a heart that pumps blood and vessels of various diameters including capillaries) transports the reactants and products of various chemical reactions in response to changes in the body's needs. (L1.5)
#7. An important kind of chemical reaction involves the interaction of oxygen with a fuel, such as hydrogen or glucose. When glucose molecules react with oxygen molecules, carbon dioxide and water molecules are produced. (L1.4)
#22. Energy is required to break bonds between atoms and energy is released when bonds form between atoms. Differences in these amounts of energy account for the net change in energy during chemical reactions.
• When a chemical reaction releases energy, it is because the amount of energy required to break bonds of reactant molecules is less than the amount of energy released when bonds of product molecules form.
• When a chemical reaction requires energy, it is because the amount of energy required to break bonds of reactant molecules is greater than the amount of energy released when bonds of product molecules form. (L2.5)
#18. Energy changes with a system and energy transfers between systems usually result in some energy being released to the surrounding environment. (L2.3)
#20. Energy is transferred from the sun to plants by light during the process of photosynthesis. Because an input of energy is required for photosynthesis to occur, the arrangement of atoms in product molecules (1 glucose and 6 oxygen molecules) must have more energy than the arrangement of the same atoms in reactant molecules (6 carbon dioxide and 6 water molecules). (L2.4)
#21. Different arrangements of atoms are associated with different amounts of energy. A change in the arrangement of atoms either releases energy or requires an input of energy. (L2.4)
#10. Within body cells, the monomers can react with one another to form polymer molecules that become part of the animal's body. Water molecules are also produced during polymer formation. (L1.5)
#8. Plants take in carbon dioxide and water molecules from their surroundings and use them to make glucose and oxygen molecules. Glucose molecules remain inside the plant and oxygen molecules are released to the air surrounding plant. The increase in mass that accompanies plant growth is because more atoms enter the plant system than leave the plant system. (L1.4)
#19. Some chemical reactions require an input of energy. A decrease in temperature or motion in surrounding systems, the decrease in charge of a battery connected to the system in a complete circuit, or the absorption of light or sound is evidence that energy was taken in during the chemical reaction and indicates that the chemical reaction system now has more energy. (L2.4)
#17. Some chemical reactions give off energy. An increase in temperature or motion in surrounding systems or the detection of light or sound is evidence that energy was given off during the chemical reaction and indicates that the chemical reaction system has less energy. (L2.3)
#9. The large carbon-based molecules that make up an animal's food must be broken down so that cells can use them. These molecules, mainly protein, carbohydrate, and fat polymers, are broken down in the digestive system into smaller carbon-based molecules (monomers) such as amino acids, glucose, and fatty acids, during chemical reactions with water molecules. (L1.5)
#6. Atoms are neither created nor destroyed during chemical reactions, so the total number of each type of atom remains the same. If the mass of a system changes, it is because atoms have either left the system (and are now in another system) or entered the system (from another system). (L1.4)
#16. A chemical reaction system is defined by the atoms that make up the reacting molecules. Defining the system this way keeps the amount of matter constant and allows us to focus on inputs and outputs of energy. (L2.3)
#15. When two systems interact, and one system decreases in energy while the other system increases in energy, we say that energy was transferred from the first system to the second. (L2.1)
#3. The polymers that make up animal body structures are made up of the same types of monomers as the food an animal eats but differ in the specific monomer composition and arrangement. Hence, they are different molecules. (L1.2)
#4. During chemical reactions, atoms that make up molecules of the starting substances (called reactants) disconnect from one another and connect to other atoms to form the molecules of the ending substances (called products). Because the arrangement of atoms in the products is different from the arrangement of atoms in the reactants, the products of a chemical reaction have different properties from the reactants. (L1.3)
#14. When the energy of one system increases, the energy of another system or systems decreases. Likewise, when the energy of one system decreases, the energy of another system or systems increases. (L2.1)
#13. Changes in temperature or motion and the production or absorption of light or sound are indicators of energy changes. (L2.1)
#2. Proteins carry out a variety of body functions, including making up body structures and speeding up chemical reactions (which would otherwise occur too slowly to maintain life). (L1.2)
#5. Systems are defined by placing boundaries around collections of interrelated things to make them easier to study. Regardless of where the boundaries are placed, a system still interacts with its surrounding environment. Therefore, when studying a system, it is important to keep track of what enters (inputs) or leaves the system (outputs). (L1.4)
#12. Models are useful for representing systems and their interactions—such as inputs, outputs, and processes—and energy and matter flows within and between systems. (L1.5, & 2.2)
#1. Multicellular organisms, such as humans, have a hierarchical structural organization, in which any one system is made up of numerous components and can itself be considered a component of the next level. (L1.2)
Energy
Matter

- Plan and conduct an investigation to provide evidence that feedback mechanisms maintain homeostasis. (HS-LS1-3)

Online resources available at *www.nsta.org/growthandactivity* provide specific details about which parts of science ideas and practices are targeted and describe and reference specific activities that contribute to each performance expectation.

Design of the Unit

The *MEGA* unit was developed by a team of scientists, educators, and curriculum developers at Project 2061 of the American Association for the Advancement of Science (AAAS) in partnership with media designers at the University of Utah's Genetic Sciences Learning Center. Development of the unit was guided by learning theory and research evidence showing that students' science understanding develops from (a) having a wide range of experiences with the natural world that are explainable by a coherent set of science ideas, (b) observing phenomena that are explicitly targeted to the science ideas and that address common student misconceptions, and (c) having an opportunity to interpret and make sense of what they experience in terms of those ideas. To help students generate conceptual understanding, however, teachers must guide them in generating ideas that are applicable across instances and in seeing how those generalizations are also useful for explaining other related phenomena. Consistent with these views, the *MEGA* unit supports student learning through lessons that focus on:

- a coherent set of ideas and connections among those ideas,
- preexisting ideas—both troublesome and helpful—that students are likely to hold,
- a variety of relevant and engaging phenomena, and
- activities that foster students' sense making and their ability to relate phenomena to the science ideas that explain them.

A Coherent Set of Ideas

The *MEGA* unit's development started by identifying a set of relevant disciplinary core ideas and crosscutting concepts in the *Framework* (National Research Council 2012) and then unpacking them into 30 discrete science ideas that are organized into a coherent content storyline for the unit.

Figure 1 represents the content storyline for the *MEGA* unit, mapping the progression of science ideas about matter on the left and energy on the right (see the Extras page at *www.nsta.org/growthandactivity* for a full-size version of Figure 1). Text boxes present the science ideas, and arrows indicate which science ideas contribute to other science ideas. Text boxes with thick borders show the life science ideas that make up the animal growth and activity storyline (Science Ideas #1–3, #9–11, and #24–30), the plant growth storyline (Science Ideas #8 and #20), and ideas about systems and models (Science Ideas #5 and #12) that students would need to make sense of phenomena involving matter and energy changes. These science ideas are based on high school core life science ideas and the crosscutting concept of systems and system models found in *NGSS* and the *Framework*.

Developers also integrated into the *MEGA* unit's content storyline several physical science ideas about matter and energy and additional ideas about systems drawn from *NGSS* middle and high school physical science core ideas and the crosscutting concept of energy and matter. Science Ideas #4, #6, and #7 show the prerequisite middle school matter ideas, and Science Ideas #13–19 and #21–23 show the prerequisite middle and high school energy ideas.

Students' Preexisting Ideas

Both the learning research literature and Project 2061's distractor-driven assessment items provided insights into misconceptions students might have when starting the *MEGA* unit. For example, *Making Sense of Secondary Science* (Driver et al. 1994) emphasizes the importance of instruction that takes account of students' ideas, describes studies that have investigated students' ideas through interviews, and reports on their findings. Life science sections on Cells, Food, Photosynthesis, Growth, Nutrition, and Respiration and physical science sections on Conservation of Matter, Mass, The Gaseous State, Combustion, Interaction, and Energy and Chemical Change provided valuable background information on naïve ideas students might bring to class. *Benchmarks for Science Literacy* (AAAS 1993) summarizes the published research on student misconceptions and learning difficulties on several topics relevant to the *MEGA* unit: Atoms and Molecules, Conservation of Matter, Chemical Reactions, Energy Transformations, Flow of Matter and Energy, Basic Functions, Reasoning, Models, and Detecting Flaws in Arguments. In addition, Project 2061's assessment items have identified misconceptions on these topics that are held by a national sample of middle and high school students (accessible at *www.assessment.aaas.org*).

Each misconception listed below is accompanied by the percentage of high school students who chose it out of four possible answer choices. As noted on the assessment website, the percentage of students holding these misconceptions in any given classroom could be higher or lower than the national sample. Different percentages for the same misconception result from differences in the appeal of distractors.

Food

- Food is a source of energy but not of building materials. (15%)
- Animals cannot store molecules from food in their bodies. (20%)
- Molecules from food are not stored in the fat tissue of animals. (20%)

Human Body Systems

- Blood does not carry simple sugar molecules to cells of the body. (31%)
- Capillaries are found only in internal organs, such as the lungs and intestines. (28%)
- Capillaries are only found in the extremities, such as hands and feet. (23%)
- Blood does not carry oxygen to the cells of the body. (17%)
- Simple sugars have to be broken down into smaller molecules before they can enter the cells of the body. (47%)
- Amino acids have to be broken down into smaller molecules before they can enter the cells of the body. (34%)
- Proteins do not have to be broken down before they can enter the cells of the body. (26%)

Because students holding these misconceptions might have difficulty following the life science content storyline, the *MEGA* unit includes activities that are designed to take account of them.

Not only did the pilot tests of the unit reveal students' misconceptions, they also showed that most students lacked critical prerequisite knowledge of chemical reactions and models for making sense of matter changes even in simple physical systems. Most students did not know that (a) new substances form during chemical reactions because atoms of reactant molecules rearrange to form products and (b) mass is conserved during chemical reactions because atoms are conserved. The following list illustrates both naïve and incorrect ideas identified by Project 2061's assessment items:

Chemical Reactions

- A chemical change is irreversible. (34%)
- The atoms of the reactants of a chemical reaction are transformed into other atoms. (40%)
- The products of a chemical reaction are the same substances as the reactants but with different properties. (19%)
- New atoms are created during a chemical reaction. (34%)
- Atoms can be destroyed during a chemical reaction. (21%)

Conservation of Mass

- Mass is not conserved in processes in which gases take part. If a gas is produced in a chemical reaction in a closed system, the mass decreases. (50% and 29%, depending on other answer choices)
- In a closed system, the total mass increases during a precipitation reaction. (49%)
- During biological decomposition in a closed system, the total mass of the system decreases. (32%)
- In a closed system, mass decreases after a solid has dissolved in a liquid. (20%)
- The mass of a closed system will increase if a new kind of atom is formed in the system. (19%)
- The number of different kinds of molecules, not the number of atoms, is conserved. (18%)

Prerequisite ideas about energy were also poorly understood. While more than half of middle and high school students understood that changes in motion, temperature, and light are all indicators of energy changes, fewer students (16–41% of middle school students and 18–49% of high school students, depending on the item) were able to apply that knowledge to chemical reactions (AAAS n.d.).

- Energy can be created. (28%)
- Energy can be destroyed. (19%)
- An object has energy in it that is used up as the object moves. (29%)
- Bond making requires energy. (52%)
- Energy is released when bonds break. (43%)

To address these gaps in students' knowledge, the *MEGA* unit's content storyline includes a set of middle and high school physical science ideas (see Figure 1) that are necessary if students are to build an understanding of matter and energy changes in living systems in terms of atoms and molecules.

Pre-test data confirmed that high school students starting the *MEGA* unit had a poor understanding of atom rearrangement and conservation, with a little over a third of them responding correctly to items aligned to those ideas. Only about 40% responded correctly to items assessing energy conservation.

The *MEGA* unit also takes into account students' readiness to carry out high school science practices of data analysis, modeling, and explanation. *NGSS* expects middle school students to be able to (a) analyze and interpret data from simple tables and graphs and draw conclusions about what the data do and do not show; (b) use physical models and drawings (though not chemical formulas or equations) to represent atom rearrangement and conservation during chemical reactions; and (c) construct valid explanations of phenomena involving changes in matter that use evidence, science ideas, and models to support claims.

Ideally, middle school students would have engaged in observing, modeling, and explaining phenomena involving chemical reactions in simple physical systems, such as those explicitly targeted in the eight-week *Toward High School Biology* (*THSB*) unit (AAAS 2017) that is a precursor to the *MEGA* unit. Pre-test data involving high school students who had not used the middle school *THSB* unit showed that students were struggling with using models, evaluating experimental designs, and drawing conclusions from experimental data, with only about 35% of students answering correctly on items assessing these practices. Classroom observations revealed that students were not able or inclined to reason with LEGO or ball-and-stick models to show how atoms disconnect from one another and connect to other atoms to form different molecules without the creation or destruction of any atoms. Nor did they demonstrate an ability to use models and science ideas about atom rearrangement and conservation to explain phenomena.

Phenomena

The *MEGA* unit is designed to provide students with opportunities to experience phenomena directly and, for phenomena that can't be observed directly, to examine data collected by scientists that can serve as evidence for the science ideas and for their explanatory power. The phenomena and data include numerous examples from both physical and life science contexts, with simple physical science phenomena typically used first to help students generate science ideas that are then applied to more complex biological phenomena. Table 1 lists science ideas found in the overarching goal of the unit (p. vi), examples of relevant physical and life science phenomena that are used in the *MEGA* unit to support those ideas, and the lessons (e.g., L1.3) in which they are used.

Each lesson includes tasks and questions that help students (1) make sense of observable phenomena and data using models of underlying matter and energy changes within systems and transfers between systems and (2) apply science ideas developed in simpler physical systems to living systems. By the end of the unit, students should be able to explain the growth and activity of all living things in terms of changes in the arrangement of atoms that result in energy changes within systems and energy transfers between systems and to recognize that the changes in matter and energy that they observe do not violate conservation principles.

Table 1. Life and Physical Science Phenomena in *MEGA* Unit

To build body structures for growth and repair, living organisms produce polymers (and water molecules) during energy requiring chemical reactions between monomers. Animals obtain monomers from their food, whereas plants make monomers from inorganic substances during energy-requiring reactions that use energy from the sun.	
Life Science Phenomena	**Physical Science Phenomena**
• Food is needed for an infant and a body builder to grow, a wound to heal, and cells to divide in culture. L1.1 • Common breakfast foods contain protein, fat, and carbohydrate polymers that are made up of amino acid, fatty acid, and glucose monomers. L1.2 • A plant gets bigger and develops more body structures as it grows. L1.4 • Isotopic labeling experiments show that the carbon atoms of glucose that plants use to build body structures for growth come from carbon dioxide in the air. L1.4 • Experiments measuring levels of polymers and monomers in digestive, circulatory, and muscular systems over time show that the protein, fat, and carbohydrate polymers used to build body structures are synthesized (along with water molecules) from amino acid, fatty acid, and glucose monomers that are digestion products of protein, fat, and carbohydrate polymers found in food. L1.5 • In the presence of light, an aquatic plant produces oxygen (based on splint test) and uses carbon dioxide (based on limewater test). L1.4 • An aquatic plant produces more oxygen when grown under a 100 w light bulb than under a 60 w light bulb. L2.4	• When a candle burns in air, the mass of the candle decreases. L1.3 • When a candle burns in air, two new substances are produced—a gas that turns limewater cloudy (CO_2) and a liquid that turns cobalt chloride paper from blue to pink (H_2O). L1.3 • When a battery is connected to water in a complete circuit, the mass of the water decreases and two new substances are produced—a gas that ignites a glowing splint (O_2) and a gas that makes a burning splint pop (H_2). L1.4 • When glucose ($C_6H_{12}O_6$) reacts with oxygen (O_2) in an open system, the mass decreases and two new substances are produced—a gas that turns limewater cloudy (CO_2) and a liquid that turns cobalt chloride paper from blue to pink (H_2O). L1.4 • When a moving ball collides with a stationary ball, a sound is heard, the stationary ball starts moving, and the moving ball stops. L2.1 • When a flask of hot water is immersed in a beaker of cold water, the temperature of the cold water increases, and the temperature of the hot water decreases. L2.1 • Coffee grounds move when a loud sound starts and stop moving when the sound stops. L2.1 • A solar-powered toy car moves farther under a 100 w light bulb than under a 60 w light bulb. L2.1 • More oxygen gas (O_2) and hydrogen gas (H_2) are produced from water (H_2O) when water is connected in a complete circuit to a new 9-volt battery than to a used battery. L2.4 • When ammonium thiocyanate and barium hydroxide react, the temperature of the surrounding system decreases. L2.4

<table>
<tr><th colspan="2">The energy both plants and animals use to produce polymers comes indirectly from energy-releasing chemical reactions such as the oxidation of glucose, fatty acid, and amino acid monomers and directly from the hydrolysis of ATP to form ADP plus inorganic phosphate.</th></tr>
<tr><th>Life Science Phenomena</th><th>Physical Science Phenomena</th></tr>
<tr><td>• Body builders synthesize more protein in an arm undergoing resistance training than in the other arm not undergoing resistance training. L2.7
• When a solution of ATP was added to a system containing amino acids and other necessary molecules for building proteins, proteins were made. L2.7
• The dry cell mass of Lactococcus lactis increases without oxygen but increases more over 14 hours with oxygen than without oxygen. L 2.8
• When L. lactis bacteria were grown on glucose in the absence of oxygen, the glucose concentration decreased, and the amount of lactic acid produced increased. L2.8</td><td>• The energy released when candle wax ($C_{11}H_{24}$) reacts with O_2 in a beaker to produce CO_2 + H_2O warms the beaker and surrounding air. L2.3
• The energy released when glucose ($C_6H_{12}O_6$) reacts with O_2 in a bottle to produce CO_2 + H_2O increases the motion of the bottle. L2.3
• When hydrogen and oxygen gas react in a bottle, light and sound are emitted, and the bottle's motion increases. L2.3</td></tr>
<tr><th colspan="2">The same energy-releasing chemical reactions provide energy for motion.</th></tr>
<tr><th>Life Science Phenomena</th><th>Physical Science Phenomena</th></tr>
<tr><td>• Limewater turns cloudy faster after 5 minutes of vigorous exercise than after 5 minutes at rest. L2.7
• The amount of oxygen taken up by thigh muscle increases with exercise time. L2.7
• Thigh muscle glucose uptake increases with time and exercise intensity. L2.7
• Isolated muscle fibers do not shorten when glucose + oxygen is added but do shorten when a solution of ATP is added. L2.7
• As the intensity of exercise increases, the concentration of glucose in the blood decreases, and the concentration of lactic acid increases. L 2.8</td><td>The physical science phenomena in the previous row that were used to develop the role of energy-releasing chemical reactions in the growth of organisms were also used to develop their role in providing energy for motion.</td></tr>
</table>

While much of the energy released can be used for growth and motion, a lot of the energy is transferred as heat to the surroundings.	
Life Science Phenomena	**Physical Science Phenomena**
• Thigh muscle temperature increases with exercise time. L2.7	• When a person lands on one end of a seesaw, lifting a person up on the other end, the second person doesn't rise as high as the first person started. L2.6 • A battery warms up as it provides energy to produce hydrogen and oxygen from water. L2.6 • When gasoline reacts with oxygen in a car engine, the hood of the car warms up. L2.6
Feedback mechanisms maintain a living system's internal conditions within certain limits and mediate behaviors, allowing it to remain alive and functional even as conditions change within some range.	
Life Science Phenomena	**Physical Science Phenomena**
• Despite predicted increases in muscle temperature, based on extrapolating from the muscle temperature increase observed in 3 minutes of exercise, the temperature of runners decreased by 1.5 °F after a 4-mile run. L2.9 • Despite predicted decreases in blood glucose and O_2 and increases in blood CO_2 and H_2O, based on an understanding of reactants and products of cellular respiration, data show that measured levels of these substances stay within a fairly narrow range. L2.9	See the Extras page at *www.nsta.org/growthandactivity* for a lesson on feedback in a simple physical system that could be used with students who may have difficulty with the idea of negative feedback.

Science Practices and Sense Making

Using scientific practices to help make sense of phenomena is central to the design of the *MEGA* unit. Students (a) analyze data they have collected or data from published scientific studies and reflect on what conclusions can be drawn from the experiments or on what additional data they could collect to test their predictions of phenomena; (b) use a variety of models to get ideas about the underlying causes of phenomena; and (c) use evidence, science ideas, and models to explain phenomena.

Data analysis. Students analyze data in nearly every lesson and use findings to support claims. For the matter story, firsthand observations; patterns in data; and inferences from tests of various substances provide evidence that chemical reactions have occurred, what the reactants and products are, and where in the body the reactions occur. For the energy story, students collect and analyze data about changes in temperature and motion in simple physical systems and then use changes in temperature and motion as indicators of energy changes associated with chemical reactions. Calorimetry is introduced to show how energy changes in chemical reactions are measured, enabling comparison of the amount of energy released when various fuels react with oxygen, the energy requirements of important biological reactions, and considerations about which energy-releasing reactions could drive the energy-requiring reactions and processes. Each chapter culminates in a lesson where students use findings as evidence to construct and revise a model.

Modeling. Students use data to construct and revise models. For the matter story, data about various body systems are used to construct hierarchical models of the components, revealing that each system is composed of organs composed of cells that are composed of polymers composed of monomers that are composed mainly of carbon, hydrogen, oxygen, and sometimes nitrogen atoms. A comparison of polymers making up body cells to polymers available in food suggests the kinds of chemical reactions that need to occur, and data from scientific papers provide evidence for which reactions do occur and where, allowing students to revise models of how food contributes to growth.

Students use **model-building activities** and **chemical equations** to model atom rearrangement and conservation during chemical reactions that lead to the production of new substances in closed and open systems. Students manipulate LEGO or ball-and-stick models to convert molecules of reactants into molecules of products and weigh models in sealed versus open baggies to represent mass conservation and changes in measured mass. They use word equations with chemical names (e.g., glucose reacts with oxygen to form carbon dioxide and water) and equations with chemical symbols and formulas ($C_6H_{12}O_6 + 6\ O_2 \rightarrow 6\ CO_2 + 6\ H_2O$).

Energy changes within systems are represented with bar graphs and system boxes. In bar graphs, shown in Figure 2, the difference in the heights of the energy bars for the reactants and the products indicates whether the chemical reaction releases energy or requires a net input of energy. In a system box, shown in Figure 3, the higher-energy state is written on the line at the top of the box and the lower-energy state is written on the line at the bottom of the box. A curved vertical arrow within a system box that connects higher-energy state and lower-energy state is used to represent the direction of the energy change. For example, in a system box that represents higher-energy reactants interacting to form lower-energy products, an arrow pointing downward would indicate that energy is given off during the chemical reaction or process. An arrow pointing upward would indicate that energy is required for the chemical reaction or process to occur, such as a tennis player going from being at rest to moving.

Figure 2. Example of a Bar Graph

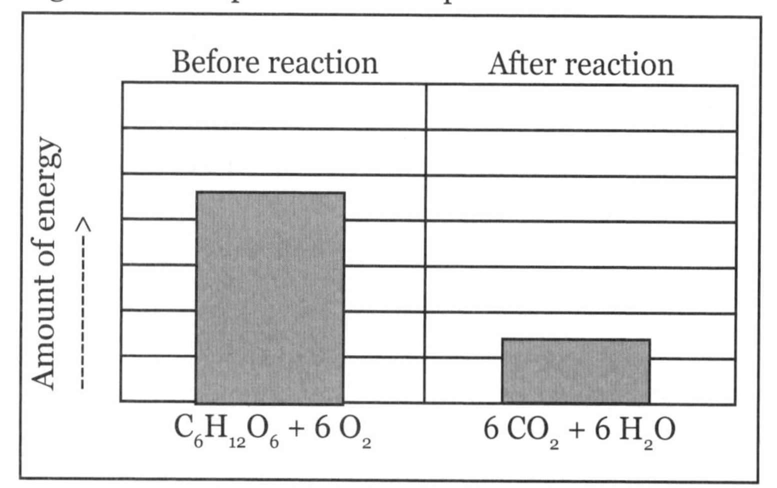

Figure 3. Example of a System Box

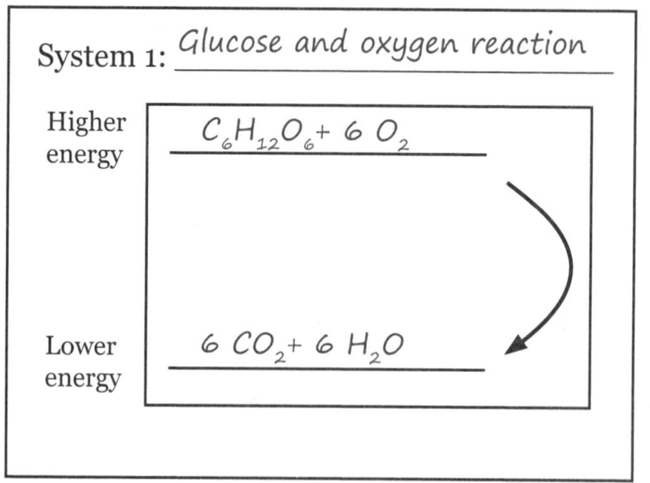

Energy transfer from one system to another system is represented with an energy transfer model, shown in Figure 4, that includes two or more system boxes and a thick arrow connecting pairs of boxes. The direction of the arrow between system boxes indicates the direction of energy transfer, which is always from the energy-releasing chemical reaction or process to the energy-requiring chemical reaction or process. Thick arrows are also used to represent energy transferred as heat from a system to its surroundings.

Figure 4. Example of an Energy Transfer Model

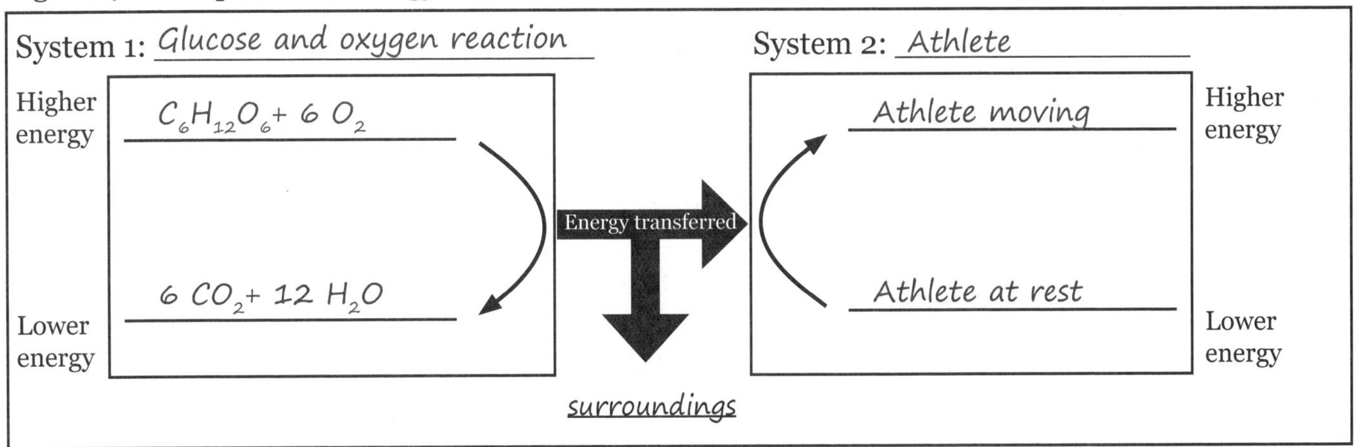

Constructing explanations. The *MEGA* unit engages students in constructing scientific explanations of phenomena, building on the support provided for this practice in the precursor middle school *THSB* unit, which uses the cognitive apprenticeship model (Brown, Collins, and Duguid 1989; Collins, Brown, and Newman 1989) to support students in constructing explanations. Teachers can refer to *THSB* for guidance on helping students develop their ability to construct evidence-based explanations, but a brief overview of how the *THSB* unit addresses this key science practice follows.

In the beginning of the *THSB* unit, students are provided with an example of a valid explanation, and the essential elements of an explanation—claim, relevant science ideas, and evidence that will be linked to the claim using science ideas—are defined. The early *THSB* lessons also establish Explanation Quality Criteria that students can use in judging an explanation's quality and introduce a template to remind students about the essential elements and help them organize their thinking and writing. After students experience modeling activities, a new element is added to the explanation template—models/modeling as tools for thinking about how something might work—and from then on students are expected to include appropriate models in their reasoning. The cognitive apprenticeship cycle is then repeated using the full explanation template. As the unit proceeds, students are reminded to use ideas about atoms in their explanations and to be sure that their explanations meet the Explanation Quality Criteria. The explanation template is faded in the *THSB* Student Edition but is available in the *THSB* Teacher Edition if the teacher feels the students still need it. At the end of the unit, the scaffolding is completely removed except for a reminder that students can use science ideas from anywhere in the unit to support their claims. Strategies used in the *THSB* unit may be helpful if students have had little or no prior experience constructing explanations. The *THSB* Teacher Edition and web-based teacher resources support teachers in helping students construct evidence-based explanations. The Teacher Edition provides ideal responses for each explanation task and for each element of the explanation template. Roseman, Herrmann-Abell, and Koppal (2017) include examples of rubrics for explanation tasks recommended for use as embedded assessments.

Nature of Science

Students have opportunities throughout the unit to experience how science works. They analyze and interpret data they collect as well as data from published studies, and they use data and reasoning from models to develop evidence-based explanations in ways that are consistent with what is expected of scientists. Students are repeatedly asked to list examples of phenomena they have observed in class that are consistent with presented science ideas and are told that the reason they can legitimately use the science ideas to support explanations of similar phenomena is that the science ideas are consistent with a wide range of observations and data—not just the small number they examined in class. Lesson 2.7 briefly recounts the discovery of the role of the ATP + $H_2O \rightarrow$ ADP+P_i reaction in coupling cellular respiration to muscle contraction.

Nonetheless, the *NGSS* note that while integrating science practices, core ideas, and crosscutting concepts "sets the stage for teaching and learning about the nature of science" (NGSS Lead States 2013, p. 97, Appendix H), engaging in such activities is not sufficient. Students would also need to reflect on their model-building, data analysis, and explanation activities and to compare their work to the work of scientists, which would likely involve examination of historical case studies. While the history of the discovery of the role of ATP in muscle contraction is fascinating, partly because the discovery involved scientists working in isolation during World War II who were unable to share their work or read about the work of others for several years, given the time that these additional activities would require and the risk they might pose to the coherence of the content storyline, we reluctantly decided not to focus more instructional time on the nature of science learning goals. However, doing so would be a way to increase the sophistication of the unit for students who readily grasp the targeted ideas.

Unique Aspects of the Unit

Understanding some of the thinking that went into the design of the *MEGA* unit and what the unit is intended to do can help to support students in achieving the unit's learning goals. This section briefly describes the development team's rationale for decisions about how the science ideas (drawn from disciplinary core ideas and crosscutting concepts) would be integrated in the unit and what vocabulary students would be expected to know and use.

Integrating Physical and Life Science Ideas and Crosscutting Concepts

The *MEGA* unit takes an approach to teaching about matter and energy changes that resembles the approach the *THSB* unit takes to teaching about matter changes, which studies have shown to be successful in improving students' three-dimensional learning (Herrmann-Abell, Koppal, and Roseman 2016; Roseman, Herrmann-Abell, and Koppal 2017). In contrast to traditional biology units, the *MEGA* unit targets ideas about matter and energy changes in chemical reactions in both simple physical systems and complex biological systems in the same unit. It introduces difficult ideas first in simple physical systems that allow firsthand observations before having students apply them to complex biological systems that require inferences from data. Unit activities use similar models to represent matter and energy changes within systems and energy transfers between systems in both simple physical and complex biological systems. This integration of physical and life science content enhances the explanatory power of the science ideas and is consistent with recommendations of *NGSS* in general and in particular with its crosscutting concepts related to matter and energy and systems and system models. Content support is provided in both the Teacher Edition and in professional development to help biology teachers become confident teaching all aspects of the unit.

Matter. The matter story in Chapter 1 engages students in considering how protein, carbohydrate, and fat polymers from food become protein, carbohydrate, and fat/lipid polymers making up their body structures. The molecular structures of these polymers appear quite complex, even after students realize that they are made up of similar monomers. So, the chapter devotes two lessons to having students use ideas about systems and various models to represent atom rearrangement and conservation to explain how products form from reactants in simple physical systems. Students are then able to model the processes involved in converting polymers from food to polymers needed to grow and repair body structures and to appreciate the roles of various body systems in these processes.

Energy. Once students can explain how changes in matter during chemical reactions contribute to biological growth and repair, they examine evidence of energy changes in the same reactions and try to explain where the energy comes from or where it goes in each case. As with the matter story, students generate energy ideas in simpler physical systems and apply them to more complex biological systems. By the end of Chapter 2, students are modeling energy transfers from the cellular respiration system to the system of proteins involved in muscle contraction, to the system that makes polymers for growth and repair, and to the surroundings as heat.

Chemical reactions. The examples of chemical reactions students observe were carefully chosen to accomplish several purposes:

- Provide phenomena in simple systems to enable students to review middle school prerequisite science ideas about BOTH matter and energy, such as the energy-releasing reaction between hydrogen and oxygen molecules to form water molecules and the energy-requiring reaction between water molecules to form hydrogen and oxygen molecules.
- Support high school science ideas about BOTH matter and energy, mainly energy-releasing reactions between carbon-based molecules and oxygen to produce carbon dioxide and water, the energy-requiring synthesis of glucose monomers and oxygen from carbon dioxide and water, and various energy-requiring reactions between monomers to produce polymers for building body structures (plus water molecules).
- Introduce students to tests of substances in simple systems that can be used to provide evidence of the production or use of those same substances in more complex systems (e.g., the combustion of candle wax can be carried out in the classroom and used to demonstrate the results of simple tests for carbon dioxide and water that later provide evidence that a similar reaction—oxidation of glucose to produce carbon dioxide and water during cellular respiration—occurs faster when muscles exercise).
- Illustrate that chemical reactions in simple physical systems can provide insights into reactions in complex biological systems.

Table 2 lists the chemical reactions used in the *MEGA* unit and provides an instructional rationale for the use of each.

Table 2. The Role of Chemical Reactions in *MEGA*

Reaction Equation	Rationale for Use of Reaction
Candle wax and oxygen → carbon dioxide and water	Energy-releasing reaction that allows collection and testing of reactants and products
Ammonium thiocyanate and barium hydroxide → water, ammonia, and barium thiocyanate	Energy-requiring reaction with product whose identity can be determined by odor
Magnesium and oxygen → magnesium oxide	Atom rearrangement easily modeled
Water → hydrogen and oxygen	Simple tests for reactants and products, atom rearrangement easily modeled
Glucose and oxygen → carbon dioxide and water	Simple tests for reactants and products, atom rearrangement easily modeled

Carbon dioxide and water → glucose and oxygen	Chemical reaction carried out by plants provides life science example
Protein polymer and water → amino acid monomers	Digestion example
Amino acid monomers → protein polymer and water	Biosynthesis example
Carbohydrate polymer (starch) and water → glucose monomers	Digestion example
Glucose monomers → carbohydrate polymer (glycogen) and water	Biosynthesis example
Fat polymer (triglyceride) and water → monomers → fat or lipid polymer and water	Digestion and biosynthesis example

Systems. The unit develops ideas about systems and the usefulness of "systems thinking" to understanding human body systems at several levels of biological organization and changes in matter and energy within and between systems. Students examine hierarchical levels of structural organization of the human body. They use what they learn about the importance of paying attention to inputs, outputs, and boundaries of systems to understand how interactions between molecules within cells lead to matter and energy changes in organs and body systems that in turn lead to observable changes in matter and energy in the human organism as a whole. The unit stops short of extending systems thinking and ideas about matter and energy changes to ecosystems, but this could be done in a subsequent unit on ecosystems.

Human body systems. The unit focuses on components of five human body systems that are needed to tell the basic story of where and how ingested food is used to build and repair body structures and the contributions of different body systems to maintaining homeostasis, even during intense exercise. For example,

- The small intestine of the **digestive system** is where food polymers are digested to monomers—e.g., carbohydrates to glucose, fats to fatty acids and glycerol, proteins to amino acids—that are absorbed into blood vessels of the **circulatory system.**
- Amino acids are taken up by muscle cells of the **muscular system** and used to build muscle proteins.
- Glucose monomers are taken up by cells of the liver (**digestive system**) and muscle (**muscular system**) and used to build glycogen polymers that store glucose for later use in cellular respiration.
- Fatty acids are taken up by fat cells of the **integumentary system.** Fatty acids are used to build phospholipid components of cell membranes and used to build triglycerides that store fatty acids for later use in cellular respiration.
- The **respiratory system** brings oxygen needed for cellular respiration to blood vessels of the **circulatory system** and removes carbon dioxide produced during cellular respiration in all body cells. During exercise, breathing becomes faster and deeper and the heart beats faster to increase the supply of oxygen to cells and increase the removal of carbon dioxide.
- Increased needs for glucose during exercise result in breakdown of glycogen to glucose in both

liver (**digestive system**) and muscle (**muscular system**), thereby maintaining the level of blood glucose and muscle glucose needed for cellular respiration.

• Muscle cells of the **muscular system** increase the rate of cellular respiration to provide increased energy for muscle contraction during exercise. Some of the energy from cellular respiration increases the temperature of the cells' surroundings.

• To counteract temperature increases, surface capillaries of the circulatory system dilate to increase heat dissipation by the skin of the **integumentary system.** Evaporation of water from the skin surface transfers energy from the body to its surrounding environment.

Regulation/homeostasis. The unit's final project engages students in predicting the effect of intense exercise on concentrations of the reactants and products of cellular respiration and on body temperature, designing experiments to test their predictions, and then consulting reference materials to explain why the data they collect or examine do not match their predictions. Although students are not expected to develop a sophisticated understanding of how our bodies maintain concentrations within a safe range, their explanations should describe the contributions of various body systems and their components.

Vocabulary

We have intentionally limited the number of technical terms to those that students need to communicate about their experiences. As a result, the unit avoids several terms typically used in middle and early high school classrooms (e.g., *element, compound, ion*) while using other terms that are less typical (e.g., *monomer, polymer*). The terms *element*, *compound*, and *ion* are not needed for students to communicate their ideas about atom rearrangement and conservation, although students do need to correctly use the terms *atom* and *molecule* to do so. The terms *monomer* and *polymer* are used instead of alternatives like *small molecule* and *large molecule* because they convey more precisely the common components of different carbohydrates and proteins and how they can be assembled from or disassembled to a small number of building blocks. We also leave out some details that are more likely to confuse than to help students develop and communicate their ideas. For example, solutions are referred to as *liquids* and ionic compounds are referred to as *molecules.*

The unit expects students to understand and correctly use the following terms in **bold:**

Atoms and **molecules.** While students may have already been taught definitions of these terms, many do not consistently distinguish them in their talking or writing or in their use of models. It is essential for students to know that atoms make up molecules, not the other way around. The unit expects students to know that an individual atom is represented by a ball or chemical symbol, that different colored balls or symbols represent different types of atoms, and that two or more balls or chemical symbols linked together represent molecules.

Monomers and **polymers.** Rather than relying on vague descriptions like "small and large molecules," which provide no information about how the small and large molecules are related, we opted to introduce the terms *monomer* and *polymer*. Using these terms in the context of chemical reactions involved in nylon formation, protein formation, glycogen formation, and lipid formation highlights their similarities (e.g., all involve formation of polymers and water molecules from monomers) and eliminates the need for students to know the names of most of the specific monomers involved (e.g., specific fatty acids, specific amino acids). However, students are expected to know that **glucose** is the monomer used to build **carbohydrate polymers** such as starch and glycogen and **amino acids** are the monomers used to build **protein polymers.** The term glucose is singled out so that students can distinguish its formation (during **photosynthesis**) from its use to make polymers needed to build plant body structures for growth. Because the term *carbohydrates* refers to both polymers and monomers, accuracy requires that we use the term *carbohydrate polymers.* Students are expected to know that proteins (such as actin and myosin, which are involved in muscle contraction, and amylase, which is involved in carbohydrate digestion), fats, and lipids (such as phospholipids, which are used to build cell membranes) are **polymers.**

Chemical reactions. Students are expected to know that chemical reactions produce **products** from **reactants** and that products have different **characteristic properties** than the reactants because they are made up of different molecules, but that the molecules making up the products are made from the atoms of the reactants. (Irreversibility is not a criterion for distinguishing chemical reactions from other changes—this is a common misconception.) Students are expected to know that **photosynthesis; aerobic** and **anaerobic cellular respiration;** and **digestion** and **synthesis** of carbohydrate, fat, and protein polymers involve chemical reactions.

Mass conservation and **measured mass.** Students are expected to know that while the **total mass** of the reactants and products does not change during a chemical reaction (total mass is **conserved**), the **measured mass** may change if one or more substances enter or leave the **system.** The term measured mass is introduced to help students understand why conservation principles aren't violated during biological growth; that is, the added mass must have come from somewhere outside the system being measured (the plant or animal's body). As a bridge to calculating the mass of the glucose + oxygen chemical reaction system before and after the reaction, students count and weigh LEGO models of the hydrogen + oxygen chemical reaction system before and after the reaction. This helps students visualize that both mass and atoms (though not necessarily molecules) are conserved during chemical reactions.

Chemical bonds. The *MEGA* unit expects students to learn that **chemical bonds** connect atoms making up molecules but not types of bonds or parts of atoms that are involved in bonding. Although most of the chemical reactions introduced involve breaking and forming covalent bonds, students are not expected to know this term or to know that each bond involves an electron pair. For the matter story, students learn that bonds between reactant molecules must break and new bonds must form in order to produce product molecules. For the energy story, students are expected to learn that **bond-breaking always requires energy** and **bond-forming always releases energy,** contrary to the incorrect idea in some textbooks that bond-breaking releases energy.

Energy changes and **conservation.** The unit is consistent with the *NGSS* approach to energy in focusing on **energy changes** within **systems** and **energy transfer** between systems rather than focusing on naming forms of energy and energy transformation. Even though the *NGSS* core ideas occasionally talk about chemical and mechanical energy, the *MEGA* unit does not expect students to identify any forms of energy. Students are expected to recognize four indicators of energy changes: changes in temperature, changes in motion, appearance or absorption of light, and production and absorption of sound. Students are expected to know that a decrease in energy somewhere is always accompanied by an increase in energy elsewhere, but they are not expected to show quantitatively that those two amounts must be the same.

Rates of chemical reactions. The *MEGA* unit does not focus on rates of chemical reactions or factors affecting rates. Understanding why, for example, all reactant molecules don't react at once or why the reaction doesn't stop once the "higher-energy" molecules have reacted requires a mental model of the Boltzmann distribution curve, how it shifts with temperature, and why remaining molecules redistribute. These ideas go far beyond *NGSS* expectations.

Enzymes. The unit does not attempt to explain how enzymes or other types of catalysts speed up chemical reactions, but it does include enzymes as examples of protein polymers that carry out functions of cells.

Systems and **models.** *Benchmarks for Science Literacy* (AAAS 1993) characterizes the role of systems and models and other common themes as tools for thinking across disciplines and clarifies what the focus of instruction should be.

> *The main goal of having students learn about systems is not for them to be able to talk about systems in abstract terms but to enhance their ability and inclination to attend to various components of particular systems in attempting to understand how the system as a whole works.* (p. 262)

In *MEGA* Chapter 1, students begin to think about systems by examining components of various human body systems (organs, cells, substances, polymers, monomers, atoms). Next, they examine simple chemical reaction systems as contexts for learning to use ideas about system inputs and outputs to keep track of substances and their atomic/molecular components. Finally, they use their knowledge of body systems and chemical reactions to model how polymers from food are digested to monomers, how monomers get to where they are used, and how the monomers are converted to polymers that become part of body structures. In Chapter 2, students apply ideas about system inputs and outputs to model energy changes within systems and energy transfer between systems. The culminating lesson engages students in considering the role of human body systems in helping to maintain levels of glucose, oxygen, carbon dioxide, and water in muscle cells so that cellular respiration can increase in response to exercise needs and in helping to ensure that energy released as heat is transferred from muscle cells to the body's external environment.

Science practices. The unit expects students to understand and correctly use the following terms in **bold** related to making sense of phenomena:

> **Models.** Students should know that a model is a simplified version of some thing or process that we hope can help us understand it better. Whether models are physical, mathematical, or conceptual, their value lies in suggesting how things either do work or might work. When a model does not mimic the phenomenon well, the nature of the discrepancy is a clue to how the model can be improved.
>
> **Evidence.** Students are expected to know that **data** from observations and measurements can be used as evidence for a claim if the data are relevant to the claim and can be confirmed by others. Students are also expected to know that a **model** can be useful for suggesting how the thing being modeled works, but there is no guarantee that ideas suggested by models are correct if they are based on the models alone. Consequently, while ideas from models should be consistent with claims, they do not provide evidence to support claims.
>
> **Explanation.** As noted in the *NGSS*, "the goal of science is to construct explanations for the causes of phenomena" (NGSS Lead States 2013, p. 60, Appendix F). By the end of the unit,

students should be able to explain changes in matter and mass during biological growth in terms of the rearrangement and conservation of atoms and to explain how living things obtain energy needed for growth and motion in terms of energy transfer from energy-releasing chemical reactions to energy-requiring chemical reactions or processes.

Structure of the Unit

The unit consists of an introductory lesson and 14 additional lessons organized into two chapters. Chapter 1, which consists of five lessons, is designed to help students develop an understanding of the role of atom rearrangement and conservation in chemical reactions that contribute to biological growth, starting with chemical reactions in simple physical systems that involve small molecules and then applying what they learn to chemical reactions in living systems that involve converting polymers from food into polymers used to build body structures. Chapter 2, which consists of nine lessons, is designed to help students develop an understanding of energy changes during chemical reactions, energy transfer, and conservation, first in simple physical systems and then in complex biological systems involved in the motion and growth in living organisms. The last lesson ties together the matter and energy stories by having students consider how the body maintains a fairly constant internal environment during exercise despite chemical reactions that involve changes in matter and energy.

Both Chapters 1 and 2 begin by engaging students with familiar phenomena that will drive learning over the chapter—starting with observable phenomena, such as bodybuilding and wound healing in Chapter 1 and athletes performing amazing feats in Chapter 2, followed by examining data that provide evidence of phenomena that can't be observed directly. Students then construct and use models to make sense of the phenomena (e.g., how various body systems contribute to converting food to biopolymers or how energy released during cellular respiration transfers energy to move muscles), and they revise their models to take account of new data. After generating important science ideas based on their observations and data, students then try to apply them in new contexts.

Each lesson of the *MEGA* unit consists of carefully sequenced activities designed to (1) draw upon students' prior knowledge and experiences relevant to classroom activities, (2) support students as they investigate and make sense of phenomena and data, (3) guide students in modeling and explaining phenomena in terms of underlying molecular mechanisms, (4) provide opportunities for students to apply or extend science ideas and practices to new phenomena, and (5) help students synthesize their ideas and reflect on changes in their thinking.

Section Headings and Purposes

Each lesson is designed with particular features that are denoted by their section headings.

What do we know and what are we trying to find out?

This introductory section situates the lesson in the content storyline by making links between science ideas of previous lesson(s) and the key question of the current lesson. This section does not "give away the answers" but provides students with some sense of what they will be working toward understanding in the lesson.

Key Question

Each lesson begins with and returns to a key question that the lesson is designed to answer. The key question aligns with the lesson's main learning goal and frames the students' inquiry. The key question is posed at the beginning of the lesson to give each student a chance to express his or her initial ideas and to give teachers a sense of the range of ideas students hold. At the end of the lesson, the question is posed again and student responses provide a way to monitor their progress.

Activity

Each lesson includes activities designed to engage students with phenomena and representations relevant to the learning goals. Some phenomena, particularly those in nonliving systems, can be observed directly and others, particularly those occurring in the bodies of living organisms, require inferences from data. Activities also engage students in modeling invisible aspects of the phenomena, particularly atom rearrangement and conservation and energy changes and transfer. Each activity includes questions to focus and guide students in observing and interpreting the phenomena. Activities are structured to encourage discussion by having students work either with a partner or a small group.

Participating in a whole-class discussion at the end of the activity can help students reach consensus on their observations of phenomena, data, and models and on the science ideas that emerge from their interpretations of them.

Science Ideas

At critical points in the unit, particularly after students have developed ideas based on phenomena, data, or models, students are asked to generalize across their experiences. Only then are relevant science ideas introduced to students as generalizations about how the world works based on a wider range of observations and data. Students have opportunities to compare the ideas they are developing with established science ideas and find examples from their work that support the science ideas.

Science ideas are used in writing explanations, so they will often precede an activity in which explanations are evaluated or developed. Or, they may precede a Pulling It Together section in which students are expected to construct an explanation of related phenomena.

Pulling It Together

These questions provide opportunities for students to individually (1) revisit and answer the lesson key question or related questions to summarize their current understanding or new learning, (2) use and apply the ideas they are developing to a new context or phenomenon, or (3) begin to link the ideas to the next lesson(s). The linking question elicits students' ideas and predictions about the key question of the next lesson or about how an idea developed in nonliving contexts might apply to growth or activity in living things.

Closure and Link

This feature, included only in the Teacher Edition, is intended to summarize what students have learned in the lesson or chapter and what unanswered questions they will tackle in the next lesson or chapter. Teacher notes outline the main points that should surface in a teacher-led discussion and questions that should grow out of what students have just learned.

References

American Association for the Advancement of Science (AAAS). 1993. *Benchmarks for science literacy*. New York: Oxford University Press.

American Association for the Advancement of Science (AAAS). 2017. *Toward high school biology: Understanding growth in living things*. Arlington, VA: NSTA Press.

Brown, J., A. Collins, and P. Duguid. 1989. Situated cognition and the culture of learning. *Educational Researcher* 18 (4): 32–42.

Collins, A., J. S. Brown, and S. Newman. 1989. *Cognitive apprenticeship: Teaching the crafts of reading, writing, and mathematics*. In *Knowing, learning, and instruction: Essays in honor of Robert Glaser*, ed. L. B. Resnick, 453–494. Hillsdale, NJ: Erlbaum.

Driver, R., A. Squires, P. Rushworth, and V. Wood-Robinson. 1994. *Making sense of secondary science*. London: Routledge.

Herrmann-Abell, C. F., M. Koppal, and J. E. Roseman. 2016. Toward high school biology: Helping middle school students understand chemical reactions and conservation of mass in nonliving and living systems. *CBE-Life Sciences Education* 15 (4). DOI:10.1187/cbe.16-03-0112.

NGSS Lead States. 2013. *Next Generation Science Standards: For states, by states*. Washington, DC: National Academies Press. *www.nextgenscience.org/next-generation-science-standards*.

NRC (National Research Council). 2012. *A framework for K–12 science education: Practices, crosscutting concepts, and core ideas*. Washington, DC: National Academies Press.

Roseman, J. E., C. F. Herrmann-Abell, and M. Koppal. 2017. Designing for the *Next Generation Science Standards:* Educative curriculum materials and measures of teacher knowledge. *Journal of Science Teacher Education* 28 (1): 111–141.

The following page has a blank explanation chart that students can use as a writing scaffold. For an example of how to develop the practice of explanation writing, see Chapter 1 of the Teacher Edition of the THSB *unit.*

Question	
Claim	
Science Ideas	
Evidence	
Models	

Explanation:

Getting Started

Matter and Energy Changes in Our Bodies

Teacher Edition

Lesson Guide
Getting Started: Matter and Energy Changes in Our Bodies

Focus: This introduction frames the unit and sets out the key question that the unit will help students answer.

Unit Key Question: How do living things use food as a source of matter for building and repairing their body structures and as a source of energy for carrying out a wide range of body functions?

Target Idea(s) Addressed

None

Science Practice(s) Addressed

None

Materials

None

Advance Preparation

None

Getting Started: Matter and Energy Changes in Our Bodies

What do we know and what are we trying to find out?

Our bodies are capable of some amazing feats, from setting an Olympic record to just staying alive while we sleep! Think about it: Every second of every day, our bodies are carrying out all kinds of activities that enable us to grow, work, play, think, create, communicate, and rest and repair ourselves. But what makes all of these activities possible? Where does the "stuff" come from to build an athlete's body or to repair a broken leg or heal a wound? Where does an athlete get the energy needed to run a marathon? How do our own bodies get the energy they need to keep us alive and functioning properly and to build and repair body structures while we sleep?

Ever since you were a child, you've probably been told that eating the right food will help you grow and give you lots of energy. In fact, it is estimated that American adults eat about 2,000 pounds of food each year! Some athletes eat much more than that when they are preparing for or participating in competitions. It's clear that something happened to all that food once it was consumed (we don't look like the food we eat, for example), but what? How does food provide building material and serve as fuel that keeps our bodies functioning?

To answer these questions, we need to know more about matter that makes up our body structures and the changes in matter and energy that take place inside us. In doing so, we will see that the same scientific ideas that explain matter and energy changes in simpler nonliving systems, such as a car or a battery, can also be used to explain matter and energy changes in more complex living systems, such as our bodies.

Teacher Talk and Actions

In this unit, we will observe examples of growth and motion and apply important science ideas about matter and energy to explain our observations:

- In Chapter 1 we will examine changes in matter that take place in our bodies as we consume food and grow—from the observable macroscopic level down to the invisible molecular level—and construct models to help us think about and explain where and how these changes occur.
- In Chapter 2 we will examine and model how energy changes in both simple physical systems and complex biological systems can make something happen. We will look at some examples, such as how a chemical reaction in a battery makes a toy car move, and then use what we learn to model and explain related phenomena in living organisms, such as how chemical reactions in our muscles make our bodies move and grow.
- The unit's final lesson brings together what we've learned to investigate how the body maintains a nearly constant internal environment during exercise despite changes in matter and energy.

By the end of the unit, you should be able to answer the Unit Key Question below. For now, brainstorm some ideas for answering the question with your class. We will return to this question at the end of the unit.

Unit Key Question: How do living things use food as a source of matter for building and repairing their body structures and as a source of energy for carrying out a wide range of body functions?

Teacher Talk and Actions

Chapter 1

Making Sense of Matter Changes Involved in Human Growth

Teacher Edition

Chapter 1 Overview

Chapter 1 tells the basic story of how our bodies convert molecules from food into molecules that make up our body structures, including the role of various body systems in carrying out the chemical reactions involved and moving reactants and products from place to place. Although students may be familiar with names of organs of various body systems, they are probably unaware of the molecules that make up the organs, giving them their structure and carrying out their functions. All these molecules need to be produced if humans are to grow and repair. The story is both abstract and complex, requiring (a) an understanding of the body from the observable macroscopic level to the invisible molecular level, (b) the ability to use a model of atom rearrangement during chemical reactions to explain how body molecules can be made from food molecules that often have very different properties, and (c) the ability and inclination to consider the inputs and outputs of a system when trying to figure out what the reactants and products are, where they come from and where they go, and which atoms rearrange during the chemical reaction.

Students make sense of a variety of phenomena related to growth and repair. For the idea "To grow and repair, living organisms need to build body structures that are different from polymers in their food," students observe, model, and explain how

- protein polymers making up human muscles, skin, and blood differ from proteins making up eggs, milk, soybeans, and wheat;
- glycogen polymers in liver and muscle differ from starch; and
- fat polymers that make up body fat and cell membranes differ from fat polymers in butter.

For the idea "During chemical reactions in both simple physical and complex living systems, atoms of reactants rearrange to form products; mass is conserved because atoms are conserved," students observe, model, and explain how new substances can be produced and mass can change when

- both a candle and a marshmallow react with oxygen to form carbon dioxide and water;
- ammonium thiocyanate reacts with barium hydroxide to form ammonia, water, and barium thiocyanate;
- water molecules react to form hydrogen molecules and oxygen molecules; and
- an aquatic plant produces glucose and oxygen from carbon dioxide and water.

For the idea "In the digestive system protein, carbohydrate, and fat polymers react with water to form monomers that are transported by the circulatory system to components of other body systems where the monomers react to form body polymers (and water)," students observe, model, and explain the following:

- As the amount of ingested protein decreases in the human gut, amino acid levels increase first in the gut and then in surrounding blood vessels.

- Shortly after humans were injected with the amino acid ^{14}C-phenylalanine, the ^{14}C atoms were found to be part of myosin protein.
- Two hours after eating ^{14}C-starch, the level of ^{14}C-glucose in human blood increases.
- The concentration of glycogen in rat liver increases from 10 minutes to 30 minutes after glucose infusion.
- An hour after rats are injected with ^{14}C-glucose, ^{14}C-glycogen is found in liver and muscle.

Lesson Overviews

Lesson 1.1 frames the matter story, the focus of Chapter 1, by engaging students in considering matter changes involving growth (videos of bodybuilding, a growing child, wound healing, cells reproducing in culture) and motion (athletes competing in an event). This provides an opportunity for students to express their ideas and for teachers to find out their students' initial ideas.

Lesson 1.2 starts with what students know about body systems—mainly external macroscopic components that can be observed directly (e.g., skin and hair), proceeds to internal organs that are observable in photographs (e.g., muscles, intestines, heart, lungs), zooms in to microscopic cells that make up some of the organs, then to the submicroscopic macromolecules (polymers) that make up cells and the types of monomers that compose them, and finally to the mainly 4–5 types of atoms that make up the monomers. The goal of the lesson is to provide evidence that polymers from food (e.g., ovalbumin from egg white, casein from milk) are different from polymers making up human body structures (e.g., hemoglobin for red blood cells, keratin for hair, collagen for skin, actin and myosin for muscles), which motivates the question of how our bodies accomplish the necessary conversions. The lesson concludes by noting that the next two lessons will give students a chance to observe and model chemical reactions in simple systems to prepare them to explain how our complex body systems convert polymers from food into body structures in Lesson 1.5.

Lesson 1.3 reviews the prerequisite middle school idea that new substances are produced during chemical reactions because atoms of molecules of reactants rearrange to form products, using examples involving molecules reacting in simple systems where the production of a new substance can be readily detected to provide evidence that a chemical reaction has occurred (e.g., barium hydroxide + sodium thiocyanate produces ammonia that can be detected by its odor, a burning candle produces carbon dioxide and water that can be detected with limewater and cobalt chloride tests, respectively) and the conversion of reactants to products can be easily modeled with LEGOs, ball-and-stick models, and chemical formulas. This serves as a review for students who have used the *THSB* unit in middle school. Students who have not used *THSB* or a comparable unit may need additional experiences with observing and modeling chemical reactions before they are ready to apply this knowledge to more complex changes in matter in living organisms. Even though the lesson uses examples from physical systems, a Pulling It Together question foreshadows the reaction that living cells use to obtain energy by asking students to consider similarities between reactants and products of a candle burning and what happens in our bodies.

It is not a goal of this lesson for students to learn to use chemical formulas to balance equations. For many students, the distinction between subscripts and coefficients is abstract and unnecessary and may distract them from learning the basic idea that atoms can rearrange but are conserved during chemical reactions. Until students can easily relate the chemical formulas to physical models, it's best to stick to physical models. Students should be encouraged to use their LEGO models to visualize atoms disconnecting from each other and connecting to different atoms and to notice that, as with their models, atoms aren't created or destroyed in the process.

Lesson 1.4 builds on the middle school idea that atom conservation explains mass conservation and extends it to having students consider the usefulness of keeping track of a system's inputs and outputs for explaining why a change in the mass of a system does not violate conservation principles. As in the previous lesson, simple physical systems are used that clearly show gases leaving a system (when water decomposes, hydrogen and oxygen gases produced leave the system) and students use LEGO models to show why the decrease in the mass of the system can be accounted for by the increase in mass of the surroundings. Ideas about system inputs and outputs are then used to help students explain why conservation principles are not violated when a marshmallow burns in air or when an aquatic plant produces bubbles of oxygen (and glucose) from water and dissolved carbon dioxide. The lesson also provides a framework for constructing valid explanations that will be a review for students who experienced *THSB* but not necessarily for students who used other middle school science curriculum materials. Students who have not previously learned to construct explanations will likely need instruction and additional opportunities to practice constructing explanations of related phenomena and to receive feedback.

Lesson 1.5 brings students back to the central question of the chapter, asking them to apply what they have learned about atom rearrangement and conservation in simple systems to consider how our body systems work together to convert polymers from food into polymers making up our body structures. Students draft initial models of the processes involved and the location of each process in the body and then use experimental data from published papers to revise their models. Because the link between reactants and products of chemical reactions occurring in the body cannot be observed directly, students learn to identify reactants and products by drawing inferences from data, some of which use isotopically labeled reactants, so the fate of labeled atoms can be traced. (Students who have used *THSB* have already learned about the value of the isotopic labeling technique in linking reactants to products and have used yellow labels on physical models to show how carbon atoms from CO_2 could become part of glucose and then cellulose.) The lesson concludes the basic story of how matter changes during chemical reactions contribute to the growth of living organisms.

Lesson Guide
Chapter 1, Lesson 1.1
Matter Changes in the Human Body

Focus: This lesson frames the chapter by engaging students in considering matter changes in videos involving growth and repair. Students will have several opportunities to revise and expand their ideas throughout the chapter.

Key Question: How can we detect changes in the matter that makes up the human body?

Target Idea(s) Addressed

None

Science Practice(s) Addressed

Obtaining, evaluating, and communicating information

Materials

Video 1: *Bodybuilder,* Video 2: *Child Growing,* Video 3: *Wound Healing,* Video 4: *Cells Reproducing in Culture*

Advance Preparation

Observe videos and decide how you plan to show them (e.g., all or only parts).

Phenomena, Data, or Models	Intended Observations	Purpose	Rationale or Notes
Activity 1 Time-lapse photos of bodybuilder, child growing, wound healing, and cells dividing in culture	• The body of the bodybuilder is getting visibly larger and more massive. Muscles are probably increasing in size. • The infant is increasing in size and mass. External body structures are increasing in size/mass (e.g., more skin, hair, teeth). Internal body structures are probably increasing in size as well (e.g., muscles, bones, heart, stomach). • New material is appearing at the site of a wound as it heals. • Cells in culture are reproducing. Prior to dividing, each cell increases in mass. (Students should observe that the cells aren't merely dividing, they are also growing prior to dividing so that the total mass of all the cells is increasing.)	Give students the opportunity to observe phenomena involving changes in matter (all involving increases in matter). Begin to develop students' abilities to carefully observe phenomena from the perspective of matter changes and to distinguish evidence from their own observations from inferences.	

Lesson 1.1—Matter Changes in the Human Body

What do we know and what are we trying to find out?

As you probably learned in middle school, all matter—including the matter making up your own body—is made up of atoms that are usually connected into molecules. Substances differ from one another because they are made up of different molecules. For example, a rusty bicycle does not look like a new bicycle because rust is made up of different molecules than the original shiny bicycle. Tarnished silver jewelry looks black because the black tarnish is made up of different molecules than the original silver.

Such changes are easy to observe because they involve color changes and occur in plain view. What about changes in living things? Consider changes that have taken place on the outside of your body as you've grown from a small baby to the young person that you are today. In addition to your being bigger and having teeth, the shape of your nose has probably changed and your hair color may be different. You can see and measure these changes easily. But what about changes that may be going on inside your body or in parts of your body that are too small to be visible? In this lesson, you will watch videos showing an athlete preparing for competition, a child growing, a wound healing, and cells reproducing in culture. As you watch, look for clues that can help you determine whether a change in matter is taking place.

Answer the Key Question to the best of your knowledge. Be prepared to share your ideas with the class.

> **Key Question: How can we detect changes in the matter that makes up the human body?**

Teacher Talk and Actions

Activity 1: Observing the Human Body

In this activity, you will observe changes that occur in human bodies as they carry out various activities.

Procedures and Questions

1. Observe the videos. In Table 1.1 below, record your observations in the second column (Observations). Watch the videos again if you need to. Each time you view the video add any additional observations you make to the table.

Table 1.1. Observable Changes in the Human Body

Video	Observations	More, Less, or the Same Number of Atoms?
Video 1 – Bodybuilder	The bodybuilder's body is getting bigger. He appears to have more muscles, but we can't actually see the muscles under his skin.	More
Video 2 – Child Growing	The child is getting taller and has more hair and teeth. We might infer that his muscles and other internal body parts are increasing in size/mass.	More
Video 3 – Wound Healing	The wound heals. New skin/scar tissue formed to repair the cut.	More
Video 4 – Cells Reproducing in Culture	Cells reproduce (grow and then divide) to form more cells.	More

Teacher Talk and Actions

The purpose of Video 4 is to encourage students to think about the relationship between body growth and cell growth. The following questions may be helpful in guiding the discussion:

- Are the cells growing? How do you know? (Students may reason that cells must grow before they divide. If not, then the cells would get smaller and smaller as more and more divisions occur.)
- Where does the matter for the cell growth come from?
- Do you think cell growth and division are contributing to the growth of the bodybuilder? Explain.
- Do you think cell growth and division are contributing to the repair of the wound? Explain.

2. We know that all matter is made up of atoms and molecules, including our bodies and the cells that make them up. Predict whether the number of atoms increases, decreases, or stays the same in each video. Enter your prediction in the third column of the table and answer the questions below.

a. In which videos did you predict the number of atoms in the body would increase? Where do you think the atoms came from?

I think the additional atoms in the bodybuilder and child growing videos came from the food the person ate. (Scar tissue could also have come from food the person ate.)

I'm not sure where the atoms came from for cells reproducing. Perhaps there were atoms in the surrounding medium that provided food for the cells.

b. In which videos did you predict the number of atoms in the body would decrease? Where do you think the atoms went?

Mass doesn't seem to be decreasing in any of the videos, so I didn't predict that the number of atoms would be decreasing in any of the videos.

c. In which videos did you predict the number of atoms would not change? Why do you think so?

Maybe the number of atoms in the body didn't change as scar tissue formed if the atoms used to form the scar tissue came from another part of the body. The video doesn't provide evidence for this.

d. How could you test your predictions?

We could measure the mass of each person or the cells before and after. Because atoms have mass, increases or decreases in atoms would be reflected in the mass.

Teacher Talk and Actions

Question for advanced students:

If you wanted to make a video in which the number of atoms in the body would decrease, what could you video and why?

I could video an athlete running a race because the athlete exhales CO_2 and breathes in O_2, and each CO_2 has a C atom that makes it heavier than O_2.

This question provides an opportunity for students to think about weight loss and for you to learn whether they think that gases have mass. Students are not expected to know the answer, which they will learn about in Chapter 2.

Pulling It Together

Discuss these questions as a class. Write any notes you have in the space below.

1. Based on Activity 1, what are some of the changes in matter that take place in our bodies that keep us alive and functioning properly?

Wound healing, building muscles, and repairing/replacing diseased cells/tissues

2. What do we still need to know to help us answer Question 1?

We don't know how our bodies make new skin or muscle.

Teacher Talk and Actions

Closure and link to subsequent lessons:

In this lesson, we observed examples of people growing and repairing their bodies and cells dividing. We talked about some changes in matter that we observed, but some of the videos didn't provide evidence that the amount of mass was changing. Because the human body is so complex, it can be challenging to determine how matter changes and where the matter we use to build and repair our bodies comes from. We will start by looking at the matter that makes up our bodies, how it is organized, and how it compares to the food we eat. Doing so will give us ideas about changes that must occur for our bodies to convert molecules making up our food to molecules making up our body structures.

Lesson Guide
Chapter 1, Lesson 1.2
Comparing Our Bodies to Our Food

Focus: This lesson begins the story of how matter changes in living organisms, extending what students learned about body systems in middle school by zooming in to the atoms and molecules that make up body structures, comparing those molecules to the molecules making up their food, and concluding with the question of how molecules from food could be turned into molecules making up their body structures.

Key Question: What components of human body systems are involved in growth and repair? How do molecules of body systems compare to molecules of our food?

Target Idea(s) Addressed

Multicellular organisms, such as humans, have a hierarchical structural organization, in which any one system is made up of numerous components and can itself be considered a component of the next level. (Science Idea #1)

Proteins carry out a variety of body functions, including making up body structures and speeding up chemical reactions (which would otherwise occur too slowly to maintain life). (Science Idea #2)

The polymers that make up animal body structures are made up of the same sets of monomers as the food an animal eats but differ in the specific monomer composition and arrangement. Hence, they are different molecules. (Science Idea #3)

Science Practice(s) Addressed

Analyzing and interpreting data
Developing and using models

Materials

Activity 1: *Per group of students:* Card pack for integumentary system; card pack for a single body system (circulatory, respiratory, muscular, and digestive); Amino Acid Key; optional: 11 × 17 paper for organizing cards into a hierarchy map; glue sticks

Activity 2: *Per group of students:* Food cards; protein cards

Advance Preparation

Organize materials in advance so that each group has materials when class begins.

Printable copies of human body system cards and the Amino Acid Key for students (along with Teacher Answer Keys) are available on the NSTA Extras page (*www.nsta.org/growthandactivity*). Each group will need a set of integumentary system cards and a set of cards for another body system.

Note: The DNA polymer and nucleotide monomer cards are not intended for use in this lesson (or in the unit). You could note in Activity 2, when students are examining the amino acid sequences of example proteins, that they will learn in another unit how proteins are made to have their correct sequence. When teaching that unit, you can give students the DNA and nucleic acids cards and have them add them to hierarchy maps of all body systems.

Phenomena, Data, or Models	Intended Observations	Purpose	Rationale or Notes
Activity 1 Data cards on components of five human body systems: integumentary, digestive, muscular, circulatory, and respiratory (each group will examine at least two)	• The human organism is made up of different organ systems, including, integumentary, digestive, circulatory, muscular, respiratory. • Human body systems are made up of various organs. • Organs making up all body systems are themselves made up of cells that vary in shape and appearance. • Cells making up organs are themselves made up of polymers, including protein, fat/lipid, and carbohydrate polymers. These polymers will need to be produced when living things grow and repair body structures. • Human body polymers are made of monomers: proteins, of amino acid monomers; glycogen, of glucose monomers; and fats/lipids, of fatty acids and a "backbone" monomer. • Glucose and fatty acid monomers are made up of C, H, and O atoms. Amino acid monomers are made up of C, H, O, N, and some S atoms. • Amino acid monomers have parts in common (NH_2, COOH, and H that are all bonded to a central carbon) but differ in the fourth group bonded to the central carbon.	Situate students' prior knowledge of human body systems in a hierarchy that zooms down to the molecular level. Apply ideas about systems and system components to body systems. Focus students on polymers that make up body structures. Familiarize students with the basic structure of amino acids, which they will use to make sense of protein sequence data in the next activity.	Targets the misconceptions that (a) cells are "in" the body rather than making up the body, (b) all cells are pretty much the same shape, and (c) atoms/molecules are in the body rather than making up the body

Phenomena, Data, or Models	Intended Observations	Purpose	Rationale or Notes
Activity 2 Food cards providing nutritional data of common foods	• Common breakfast foods contain protein, fat, and carbohydrate polymers, though different foods contain different amounts of these polymers.	Provide evidence that foods are made up of the same types of polymers that make up their body structures.	Targets the misconception that all proteins are the same
Data cards with the name, source, total number of amino acids, and partial amino acid sequence of 10 proteins—6 proteins from the human body and 4 proteins from the bodies of other organisms (both plants and animals that serve as food for humans)	• Amino acids that are linked together in proteins have fewer H and OH atoms than do amino acids that are not linked together in proteins. • Human body proteins vary in both the total number of amino acids and in their amino acid sequence (e.g., the first five amino acids of collagen are mhpgl, whereas the first five amino acids of myosin are msass).	Familiarize students with a few proteins to provide evidence that proteins differ from one another in their structure and function.	

Lesson 1.2—Comparing Our Bodies to Our Food

What do we know and what are we trying to find out?

In the previous activity, you observed some changes in the bodies of both elite athletes and ordinary people—as well as changes in a single cell—as they grew and repaired themselves. But what exactly is happening as a small baby grows into a much larger adult or as a cell divides? And where and how do these changes take place?

As you probably know, the human body is composed of multiple organs and other body structures (for example, tissues and cells) that form distinct systems that carry out specific functions (for example, digestion and movement). In this lesson we will consider some of these human body systems and their components to see what they are made of and how they are involved in the changes in matter that take place as our bodies grow and repair themselves. Then we will compare molecules that enter our bodies as food to molecules that make up all our body structures.

Answer the Key Questions to the best of your knowledge. Be prepared to share your ideas with the class.

> **Key Questions: What components of human body systems are involved in growth and repair? How do molecules of body systems compare to molecules of our food?**

Teacher Talk and Actions

Activity 1: Exploring Human Body Systems

Materials

For each team of students
Card pack for integumentary system
Card pack for another human body system (muscular, circulatory, respiratory, digestive)
Amino Acid Key
Glue sticks
Optional: 11 × 17 paper for organizing cards into hierarchy map

In this activity, we will examine various human body systems to get a better sense of what they include, how they are organized, and what they are made of. We will start by looking first at macroscopic components of each system that can be seen easily (for example, parts of the respiratory system such as the mouth and nose). We will then zoom in on invisible components of each system (for example, protein molecules and atoms). Doing so can help us figure out what the various components of these body systems are made of and what our bodies must produce in order to grow and repair themselves after injuries.

Procedures and Questions

1. Examine the cards for the integumentary system. With your class, discuss the kinds of information that are provided on each card and answer the following questions.

a. Why do you think all the components shown on the cards are part of the integumentary system?

Perhaps they all help to protect the body in some way.

b. Can you think of any components of the system that are not included in the set of cards? If so, why do you think they are not included?

The cards show only a few proteins; there are probably others.

c. What is the largest component of the integumentary system shown on a card?

Skin is the organ that covers nearly all the body.

d. What is the smallest component shown?

Atoms (mainly carbon, hydrogen, oxygen, and nitrogen atoms)

e. What kinds of information are provided on the cards?

Each card contains information about the function of that component.

Teacher Talk and Actions

By the end of this activity, students should recognize that (a) protein polymers are components of all body systems, though different body systems have proteins with different names and functions, (b) lipids are components of cell membranes of cells of all body systems, and (c) some body systems also have or are made up of significant amounts of fat or carbohydrate polymers.

2. Work with your teacher to organize cards of the integumentary system into a hierarchy that zooms from the largest scale (macroscopic) to the smallest scale (atoms and molecules). Use the information on the cards to help construct the hierarchy. Draw the hierarchy below.

Hierarchy map of the human integumentary body system

Please find the answer keys to the body system hierarchies on the Extras page at www.nsta.org/growthandactivity.

Teacher Talk and Actions

You will need to save or photograph hierarchy maps of each body system. Students will need to use the hierarchy maps in Lessons 1.5 and 2.9.

3. Examine the cards in the body system card pack your teacher gave you.

a. Does your set of body system cards include the same categories of information as the set for the integumentary system?

Yes

b. List all of the different categories of information that are provided on the cards from largest to smallest.

Body system > tissues/organs > cells > polymers > monomers > atoms

4. Now organize your set of body system cards into a hierarchy, putting the largest category at the top. If another group in the class has the same body system, compare your hierarchy map to theirs, discuss any differences, and create a revised map that everyone agrees on. Draw or insert a photo of your revised hierarchy map below and label the body system that it represents.

Hierarchy map of _______________ body system

Teacher Talk and Actions

5. Examine the hierarchy maps of body systems constructed by other groups. List the categories of components that make up all of the systems from largest to smallest. Highlight the components that can be seen without magnification, underline the components that can be seen with a microscope, and circle the components that cannot be seen even with a microscope.

Body system > Tissues/Organs > Cells > Polymers > Monomers > Atoms

6. Now let's zoom in on one important component of all of the body systems: polymers. For each body system in the table below, list examples of types of polymers shown on the hierarchy maps. To help you get started, the table already includes the polymers for the integumentary body system.

Body System	Polymers That Make Up Structures and Carry Out Functions of the Body System		
	Protein Polymer	**Fat/Lipid Polymer**	**Carbohydrate Polymer**
Integumentary	*Keratin, collagen*	*Membrane phospholipid, triglyceride*	*No examples*
Muscular	*Actin, myosin, glycogen synthase enzyme*	*Membrane phospholipid*	*Glycogen*
Respiratory	*Collagen*	*Membrane phospholipid*	*No examples*
Circulatory	*Hemoglobin*	*Membrane phospholipid*	*No examples*
Digestive	*Amylase enzyme*	*Membrane phospholipid*	*Glycogen*

7. Based on the information you listed in the table above and in the hierarchy maps, which type of polymer appears to carry out the greatest variety of functions? Justify your answer, giving specific examples of functions that are listed on polymer cards.

Protein polymers make up body structures (keratin, collagen, actin, myosin), carry oxygen from lungs to body cells (hemoglobin), and speed up chemical reactions (glycogen synthase enzyme, amylase enzyme).

Teacher Talk and Actions

8. Examine the hierarchy maps again to find an example of a monomer that makes up each kind of polymer and write it in the table below.

Polymer	Monomers that make up body polymers
Protein polymers	Amino acids
Fat/lipid polymers	Fatty acids, glycerol
Carbohydrate polymers	Glucose

9. We know that protein polymers have many different functions, so there must be lots of different kinds of proteins. In fact, scientists estimate that there are about 50,000 different proteins in the human body! But the most surprising thing is that all of these proteins are mostly made up of only 20 different amino acid monomers. The amino acid monomer cards on your hierarchy maps show models of four of the monomers. The Amino Acid Key provides the name, letter code, and a model of the types and arrangement of the atoms that make up each of the 20 amino acids.

Examine the Amino Acid Key and list the atoms that are present in all 20 amino acids:

C, H, O, N

10. Examine the models of several different amino acids until you notice what they all have in common. Draw two amino acids and circle the atoms they have in common.

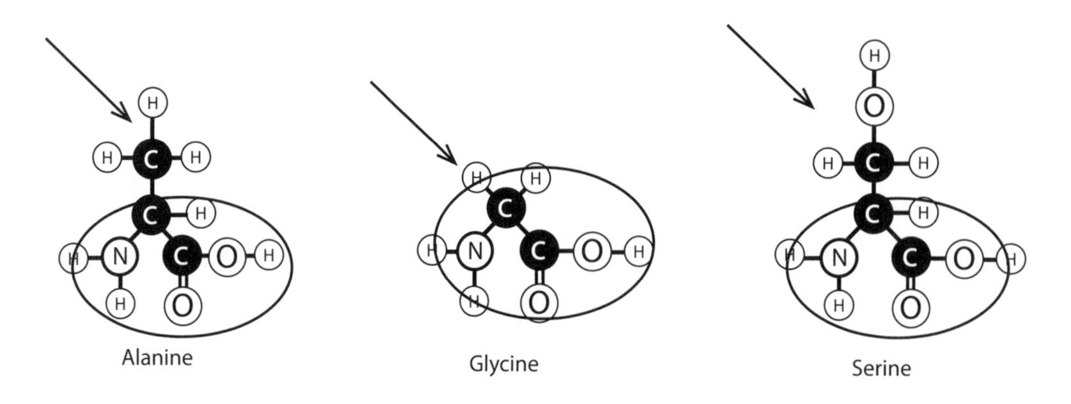

11. What part of an amino acid molecule differs from one amino acid to the next? Draw arrows to the part(s) that are different.

Teacher Talk and Actions

There may be occasions where the number of each type of bond may look like it's the same. This is only because we are treating every type of bond (such as C-H bonds) as identical, whereas in the real world not every C-H bond is the same. Bonds will behave differently in different molecules because of the types of atoms connected to the C. For example, the C-H bond in CH_2O (formaldehyde) is different from the C-H bond in CH_4 (methane).

Activity 2: Comparing Body Molecules to Food Molecules

Materials

For each team of students
Food cards
Protein cards

In the previous activity, you observed that cells that make up the structures of all body systems are composed of protein, fat, and carbohydrate polymers. A lean adult male is about 20% protein, 15% fat, and 2% carbohydrate (the rest is water). So growing must involve making lots of new polymers. We now know, too, that each type of body polymer is made up of monomer units. If our bodies had a source of the monomer units, perhaps it could make the polymers it needs.

But where do you think those monomer units could come from? Can we get them from the food we eat? Let's consider a few different breakfast meals—a bowl of fruit with yogurt, a fried egg with toast, a bowl of cereal with milk, a banana—and see what types of molecules they contain.

Procedures and Questions

1. A packet of food cards has been placed at your table. The food cards contain the following information:
    - The Nutrition Facts label highlights the type and relative amount of various molecules in a serving of the food item.
    - The Primary Food Polymer(s) identifies the one or two main polymers present in the food item and provides data about their molecular composition and structure.

Teacher Talk and Actions

2. Examine the set of food cards. What types of polymers are present in various foods?

Protein, fat, and carbohydrate polymers

3. Use data on the cards to complete Table 1.2 and respond to the questions below.

Table 1.2. Molecular Composition of Some Common Foods

Food	Types of Polymers Contained in a Serving of the Food		
	Protein (g)	**Fat (g)**	**Carbohydrates (g)**
Egg white	4	0	0
Whole milk	8	8	13
Banana	1	0	27
Wheat bread	3	2	15
Butter	0	11	0

4. We now know that a healthy breakfast contains protein, fat, and carbohydrate polymers—just like our bodies. Yet our bodies can look quite different from the food we eat at breakfast. Let's start by looking more carefully at a few example proteins present in our food and compare them to example proteins that make up our bodies. Obtain a set of protein cards. Examine the protein cards and use the data on the cards to respond to the questions below.

a. What do ovalbumin (from egg white) and myosin (from human muscle) have in common? Provide evidence.

Both ovalbumin and myosin are made up of a variety of amino acid monomers linked together in a chain (called a polymer). For example, the first 10 amino acids of both contain amino acids m, a, s.

b. How do ovalbumin and myosin differ from one another? Provide evidence.

The total number of amino acids making up each protein is different and the sequence of amino acids making up the partial sequence shown (1–25 and 61–86) is different. Ovalbumin is made up of 386 amino acids and myosin is made up of 1,937 amino acids. The first 10 amino acids of ovalbumin are gsigaasmef, whereas the first 10 amino acids of myosin are msassdaema.

c. Do your answers to 4a and 4b apply to the other proteins? Explain.

Yes, all the proteins are made up of amino acids, but the amino acid sequence and total number of amino acids vary from protein to protein.

d. Do you think the proteins in food can be turned into the proteins in the body?

Possibly. If the body could convert food polymers to monomers, maybe it could use the monomers to build body polymers.

Teacher Talk and Actions

Body Protein (organ)	Organism	Total Number of Amino Acids	Partial Amino Acid Sequence (#1–25 and #61–86)
Keratin (skin)	Human	505	1 mtcgsgfggr afscisacgp rpgrc 61 rsfgyrsggv cgpsppcitt vsvne
Collagen (lungs, skin)	Human	1678	1 mhpglwlllv tlclteelaa ageks 61 gpqgftgstg lsglkgergf pgll
Hemoglobin (red blood cells)	Human	142 (α)	1 mvlspadktn vkaawgkvga hagey 61 kkvadaltna vahvddmpna lsals
Myosin (skeletal muscle)	Human	1937	1 msassdaema vfgerapylr kseke 61 kvtvktegga tltvredqvf pmnpp
Actin (skeletal muscle)	Human	377 (α)	1 mcdedettal vcdngsglvk agfag 61 qskrgiltlk ypiehgiitn wddme
Amylase (pancreas)	Human	511 (α)	1 mkfflllfti gfcwaqyspn tqqgr 61 nenvaiynpf rpwweryqpv syklc
Ovalbumin (egg white)	Hen (chicken)	386	1 gsigaasmef cfdvfkelkv hhane 61 klpgfgdsie aqcgtsvnvh sslrd
Casein (milk)	Cow	214	1 mkllilitclv avalarpkhp ikhqg 61 sestedqame dikqmeaesi sssee
Soy	Soybeans	43	1 skwqhqqdsc rkqlqgvnlt pcekh 26 lmekiqgrgdddddddd
Glutenin (flour)	Wheat	101	1 eklgqgqqpr qwlqprqgqq gyypt 61 pyhvsaehqa aslkvakaqq laaql

Source: U.S. National Library of Medicine. Protein database. Retrieved on August 12, 2019, from *www.ncbi.nlm.nih.gov/protein.*

Science Ideas

Science ideas are accepted principles or generalizations about how the world works based on a wide range of observations and data collected and confirmed by scientists. Because these science ideas are consistent with the available evidence, you are justified in applying science ideas to other relevant observations and data.

Activities 1 and 2 were intended to help you understand two important ideas about the matter that makes up bodies of living organisms. Read each idea below. Look back through Lesson 1.2. In the space provided after each science idea, give evidence from 2–3 different examples that support the idea.

Science Idea #1: Multicellular organisms, such as humans, have a hierarchical structural organization, in which any one system is made up of numerous components and can itself be considered a component of the next level.

Evidence:

Students should fill in (specific examples) like those shown below but from the body system or systems they worked on:

Humans consist of more than five body systems that keep them alive and functioning properly. According to the data cards for the body systems we examined, each body system is made up of atoms (mainly C, H, O, N) organized into monomers (e.g., amino acids) that are linked together in polymers (e.g., proteins), that are organized into cells (e.g., epidermal cells, muscle cells), that are organized into tissues/organs (e.g., skin, muscles), that are organized into body systems (e.g., integumentary, muscular) that work together to keep the organism alive and functioning (e.g., protection, movement).

Science Idea #2: Proteins carry out a variety of body functions, including making up body structures and speeding up chemical reactions (which would otherwise occur too slowly to maintain life).

Evidence:

The cards showed us that cells of a variety of body systems have proteins that carry out specific functions (e.g., actin and myosin in the muscular system help muscle cells contract, hemoglobin in red blood cells of the circulatory system carries oxygen from the lungs to body cells, collagen in the lungs provides support for air sacs, and amylase in the small intestine helps the body digest starch). Cells of all body systems have enzymes, which help speed up chemical reactions. Cells of all body systems have protiens that act as enzymes to speed up chemical reactions.

Teacher Talk and Actions

Encourage students to provide evidence for at least two body systems.

Science Idea #3: The polymers that make up animal body structures are made up of the same sets of monomers as the food an animal eats but differ in the specific monomer composition and arrangement. Hence, they are different molecules.

Evidence:

Proteins making up human body structures (e.g., actin, myosin, hemoglobin, keratin, collagen) and proteins making up human food (e.g., albumin, gluten, soy protein, casein) are all made up of amino acid monomers. However, the proteins differ in both the total number and sequence of their amino acid components.

Teacher Talk and Actions

Pulling It Together

Work on your own to answer these questions. Be prepared for a class discussion.

1. Based on Activities 1 and 2, what have we learned that will help us begin to answer the Key Questions and what do we still need to find out? Key Questions: **What components of human body systems are involved in growth and repair? How do molecules of body systems compare to molecules of our food?**

What we know:

- *What matter makes up our bodies: Our bodies are made up of matter, specifically proteins, fats, and some carbohydrate polymers (and a lot of water molecules). For our bodies to build muscles and repair organs/cells they need to build proteins, fats, and carbohydrates.*
- *What matter makes up our food: Our food is also made up of proteins, fats, and carbohydrate polymers (and a lot of water molecules).*
- *Similarities among matter making up our bodies and our food: Proteins making up our bodies and proteins making up our food are made up of the same 20 types of amino acid monomers.*

What we still need to find out:

- *How do our bodies convert polymers from food into polymers needed for building and repairing body structures?*

2. Do you think other multicellular organisms are made up of similar polymers? Support your claim with evidence from Activities 1 and 2 and what you have observed in your daily life.

- *Muscular System cards: The meat that we eat comes from muscles of cows, chickens, turkeys, lambs, fish, and seafood such as crabs, shrimp, clams.*
- *Integumentary System cards: The turkey we eat at Thanksgiving has skin.*
- *Circulatory System: If you buy a whole chicken or turkey it comes with a heart, and sometimes you can find blood vessels and blood in body parts of cows, chickens, turkey, sheep.*
- *Digestive system: Grocery stores sell livers of chickens and cows.*
- *Food cards show that milk (from a cow) and eggs (from a chicken) have protein, and protein cards show that soybeans and wheat (both plants) have protein.*
- *Packages of chicken parts and steak sold at the grocery store usually have some fat.*
- *Food cards show that wheat bread (mostly flour from wheat plants) has a lot of carbohydrate.*

Teacher Talk and Actions

Lesson Guide
Chapter 1, Lesson 1.3
Matter Changes During Chemical Reactions

Focus: This lesson engages students in observing several chemical reactions involving small molecules reacting in simple systems where the properties of reactants and products can be observed directly and the conversion of reactants to products can be easily modeled with LEGOs, ball-and-stick models, and chemical formulas. Depending on students' prior knowledge, this lesson can serve as either a review of or as an introduction to ideas about chemical reactions.

Key Question: How can some molecules be converted to other molecules?

Target Idea(s) Addressed

During chemical reactions, atoms that make up molecules of the starting substances (called reactants) disconnect from one another and connect to other atoms to form the molecules of the ending substances (called products). Because the arrangement of atoms in the products is different from the arrangement of atoms in the reactants, the products of a chemical reaction have different properties from the reactants. (Science Idea #4)

Models are useful for representing systems and their interactions—such as inputs, outputs, and processes—and energy and matter flows within and between systems. (Science Idea #5)

Science Practice(s) Addressed

Analyzing and interpreting data
Developing and using models

Materials

Activity 1: *Per group of students:* Tea light candles; 1 liter glass beaker; LEGO kits; ball-and-stick models; cobalt chloride paper; videos: *Are New Substances Produced When a Candle Burns? (Test 1 Mass Measurements; Test 2 Cobalt Chloride Test; Test 3 Limewater Test)*

Activity 2: *TEACHER DEMO:* Ammonium thiocyanate; barium hydroxide; beaker; pH paper; *per group of students:* LEGOs (see p. 67)

Advance Preparation

1. Download videos and test them in advance. Videos can be used as a demonstration when it is not feasible for students to carry out their own tests during class time. Alternatively, the videos can be shown after students have done their own tests to be sure they make the intended observations.

2. Organize materials in advance so that each team has materials when class begins.

3. The cobalt chloride paper strips should start out blue. If they are not blue, microwave the cobalt chloride paper for 3–4 minutes on low heat to remove all the water (currently in the strips from the atmosphere).

4. Build the models of candle wax and oxygen (see p. 61) ahead of time to save class time for manipulation and instruction.

Phenomena, Data, or Models	Intended Observations	Purpose	Rationale or Notes
Activity 1 Candle wax + oxygen reaction Modeling atom rearrangement and conservation with LEGO bricks or ball-and-stick models	• When a candle burns, the mass of the candle wax decreases. •When a beaker is placed over a burning candle, the candle stops burning and two new substances can be detected—one that turns cobalt chloride paper from blue to pink and another that turns limewater cloudy. •Models of candle wax ($C_{11}H_{24}$), water (H_2O), and carbon dioxide (CO_2), and the knowledge that air is about 20% O_2, suggest that O_2 is another reactant. • LEGO bricks, ball-and-stick models, and molecular formulas could help us figure out why the candle stopped burning after it was covered by a beaker. • A candle continues to burn in air but stops burning when covered by a beaker. • Although the number of C, H, and O atoms is unchanged during the candle-burning chemical reaction, the atoms are bonded differently. • We were able to figure all this out because we carried out the chemical reaction in a closed container, where no molecules/atoms could enter or leave.	Introduce students to a combustion reaction where the production of new substances can be easily detected. Introduce students to tests for H_2O and CO_2, which they will need for subsequent lessons. Give students an opportunity to model atom rearrangement and conservation (with LEGOs, ball-and- stick models, molecular formulas, and space filling models) during chemical reactions in simple systems. Give students an opportunity to see the value of modeling to help them make sense of phenomena. Get students to start thinking that it would have been difficult to figure out what was going on if they hadn't kept track of all the starting and ending substances in the candle burning because the open system allowed gaseous products to escape.	

Phenomena, Data, or Models	Intended Observations	Purpose	Rationale or Notes
Activity 2 Barium hydroxide + ammonium thiocyanate reaction	• When two white solids are mixed, two new substances are detected—one that is a liquid that turns cobalt chloride paper from blue to pink and another that is a gas that turns pH paper blue/purple. • LEGO bricks and molecular formulas help us make sense of our observations: before the chemical reaction, molecules of $Ba(OH)_2$ and molecules of NH_4SCN are present; after the chemical reaction, molecules of H_2O and NH_3 and probably molecules of $Ba(SCN)_2$ are present. • Although the number of N, H, S, C, O, and Ba atoms is unchanged during the chemical reaction, the atoms are bonded differently. • We are able to figure all this out because we carried out the chemical reaction in a way that we could observe the production of new substances and carry out tests to provide evidence of their identities. (While a gas did escape, its odor provided evidence of its presence outside the container.)	Introduce students to a combustion reaction where the production of new substances can be easily detected. Provide an opportunity for students to use the test for H_2O and learn a new test (pH), which they will need for subsequent lessons. Give students an opportunity to model atom rearrangement and conservation (with LEGOs, ball-and-stick models, molecular formulas, and space-filling models) during a chemical reaction in a simple system. Give students an opportunity to see the value of modeling to help them make sense of phenomena. Get students to start thinking that it would have been difficult to figure out what was going on (a) if they hadn't kept track of all the starting and ending substances and (b) if the gas produced in the barium hydroxide and sodium thiocyanate reaction was odorless because the open system allowed gaseous products to escape.	

Phenomena, Data, or Models	Intended Observations	Purpose	Rationale or Notes
Pulling It Together Q2: Magnesium + oxygen reaction	• When oxygen molecules react with magnesium atoms, light is given off and the O and Mg atoms rearrange to form new molecules that contain only Mg and O atoms. • If two Mg atoms react with one O_2 molecule, the ending substances must be made up of two Mg atoms and two O atoms because the number of each type of atom remains the same during a chemical reaction.	Give the students an opportunity to use the science ideas to reason about the correct chemical reaction equation for a particular reaction.	
Pulling It Together Q3: Foreshadowing chemical reactions in which monomers from food are oxidized to produce carbon dioxide and water	Since reactants have C-C bonds but CO_2 does not, atom rearrangement must have occurred.	Give students an opportunity to try to use the idea that during chemical reactions, atoms of reactants rearrange to form products as they think about a biological system.	
Pulling It Together Q4: ATP hydrolysis reaction	When ATP reacts with H_2O, two bonds are broken and two new bonds form to produce ADP + Pi.	Familiarize students with the ATP hydrolysis reaction, which will be shown to be essential for transferring energy from the glucose + oxygen reaction to move animal muscles and to synthesize starch in plants (in L 2.7). Help students "see through" the details of molecular structure to focus on which atoms are rearranged during the reaction.	

Lesson 1.3—Matter Changes During Chemical Reactions

What do we know and what are we trying to find out?

In Lesson 1.2, we observed that the molecules that make up our body structures are different from the molecules that make up our food. Can the molecules from our food become the molecules that make up our body? Because both food and body molecules are large and exist in mixtures of several substances, it isn't easy to study whether any particular food molecule can be turned into a particular molecule that makes up our body structures.

To better understand how molecules can be changed into other molecules, it is helpful to start with pure substances—each of which is made up of a single type of molecule—and observe what happens when they interact. In this lesson, we will look at pure substances interacting in a container that is simpler than the body. We will then use models to get ideas about what happens to the molecules that make up those substances. Once we understand how substances interact to form other substances in these simpler systems, we will apply our understanding to interactions of substances in the body.

Answer the Key Question to the best of your knowledge. Be prepared to share your ideas with the class.

> **Key Question: How can some molecules be converted to other molecules?**

Teacher Talk and Actions

Activity 1: What Happens When a Candle Burns?

Materials

For each team of students
Candle
Glass beaker
LEGOs or ball-and-stick models
Cobalt chloride paper

In this activity, we will examine what happens when a candle burns to see if we can detect any changes in matter. We will try to determine the identities of the substances we start with and whether any new substances are produced. If we can show that when a candle burns, the substances we start with are converted to different ending substances, then we know that a **chemical reaction** occurred.

Procedures and Questions

Identifying the starting and ending substances

1. Examine the candle wax and describe some of its characteristic properties (i.e., solubility, odor, color, and whether it is a solid, liquid, or gas at room temperature).

The candle wax is a white, odorless solid at room temperature that is not soluble in water.

2. Your teacher will light your candle. Observe the candle burning and describe what you see.

I see a flame on the wick and liquid surrounding the wick.

3. What do you think the candle will look like if we let it burn for about 10 minutes? Explain.

The candle would get smaller because the candle wax is burning.

4. Your teacher measured the mass of the candle before burning it. If burning involves a chemical reaction and the wax is a starting substance, then what would you expect to happen to the mass of the candle as it burns? Watch either the demonstration or video. Record the masses of the candle in the table below and compare the results to your prediction above.

Mass of candle before burning	*40.20 g*
Mass of candle after burning for about 10 minutes	*38.34 g*

Do your observations and the mass data provide evidence that candle wax is a starting substance? Explain.

Yes, because a starting substance will be used during a chemical reaction, and its mass will decrease.

Teacher Talk and Actions

Safety Note
Students should wear protective eyewear and gloves during all investigations.

Optional: Show the video *Are New Substances Produced When a Candle Burns?—Test 1 Mass Measurements*.

5. If a chemical reaction occurred, then one or more new substances would be produced. To see if that happened, we'll need to carry out the reaction in a closed container so that any substances produced remain in the container rather than floating away in the air. While the candle is still burning, cover the candle with the beaker. Record your observations below.

The candle stops burning and smoke comes from the wick.

If we look carefully, we can observe that small droplets of a clear colorless liquid form on the inside of the beaker.

6. Based on your observations from Step 5, propose a molecule that could be a new substance that was produced and explain why you think so.

The substance produced could be water because water is a clear colorless liquid. The molecular formula for water is H_2O.

7. We can use the cobalt chloride test to learn more about the identity of the new substance that formed. Test the inside of the beaker with cobalt chloride test paper and describe your observations and what they indicate about an ending substance of the chemical reaction.

The cobalt chloride test paper turns from blue to pink (positive test), indicating that water (H_2O) is an ending substance (product) of the chemical reaction.

8. Do you think this ending substance is the only new substance produced during the reaction? We can get a clue from looking at a molecular model of the chemical formula for candle wax ($C_{11}H_{24}$), which tells us that every molecule of candle wax is made up of 11 carbon (C) atoms and 24 hydrogen (H) atoms.

Because atoms are not created or destroyed during chemical reactions, there must be another ending substance that contains carbon (C) atoms. The product water contains H atoms, so that could be where the H atoms of candle wax end up. The C atoms could be part of a gas such as carbon dioxide (CO_2), carbon monoxide (CO), or methane (CH_4), all of which contain C atoms.

9. If another ending substance formed, it is logical to think that it is a gas because we did not observe the formation of a solid and we already determined the identity of the clear liquid. Observe Test 3 of the video: *Are New Substances Produced When a Candle Burns?* and describe your observations and what they indicate about an ending substance of the chemical reaction.

The limewater turns cloudy (positive test), indicating that carbon dioxide (CO_2) is an ending substance (product) of the chemical reaction.

10. Now that we have identified two new substances that were formed, let's think about whether other substances might be involved in the candle-burning reaction. Use your models and what you know about the chemical formulas of the substances to write a chemical equation that includes what you now know about the reactants and products.

$C_{11}H_{24} \rightarrow CO_2 + H_2O$

Teacher Talk and Actions

Make sure the students notice the clear liquid droplets that form on the inside surface of the beaker.

The cobalt chloride paper is very sensitive to water in the air. Be sure to dry the paper before the class so that it starts out blue. You can remove most of the moisture from the paper by microwaving it for a few minutes. After removing the moisture, the paper should be used immediately as it will begin reacting with the moisture in the air.

Optional: Show the video *Are New Substances Produced When a Candle Burns?—Test 2 Cobalt Chloride Test.*

Show the video *Are New Substances Produced When a Candle Burns?—Test 3 Limewater Test.*

11. What type of atom(s) must be present in the other reactant? Explain.

The other reactant must contain O atoms, because the products H_2O and CO_2 contain O atoms and the candle wax does not. Since atoms aren't created or destroyed during a chemical reaction, all the atoms that make up molecules of products must have come from the molecules of reactants.

12. Use models to determine how many O_2 molecules are needed to react with each molecule of candle wax and represent the complete chemical equation for the reaction below. You can use LEGOs or ball-and-stick models to help you figure it out.

$C_{11}H_{24} + 17\ O_2 \rightarrow 11\ CO_2 + 12\ H_2O$

13. We can use the table below to check to be sure that all the atoms are accounted for in the chemical equation. Complete the table for the equation you have written above.

Atom	# Before reaction	# After reaction
Hydrogen	24	24
Carbon	11	11
Oxygen	34	34

14. Now that you have written the complete equation for the reaction, let's try to use the information to model why the flame went out when we covered the burning candle with the beaker. Let's start by building models to represent the number of candle wax molecules present before we lit the candle. How many molecules of $C_{11}H_{24}$ and how many molecules of O_2 should we build?

Based on the complete equation, we know that 1 molecule of $C_{11}H_{24}$ reacts with 17 molecules of O_2. We need to build enough $C_{11}H_{24}$ molecules so that some can react with O_2 to produce CO_2 and H_2O but some $C_{11}H_{24}$ molecules will remain when the reaction stops. (If students build 17 molecules of O_2, they need to build at least 2 molecules of $C_{11}H_{24}$.)

15. We can create a table like the one in Step 10 to check to be sure that all the atoms are accounted for during the chemical reaction. However, for this situation, we need to keep in mind that some of the reactant molecules are left over after the reaction has stopped.

Atom	# Before reaction	# After reaction			
		In products		In leftover reactants	
		H_2O	CO_2	$C_{11}H_{24}$	O_2
Hydrogen	48	24	0	24	0
Carbon	22	0	11	11	0
Oxygen	34	12	22	0	0

Teacher Talk and Actions

Below is an example of a LEGO model of a candle wax molecule (black bricks represent carbon atoms, and white bricks represent hydrogen atoms).

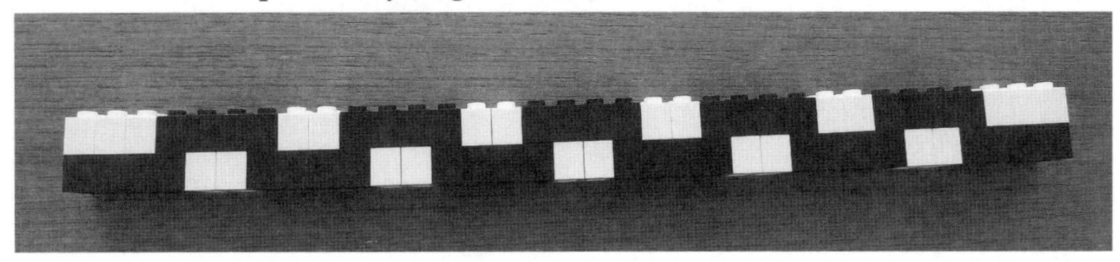

Provide students with 34 red LEGOs to make 17 O_2 molecules. They can then "react" their LEGO models to produce LEGO models of 11 CO_2 and 12 H_2O molecules.

If students use ball-and-stick models of candle wax, they will need more curved sticks to make 11 CO_2 (22 double bonds) molecules, but there are only curved sticks for the 17 double bonds of oxygen. Let students know that they can trade 1 curved stick for 1 straight stick in their kits.

Discuss the meaning of the chemical equation by asking students the following questions:

How many O_2 molecules are needed to react with 1 molecule of $C_{11}H_{24}$?

17 molecules of O_2

How many O_2 molecules would be needed to react with 2 molecules of $C_{11}H_{24}$?

34 molecules of O_2

How many O_2 molecules would be needed to react with 20 molecules of $C_{11}H_{24}$?

340 molecules of O_2

16. Some students think that the reason the candle stops burning is that the beaker "snuffs it out." Use what you have learned in this activity to explain to those students why the candle stops burning.

The candle stops burning because the beaker blocks oxygen gas (O_2) from entering. Since O_2 is a reactant of the chemical reaction, the candle stops burning when no oxygen is available to react.

17. What do you think would happen to the burning candle if you had not covered it with a beaker? Explain.

The candle would have continued to react with oxygen in the air until at least one of the reactants was used up.

Teacher Talk and Actions

The air in this room contains about 20% oxygen, which is available to react with the candle wax.

Summarize the activity and foreshadow the next activity:

In this activity, you have observed that two new substances—carbon dioxide and water—were produced when candle wax reacted with oxygen. We used the term **reactants** to refer to the starting substances of the chemical reaction and the term **products** to refer to the ending substances. We used LEGOs, ball-and-stick models, and molecular formulas to represent the atomic composition of the molecules of the reactants and products, and we used a reaction equation to represent how matter changed during the chemical reaction. Specifically, we saw that 12 H_2O and 11 CO_2 molecules formed during the chemical reaction because atoms making up $C_{11}H_{24}$ and 17 O_2 disconnected from each other and connected to other atoms to form the molecules of products. Apparently, the rearrangement of atoms resulted in the production of new substances during the chemical reaction.

In Activity 2, we'll see if we can use models and ideas about atom rearrangement to explain a very different chemical reaction.

Activity 2: What Happens When Ammonium Thiocyanate and Barium Hydroxide Interact?

Materials

For each team of students
LEGOs

In the last activity, we used our own observations and tests to detect new substances that were produced when a candle was burned inside a beaker. In this activity, we will examine the interaction between ammonium thiocyanate and barium hydroxide and use some different tests to see if any new substances form, which would indicate that a chemical reaction has occurred. We will then use models to represent what happens to the atoms.

Procedures and Questions

Describing the properties of the starting substances

1. Describe the characteristic properties of ammonium thiocyanate that you can observe (i.e., solubility, odor, color, and whether it is a solid, liquid, or gas at room temperature).

Ammonium thiocyanate is a white, odorless solid at room temperature and is soluble in water.

2. Describe the characteristic properties of barium hydroxide that you can observe (i.e., solubility, odor, color, and whether it is a solid, liquid, or gas at room temperature).

Barium hydroxide is a white, odorless solid at room temperature and is soluble in water.

Observing the interaction

3. Watch as your teacher mixes ammonium thiocyanate and barium hydroxide in a beaker. Record your observations below. Focus on observations that provide evidence that new substances are being formed.

As the solids interact, they start to look wet, indicating that a liquid is forming. There is also an odor that was not there before the reaction, suggesting that a gas is forming.

4. Your teacher will use pH paper to test how acidic or basic the gas is. Record the results of the test below.

The gas turns the pH paper blue—indicating that it is basic.

5. Scientists have isolated the liquid from other substances in the mixture and tested it with cobalt chloride paper. They found that the liquid turned the paper pink. What can you conclude from these results?

We can conclude that the liquid formed is water.

Teacher Talk and Actions

The teacher should handle all chemicals in Activity 2. Demonstrate the properties of the substances to students by showing them the color and having them smell the aroma, but do not let students directly handle the products or reactants.

Demonstrate the solubility of each starting substance by adding about 1 g to a liter of water.

The properties of barium hydroxide and ammonium thiocyanate that students can readily observe are the same.

If a student asks if this means that they are the same substance, here is a list of some other characteristic properties of these substances that proves that they are not the same substance:

Property	Barium hydroxide	Ammonium thiocyanate
Density	3.743 g/ml	1.205 g/ml
Melting point	300 °C	150 °C
Boiling point	780 °C	170 °C

Preparation:

- Pour barium hydroxide into a 400 ml beaker until it reaches the 50 ml mark.
- Pour ammonium thiocyanate into a second 400 ml beaker until it reaches the 50 ml mark.

Demonstation:

- Allow students to make observations about the starting substances.
- Pour the barium hydroxide into the beaker of ammonium thiocyanate.
- Stir with a glass stirring rod.
- Allow students to make observations. Some students may recognize that the odor is like that of ammonia. However, since ammonia is used less often as a household cleaning product, many students may be unfamiliar with its odor.
- Remind students to use the wafting technique to smell the contents of the beaker.
- Conduct a pH test by holding pH paper at the mouth of the beaker. If the pH paper turns blue, the substance is basic. If it turns red, the substance is acidic. If it turns green, the substance is neutral.

6. Did a chemical reaction occur when ammonium thiocyanate and barium hydroxide were mixed? What evidence do you have to support your answer?

Yes, we have evidence that a clear liquid and a smelly basic gas are formed. (The odor of the gas smells like ammonia, which is basic.)

Representing the chemical reaction

7. We collected evidence that at least two new substances were formed when ammonium thiocyanate interacted with barium hydroxide. Let's use models to determine if these are the only substances that form. The molecular formulas of the starting substances are listed below. Use them to build LEGO models that can help you make sense of the chemical reaction. Scientist have determined that the ratio of starting substances for this reaction is one $Ba(OH)_2$ for every two NH_4SCN. So, start by building one $Ba(OH)_2$ model and two NH_4SCN models.

Ammonium thiocyanate: 2 NH_4SCN
Barium hydroxide: $Ba(OH)_2$

8. Count the number of each type of atom and complete the "Number before reaction" column in the table in Step 13.

9. First build models of the two products we know about. How many models of product molecules were you able to build?

Two water and two ammonia molecules

10. Are there any atoms of the reactant molecules that were not used to build the product molecules? If so, what types of atoms remain?

We have one barium, two sulfur, two carbon, and two nitrogen atoms left over.

11. What does this suggest about the products that formed? Try to write the chemical formula for any other products.

Perhaps there is another product made up of the remaining atoms (one barium, two sulfur, two carbon, and two nitrogen atoms). The chemical formula might be $Ba(SCN)_2$.

12. Write a reaction equation using the chemical formulas to represent what happened during this chemical reaction.

$Ba(OH)_2 + 2\ NH_4SCN \rightarrow Ba(SCN)_2 + 2\ H_2O + 2\ NH_3$

Teacher Talk and Actions

Below is an example of a LEGO model of ammonium thiocyanate and barium hydroxide.

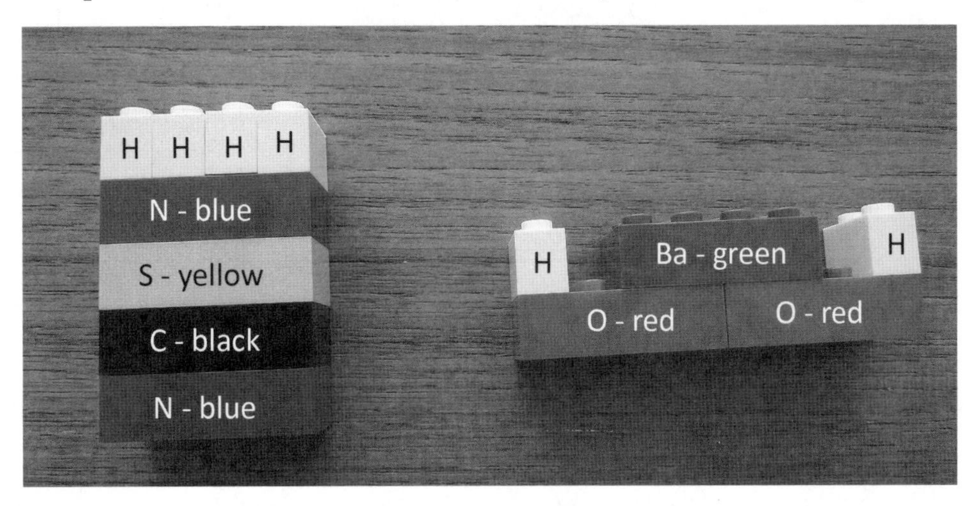

Once the class agrees that there is a third ending substance made up of 1 Ba, 2 S, 2 C, and 2 N atoms, tell them that scientists have used other tests to discover that the third ending substance is barium thiocyanate, which has a molecular formula of $Ba(SCN)_2$.

13. Count the number of each type of atom and complete the “Number after reaction” column in the table below:

Atom	Number before reaction	Number after reaction
Nitrogen	4	4
Hydrogen	10	10
Sulfur	2	2
Carbon	2	2
Oxygen	2	2
Barium	1	1

14. Using the reaction equation and your models, write a description of what happened to the atoms that make up the ammonium thiocyanate and barium hydroxide during the chemical reaction. Be sure to provide specific examples.

The atoms that were bonded together in the reactant molecules disconnected and then formed new bonds to make the product molecules. For example, the barium atom was connected to the (OH) group before the reaction and was connected to the (SCN) group after the reaction.

Teacher Talk and Actions

Activity 3: Representing Matter Changes During Chemical Reactions

In this activity, we will think about how we used different kinds of models to help us make sense of our observations of the chemical reactions that we have investigated so far in Lesson 1.3. We will focus on how models can help us visualize where new substances that we encounter every day come from.

Procedures and Questions

1. Think about the chemical reactions you observed during the last two activities. What can you generalize about chemical reactions from these activities?

During a chemical reaction, atoms that make up the molecules of the starting substances disconnect (as bonds between those atoms break) and then connect to other atoms (new bonds form) to form new molecules.

2. Look back at the modeling activities you did during this lesson. What are some similarities among the activities?

Each time we converted one set of molecules into another set of molecules by rearranging the way the atoms were connected.

Each time we used something (balls, LEGO bricks, circles with letters, letters in chemical formulas) to represent the atoms that make up the reactants and products.

3. What are some differences?

The activities involved different chemical reactions, and we used different models to represent atom rearrangement (rearranging physical models like balls or LEGO bricks or writing chemical equations or using circles or letters to represent atoms). Each activity involved models of different substances.

4. How did the models differ in the way they represented the atoms and molecules?

The LEGO models didn't use anything to represent the bonds like the sticks in the ball-and-stick models. The chemical reaction equation with the molecular formulas didn't represent the way in which the atoms were bonded (didn't show the structure of the molecules). The LEGOs used rectangles to represent the atoms whereas the balls used spheres.

5. How were the models useful in representing chemical reactions?

All the models showed that atoms of reactant molecules were arranged differently from atoms of product molecules. The models showed us that atoms of reactant molecules must have rearranged to form product molecules.

Teacher Talk and Actions

6. Do you think one type of model better represented what happened during chemical reactions? Explain.

I think the ball-and-stick models were better because they more accurately represented the structure of the molecules and helped me see what atoms were connected.

I think the LEGOs were better because they didn't have any sticks. Atoms don't actually have "sticks" that hold them together.

I think the LEGOs and ball-and-stick models both did a good job representing the reactions because they both showed what was important—that atoms disconnected and then connected in new ways.

7. How are the models you used different from the real atoms they represent? List as many differences as you can.

The models of the atoms are a lot bigger than the real atoms.

The model atoms have color but an individual atom does not.

The model atoms do not accurately represent the shape of a real atom.

Real atoms are not held together by sticks like the model atoms.

8. Was anything about the models confusing? Explain.

I didn't like the ball-and-stick models because the sticks confused me. They made me wonder about whether the bonds are matter.

I didn't like the LEGOs because they don't look like typical molecular models that I have seen before.

9. Can you think of any other questions about the world around you that using a model helped you understand?

A good example might be students' having used a sphere rotating near a light source in an earth science class to help them understand why the Sun appears to travel across the sky from east to west. Or students might have used a map to decide how to get from one place to another.

Students who have used Toward High School Biology (THSB) might recall using LEGO bricks to explain why iron rusts when exposed to oxygen in the air or why bubbles form when H_2O_2 is put on a wound.

Teacher Talk and Actions

Science Ideas

The activities in this lesson were intended to help you understand important ideas about chemical reactions and models. Read the ideas below. Look back through the lesson. In the space provided after the science idea, give evidence from your work that supports the idea.

Science Idea #4: During chemical reactions, atoms that make up molecules of the starting substances (called reactants) disconnect from one another and connect to other atoms to form the molecules of the ending substances (called products). Because the arrangement of atoms in the products is different from the arrangement of atoms in the reactants, the products of a chemical reaction have different properties from the reactants.

Evidence:

Our models showed that the arrangement of atoms in candle wax and oxygen is different from the arrangement of atoms in carbon dioxide and water. For example, wax and oxygen contain C-H and O=O bonds whereas carbon dioxide and water contains O-H and C-O bonds.

Our observations and the results of the chemical tests showed that candle wax and oxygen have different properties than water and carbon dioxide. For example, candle wax is a solid at room temperature and water is a liquid. Oxygen causes a burning splint to keep burning and carbon dioxide turns limewater cloudy.

Teacher Talk and Actions

This science idea is where we introduce the terms *products* and *reactants*. Please highlight the terms to your students, and use them consistently hereafter.

Pulling It Together

Work on your own to answer these questions. Be prepared for a class discussion.

1. Now that you have had more experiences with chemical reactions, how would you answer the Key Question: **How can some molecules be converted into other molecules?**

During chemical reactions, atoms making up molecules of reactants disconnect (bonds break) and connect to other atoms (new bonds form). For example, in the candle wax + oxygen reaction, carbon-carbon, carbon-hydrogen, and oxygen-oxygen bonds of reactant molecules were broken and C-O and O-H bonds formed to make product molecules carbon dioxide and water.

2. Early camera flashes utilized a chemical reaction between oxygen (O_2) and magnesium (Mg) because the reaction produces light with qualities that are similar to daylight.

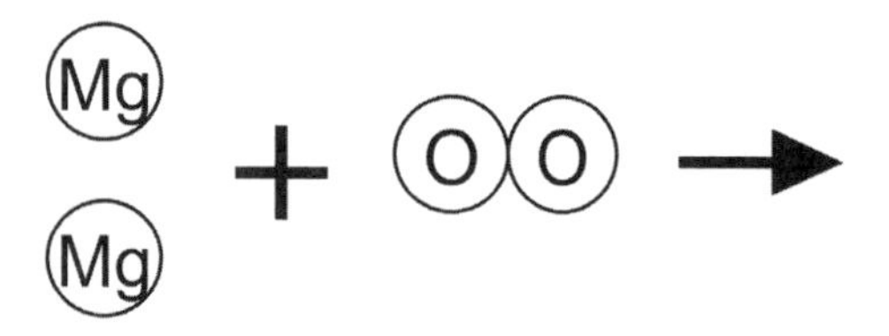

a. Which of the following models represents the products of this chemical reaction?

a. O C O + O H H

b. Mg O

c. Mg O Mg O

d. Mg O + O H H

b. Which science idea(s) did you use when selecting your answer?

I used the idea that atoms of reactant molecules rearrange to form product molecules during chemical reactions (Science Idea #4).

c. How did the idea(s) help you?

The science idea helps me to understand that the product molecules must be made up of the same atoms as the reactant molecules.

Teacher Talk and Actions

3. Now that we have investigated changes in matter that occur during chemical reactions, let's think about whether chemical reactions occur inside the human body.

You may know that we breathe in oxygen (O_2) and breathe out carbon dioxide (CO_2) and water (H_2O). And you know from Lesson 1.2 that the polymers that make up the food we eat all have C-C bonds. Does this provide evidence that a chemical reaction occurred in the body? Explain.

Yes, this does provide evidence that a chemical reaction occurred. The molecules that enter the body (oxygen and food polymers) have different properties (neither turns limewater cloudy or cobalt chloride paper pink) than the molecules that leave the body (water, which turns cobalt chloride paper pink, and carbon dioxide, which turns limewater cloudy). Because new substances formed, a chemical reaction occurred.

Looking at the models, we see that food polymers all have C-C bonds whereas neither CO_2 nor H_2O have C-C bonds. Similarly, oxygen has O=O bonds but neither CO_2 nor H_2O has O=O bonds. That means C-C bonds of food polymers and O=O bonds must break, and C-O and H-O bonds must form. Because atoms rearranged, a chemical reaction must have occurred.

4. The models below show the reaction between adenosine triphosphate (ATP) and water, which forms adenosine diphosphate (ADP) and inorganic phosphate. This reaction occurs in all living things and is essential to survival. Like all other chemical reactions, bonds break and bonds form during this reaction. On the models, place a "/" through the bonds that are broken and place a circle around the bonds that form.

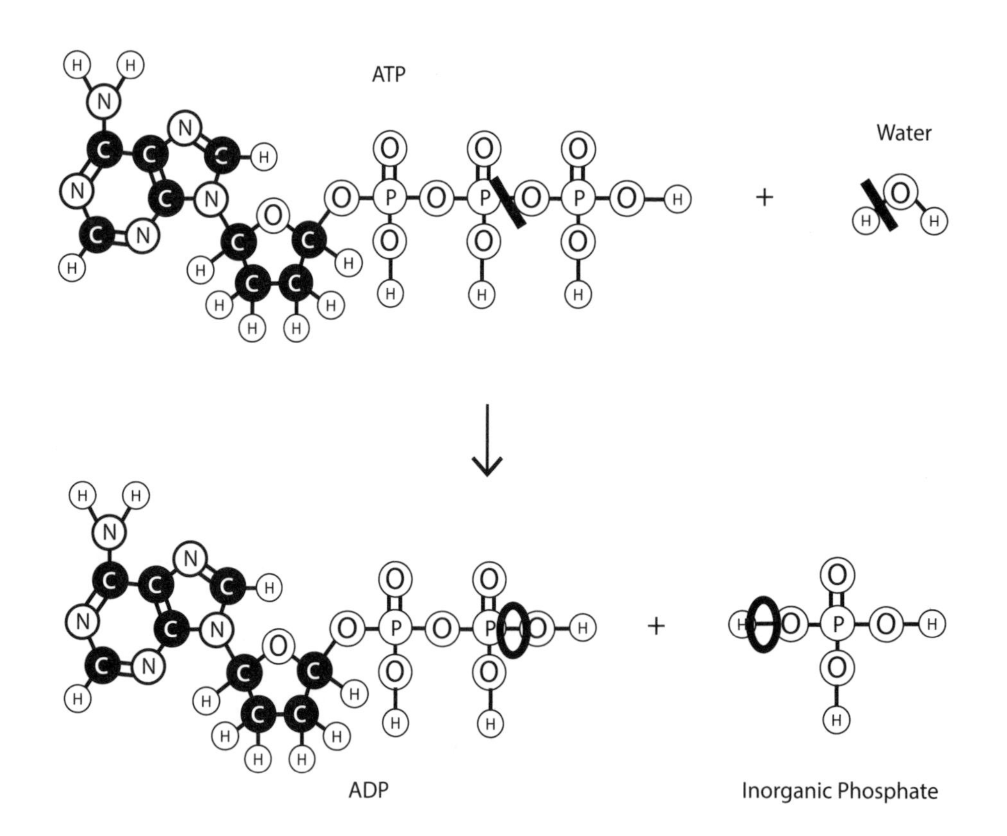

Teacher Talk and Actions

Students can color the model before answering the question. Oxygen (O) should be colored red, nitrogen (N) blue, and phosphorus (P) orange.

Lesson Guide
Chapter 1, Lesson 1.4
Chemical Reactions and Systems

Focus: This lesson introduces ideas about atom conservation and systems as useful tools for thinking about matter changes during chemical reactions.

Key Question: How can defining a system's inputs and outputs help us make sense of what is happening during a chemical reaction?

Target Idea(s) Addressed

Systems are defined by placing boundaries around collections of interrelated things to make them easier to study. Regardless of where the boundaries are placed, a system still interacts with its surrounding environment. Therefore, when studying a system, it is important to keep track of what enters (inputs) or leaves (outputs) the system. (Science Idea #5)

Atoms are neither created nor destroyed during chemical reactions, so the total number of each type of atom remains the same. If the mass of a system changes, it is because atoms have either left the system (and are now in another system) or entered the system (from another system). (Science Idea #6)

An important kind of chemical reaction involves the combination of oxygen with a fuel, such as hydrogen or glucose. When glucose molecules react with oxygen molecules, carbon dioxide and water molecules are produced. (Science Idea #7)

Plants take in carbon dioxide and water molecules from their surroundings and use them to make glucose and oxygen molecules. Glucose molecules remain inside the plant and oxygen molecules are released to the air surrounding the plant. The increase in mass that accompanies plant growth is due to more atoms entering the plant system than leaving the plant system. (Science Idea #8)

Science Practice(s) Addressed

Analyzing and interpreting data
Developing and using models
Constructing explanations

Materials

Activity 1: *Per group of students:* 9-volt battery; wires with alligator clips on one end and battery connector on the other; 2 pencil leads; petri dish; (salt) water (see advance preparation); LEGOs (see p. 87); paper plate; *Per class:* Balances; video: *Water Reaction Splint Tests*

Activity 2: *Per group of students:* Ball-and-stick models; video: *Burning Marshmallow*

Activity 3: *Per group of students:* Ball-and-stick models; *Per class:* Videos: *Plant Growing, Aquatic Plant in Light and Dark,* and *Plant Experiment With Limewater.*

Advance Preparation

Organize materials in advance so that each group has materials when class begins.

Activity 1: For a more dramatic reaction, make a salt solution in advance and use that during the class instead of water. Students do not need to know that there is salt in the water.

Phenomena, Data, or Models	Intended Observations	Purpose	Rationale or Notes
Activity 1 Water molecules react to form hydrogen and oxygen molecules Modeling atom rearrangement and conservation with LEGO bricks or ball-and-stick models	• When a battery is connected to water in a complete circuit, two new substances are produced—a gas that ignites a glowing splint and a gas that makes a burning splint pop. • The two substances that are formed leave the Petri dish system. • If we allow the chemical reaction to continue and the gases leave the Petri dish, the mass should decrease. • Using LEGOs to model the chemical reaction on a balance allows us to make sense of the decrease in mass: when some of the LEGOs representing H_2O molecules react to form gaseous products H_2 and O_2 that leave the Petri dish system and enter the surrounding air system, the Petri dish system now has fewer atoms so the mass of the Petri dish system will be less because there are fewer atoms being measured on the balance. • If we count the LEGOs representing each type of atom before the reaction and after the reaction in both the Petri dish system and the surrounding air system, we observe that (a) H and O atoms have left the Petri dish and are now in the air and (b) all the atoms are accounted for—that is, each type of atom is conserved—if we add together the atoms in the two systems.	Introduce students to the matter change in the water-splitting reaction, which will be used to illustrate an energy-requiring chemical reaction in Chapter 2. Give students an opportunity to use models to make sense of a change in measured mass during a chemical reaction. Give students an opportunity to see the value of modeling to help them make sense of phenomena. Give students an opportunity to use systems thinking to help them make sense of phenomena.	

Phenomena, Data, or Models	Intended Observations	Purpose	Rationale or Notes
Activity 2 Glucose and oxygen react to form carbon dioxide and water	• When glucose ($C_6H_{12}O_6$) reacts with oxygen (O_2), two new substances are produced—a gas that turns limewater cloudy (CO_2) and a colorless gas that condenses to a liquid that turns cobalt chloride test paper from blue to pink (H_2O). • For every glucose molecule that reacts, 12 oxygen atoms enter the marshmallow system and 6 carbon, 12 hydrogen, and 18 oxygen atoms leave. • Using molecular models and the atomic mass of the atoms involved allows us to make sense of what happens to the mass of the marshmallow system: the number of atoms in the marshmallow system, and therefore the mass of the system, decreases because the products of the reaction are gaseous substances that leave the marshmallow system. • Using molecular models and the atomic mass of the atoms involved allows us to make sense of what happens to the mass of the surrounding system: the number of atoms in the surrounding system, and therefore the mass of the system, increases because the products of the reaction are gaseous substances that leave the marshmallow system and enter the surrounding system. • We only included the molecules that participated in the reaction in our model because the molecules that don't react remain unchanged and do not help us make sense of the chemical reaction or the changes in mass.	Introduce students to the matter change in the glucose combustion reaction, which will be related to cellular respiration in Chapter 2. Give students an opportunity to use models to make sense of a change in measured mass during a chemical reaction. Give students an opportunity to see the value of modeling to help them make sense of phenomena. Give students an opportunity to use systems thinking to help them make sense of phenomena.	

Phenomena, Data, or Models	Intended Observations	Purpose	Rationale or Notes
Activity 3 Carbon dioxide and water react to form glucose and oxygen	•A plant gets bigger and develops more body structures when it grows. •When an aquatic plant is exposed to light, gas bubbles form—a gas that sustains a burning splint (O_2). •Limewater does not turn cloudy when a plant leaf is present but does turn cloudy without the leaf (indicating the absence of CO_2 around the leaf). •Labeling experiments show that carbon atoms from carbon dioxide enter the plant and then become part of glucose molecules and that oxygen atoms from water enter the plant and then become part of oxygen gas that leaves the plant. •If we model the conservation of atoms during the chemical reaction with a reaction equation and use the atomic masses of the atoms in the equation, we can predict that the mass of the plant system will increase, the mass of the surrounding system will decrease, and the mass of the chemical reaction system will stay the same.	Give students an opportunity to learn the elements of a scientific explanation and practice writing explanations using science ideas, evidence, and models.	
Pulling It Together Q3: Candle burning	The mass of a candle decreases after it has been burned for two hours.	Give students a chance to revisit the candle wax + oxygen reaction and explain why the mass of the candle decreases.	
Pulling It Together Q4: Chemical reactions in the body that convert food molecules to body molecules	In which body systems do you think the reactions occur?	Give students the opportunity to think about where the chemical reactions involved in growth are occurring.	

Lesson 1.4—Chemical Reactions and Systems

What do we know and what are we trying to find out?

In Lesson 1.3 we learned that when substances interact, the appearance of one or more new substances is evidence that a chemical reaction has occurred. To determine whether a new substance formed, we either observed the appearance of a substance with different properties directly (e.g., the drops of liquid that formed in the beaker when the candle was burned) or we used various tests to detect its presence (e.g., limewater was used to detect the presence of carbon dioxide, and cobalt chloride paper was used to detect the presence of water).

We were able to use direct observations or tests because we prevented substances from leaving the container in which the chemical reaction occurred. If we hadn't kept track of all the starting and ending substances, it would have been difficult to figure out what was going on. For example, if we burned the candle without the beaker on top of it, we would not have observed any evidence of new substances being formed because the carbon dioxide and water vapor were both colorless and odorless gases that would have disappeared into the surrounding air.

In many situations, it is difficult to keep all the substances together, so it can be difficult to figure out if a chemical reaction has occurred. In the human body, substances are moving from one organ to another or even from one body system to another. Food moves from the stomach to the small intestines, and carbon dioxide moves from blood vessels to the lungs. At one moment blood is moving through the heart, and at the next it is moving through veins in the leg. In these more complicated situations, we need another way to keep track of substances involved in chemical reactions.

In this lesson, we will take a more formal approach to keeping track of substances involved in a chemical reaction by carefully defining the system or systems in which the chemical reaction occurs. Any collection of things that have some influence on one another can be thought of as a **system** (e.g., the heart and blood vessels that make up the circulatory system work together to transport substances from one part of the body to another). In defining a system, we must include enough parts so that their relationship to one another makes sense but not so many parts that they distract us from the process we are studying. How well we define a system can make the difference between understanding and not understanding what is going on.

Answer the Key Question to the best of your knowledge. Be prepared to share your ideas with the class.

Key Question: How can attending to a system's inputs and outputs help us make sense of what is happening during a chemical reaction?

Teacher Talk and Actions

Activity 1: Defining Systems

Materials

For each team of students
9-volt battery
Wires with alligator clips on one end and battery connector on the other
2 pencil leads
Petri dish
Water
LEGOs
Paper plate
Balance

In Lesson 1.3, we observed several chemical reactions and modeled what was happening as the atoms of the molecules that made up the starting substances rearranged to form the molecules of the ending substances. In this activity, we will see if thinking about **systems** can help us understand changes in mass that happen during a chemical reaction. We will define our system so that we can focus more clearly on the substances entering the system, **inputs,** and substances leaving the system, **outputs.** We will then look for evidence that new substances have been formed and predict what will happen to the mass of our system during the reaction.

Procedures and Questions

Petri dish system

1. Fill the Petri dish with water.

2. For this reaction, we will define the system in terms of the components that are inside the Petri dish (i.e., the Petri dish system).

a. What are the **components** of the Petri dish system?

The components of the system are all the water molecules that are inside the Petri dish.

b. What are some substances or things that are not included in the Petri dish system?

Anything outside the Petri dish, such as the surrounding air and the table

c. The boundary of a system separates the components included in the system from the components excluded from it. What is the boundary of the Petri dish system?

The boundary of the Petri dish system is the wall of the dish. The wall keeps liquid water from leaving the system.

d. Can matter enter or leave the Petri dish system?

Yes, matter can flow in or out of the system because there is no lid on the dish. For example, water that evaporates would leave the system as a gas.

Teacher Talk and Actions

For this activity, each team needs 6 red LEGOs and 12 white LEGOs to build 6 LEGO water molecules.

Safety Notes

1. Wear safety goggles or glasses with side shields during the setup, hands-on, and take down segments of the activity.
2. Use caution when handling sharp tools and materials.
3. Immediately wipe up any spilled water on the floor to avoid a slip-and-fall hazard.
4. Properly clean up and dispose of waste materials.
5. Wash your hands with soap and water immediately after completing this activity.

Observing the chemical reaction: water

3. Connect the wires to the 9-volt battery.

4. Clamp one pencil lead in each alligator clip.

5. Place both pencil leads into the water in the Petri dish, making sure they do NOT touch each other.

6. Record your observations of what happens in the Petri dish. Focus on observations that provide evidence that new substances are being formed during the chemical reaction.

   When both leads are placed into the water, bubbles form around the leads. More bubbles appear to form around one lead than the other.

7. Given that a water molecule is made up of two hydrogen atoms and one oxygen atom, propose some substances that could be the products of this reaction.

   Students may propose a variety of substances here like hydrogen gas, oxygen gas, hydrogen peroxide, or ozone.

8. If we were to collect the products, we could use splint tests to gather evidence that will help us identify which substances they are. Watch the video in which a scientist uses either a burning or a glowing splint to test the contents of each test tube after the reaction and record the results of each test below.

   Glowing splint: The glowing splint reignited when placed in one test tube (a positive test for oxygen).

   Burning splint: The burning splint made a squeaky popping sound when placed in the other test tube (a positive test for hydrogen).

9. Were any new substances formed? What evidence do you have?

   Hydrogen gas and oxygen gas formed. Results of the glowing splint test provided evidence that one gas was oxygen, and results of the burning splint test provided evidence that the other gas was hydrogen.

10. How did your Petri dish system change after the chemical reaction occurred?

   The gases (hydrogen and oxygen) produced during the chemical reaction left the system.

11. What do you predict would happen to the mass of the Petri dish system if we let the reaction continue for a long time? Explain.

   The mass would decrease because the hydrogen and oxygen gases that form from the water are leaving the system (over more time the decrease in the level of water might be noticeable and the decrease in mass might be detectable).

Teacher Talk and Actions

If students suggest water vapor, ask if the formation of water vapor would be a product of a chemical reaction that starts with water. Help students conclude that the production of H_2O gas from H_2O liquid would not be evidence of a chemical reaction because atoms don't rearrange.

Use the video *Water Reaction Splint Test* here.

A glowing splint will reignite in the presence of oxygen. A burning splint will make a squeaky popping sound in the presence of hydrogen.

Keeping track of atoms

We can use models to determine whether our prediction makes sense. In our model, we'll use LEGOs to represent water molecules and a paper plate to represent the Petri dish. If we define our system as the Petri dish, then the LEGO models sitting on the paper plate represent the molecules in the system. Anything not on the paper plate is not part of the Petri dish system.

12. Construct six LEGO models of water molecules.

13. Obtain a paper plate and place the six LEGO models on the paper plate.

14. Place the paper plate with the LEGO models on the balance and record the mass in the "Before Reaction" column on Table 1.3 on the next page. This represents the mass of the Petri dish system before the reaction occurs.

15. Next, record in the "Before Reaction" column the number of hydrogen and oxygen atoms that are in the Petri dish system (i.e., how many white and red LEGOs are on the paper plate).

16. Refer back to your observations in Step 9. What are the ending substances (products) of this reaction?

17. Use four of the six LEGO models of water molecules to make as many product molecules as you can.

18. How many LEGO models of each product molecule did you make?

Two oxygen molecules and four hydrogen molecules

19. Where should you place the models of product molecules you just made? Explain.

Because the hydrogen and oxygen produced are gases that can leave the Petri dish system, the models of these molecules should be removed from the paper plate.

20. If any atoms have left the system, we need a way to specify where they go. We do so by defining a new system, the surrounding system, which is separate from the Petri dish system. Record the number of white LEGOs (H atoms) and the number of red LEGOs (O atoms) that you removed from the paper plate in the "After Reaction" column for the surrounding system.

21. Record the number of white LEGOs (H atoms) and the number of red LEGOs (O atoms) that remain on the paper plate in the "After Reaction" column for the Petri dish system.

22. Record the mass of the Petri dish system after the reaction.

23. Record the mass of the surrounding system (LEGOs that are not on the paper plate) after the reaction.

Teacher Talk and Actions

Students can respond verbally to this question about the products of the reaction. They do not have to write a response to Step 16.

Table 1.3. The Mass and Number of Atoms in Each System Before and After the Chemical Reaction

	Before Reaction			**After Reaction**		
	Petri Dish System	**Surrounding System**	**Total**	**Petri Dish System**	**Surrounding System**	**Total**
Mass (g)	24 g	0 g	24 g	8 g	16 g	24 g
	Petri dish system	**Surrounding system**	**Total**	**Petri dish system**	**Surrounding system**	**Total**
# H atoms	12	0	12	4	8	12
# O atoms	6	0	6	2	4	6

24. Use the information in the table to answer the following questions:

a. Do the models and data support our prediction about the mass of the Petri dish system? Explain.

Yes, the mass of the Petri dish system decreased during the reaction because it contains fewer atoms than it did before the reaction. Atoms making up product molecules, oxygen gas (O_2) and hydrogen gas (H_2), left the system.

b. What happened to the mass of the surrounding system during the chemical reaction? Explain.

The mass of the surrounding system increased because it contains more atoms from the gases produced (O_2 and H_2) than it did before the reaction.

c. How did keeping track of inputs and outputs to the Petri dish system help us understand why the mass of the Petri dish system decreased?

There were no inputs to keep track of in the chemical reaction, but there were outputs to the surrounding system. Keeping track of the outputs of matter showed us why the mass of the system decreased during the chemical reaction. If we weighed the outputs (i.e., the oxygen and hydrogen gas that left the system), we would have found that the gases carried away the mass that was "lost" from the system.

d. What happened to the total mass of the Petri dish system plus the surrounding system during the chemical reaction?

The total mass of the Petri dish system plus the surrounding system stayed the same because the number of each type of atom stayed the same during the reaction, regardless of where the atoms were located.

e. Was it important to include the other two LEGO water molecules in our model of the Petri dish system? Explain.

No. We really didn't need to include those models because they didn't play a role in the process we were studying.

Teacher Talk and Actions

Activity 2: Using Systems to Help Detect Changes in Mass

Materials
For each team of students Ball-and-stick models

In this activity we will investigate another reaction in which a fuel (in this case, a sugar called glucose) interacts with oxygen. We looked at a similar reaction in the last lesson in which another fuel, candle wax, interacted with oxygen. Later in this unit, we'll focus on the energy aspects of these and other reactions. But for now, we need to focus on the matter in the systems that are involved in the chemical reaction and whether we can detect any changes in the mass of the systems.

Procedures and Questions

Observing the chemical reaction: glucose and oxygen

1. Watch the video of the marshmallow reacting with oxygen as it burns. Record your observations below. Focus on observations that provide evidence that new substances are being formed.

A black coating forms on the marshmallow when it is burned.

2. Were there any reactants remaining after the reaction stopped? If so, which reactants are left over?

There is still some unburned marshmallow after the reaction stopped and there is still oxygen in the air after the reaction stopped.

3. One of the main ingredients in marshmallows is sugar (glucose). The formula for glucose is $C_6H_{12}O_6$. Below is a model of a molecule of glucose.

Glucose

4. In the video, the products of this reaction were allowed to escape into the surrounding system. If we collected the products, we would find that two products were formed: one that turns limewater cloudy, and one that turns cobalt chloride paper pink. Based on your experiences with these tests in Lesson 1.3, what can we conclude are the products?

Carbon dioxide (CO_2) and water (H_2O)

Teacher Talk and Actions

For this activity, each team needs pre-built glucose molecules (6 black, 6 red, and 12 white for each glucose) and oxygen molecules (12 red to build the 6 oxygen molecules that react with each glucose molecule).

Safety Note
Wear safety goggles or glasses with side shields during the setup, hands-on, and take down segments of the activity.

Show the video *Burning Marshmallow* here.

The black coating that forms on the marshmallow is made up of carbon-based substances that are the products of incomplete combustion. The atmosphere is only 20% oxygen so it is difficult to get complete combustion.

5. How many carbon atoms are in one molecule of glucose? Based on that number, how many carbon dioxide molecules do you predict will form during the reaction for each glucose molecule that burns?

Because there are six carbon atoms in each glucose molecule and one carbon atom in each carbon dioxide molecule, six carbon dioxide molecules should form.

6. Based on the number of hydrogen atoms in a molecule of glucose, how many water molecules do you predict will form during the reaction?

Because there are 12 hydrogen atoms in each glucose molecule and 2 hydrogen atoms in each water molecule, 6 water molecules should form.

7. Based on your predictions in Steps 5 and 6 of the number of carbon dioxide molecules and water molecules that would form during the reaction, how many oxygen molecules are required to react with one glucose molecule?

There will be 12 oxygen atoms in 6 carbon dioxide molecules and 6 oxygen atoms in 6 water molecules. Therefore, there will be a total of 18 oxygen atoms after the reaction. There are already 6 oxygen atoms from glucose, leaving 12 oxygen atoms needed from oxygen gas. So, there are 6 oxygen gas molecules needed.

8. Write a reaction equation using the chemical formulas to represent what happened during this chemical reaction.

$C_6H_{12}O_6 + 6\ O_2 \rightarrow 6\ CO_2 + 6\ H_2O$

The marshmallow system

Let's look at the marshmallow system and think about what will happen to the mass of the marshmallow system during the chemical reaction you just investigated.

9. Describe the marshmallow system during the reaction. Be sure to include a description of the boundary, components, inputs, and outputs. Remember that you only need to include parts that help you make sense of the reaction.

The exterior of the marshmallow forms the boundary of the system. The component of interest is glucose. The input is oxygen and the outputs are carbon dioxide and water.

10. Record in Table 1.4 on the following page the number of each type of atom in the marshmallow system before and after the chemical reaction occurs. You can use the equation you wrote in Step 8 or models to help you count the atoms.

11. What happens to the number of atoms that make up the marshmallow system during this chemical reaction? Explain.

The number of atoms decreases because for every glucose molecule that reacts, 12 oxygen atoms (as oxygen gas) enter the marshmallow system and 6 carbon atoms, 12 hydrogen atoms, and 18 oxygen atoms (as carbon dioxide gas and water vapor) leave the marshmallow system.

Teacher Talk and Actions

It might be helpful for some students to build models of these molecules to help them with the system analyses.

Ask students how they know that oxygen is an input and guide students to recognize that they saw the candle go out when the oxygen was gone. The glucose reaction is similar, and they could test the need for oxygen by putting the burning marshmallow under a glass beaker (closed system) and observe whether it goes out.

12. What does your answer to Step 11 mean in terms of the mass of the marshmallow system? To help you think about the answer to this question, record the number and mass of each atom before and after the reaction in Table 1.4 below. (The mass of each atom is provided in the table in atomic mass units [amu].)

Table 1.4. Atoms in the Marshmallow System

		Before the Reaction		After the Reaction	
Atom	**Mass (amu)**	**Number of Atoms**	**Mass of Atoms (amu)**	**Number of Atoms**	**Mass of Atoms (amu)**
Carbon	12	6	72	0	0
Hydrogen	1	12	12	0	0
Oxygen	16	6	96	0	0
		Total mass before	180	Total mass after	0

If you subtract the mass before the reaction from the mass after the reaction you get −180 amu. Therefore, the marshmallow system loses 180 amu for every glucose molecule reacted because during the reaction 6 carbon atoms, 12 hydrogen atoms, and 6 oxygen atoms have left the system.

13. Why didn't we include as part of the marshmallow system shown in Table 1.4 all of the atoms of the glucose molecules that make up the original marshmallow? Explain.

Only the glucose molecules that react with oxygen should be included in the system because the other glucose molecules are not involved in this reaction, and we only need to include in our model components that help us make sense of the phenomenon we are studying. The glucose molecules that do not react remain unchanged, so they do not illustrate that atoms were rearranged during the reaction and they don't help illustrate that atoms left the marshmallow system.

The surrounding system

Now let's look at the surrounding system and think about what will happen to the mass of the surrounding system during the chemical reaction.

14. Describe the system surrounding the marshmallow. Be sure to include a description of the boundary, components, inputs, and outputs. Remember that you only need to include parts that help us make sense of the reaction.

The exterior of the marshmallow forms the boundary between the marshmallow system and the surrounding system. The components of interest are the molecules involved in the reaction. During the reaction, the output from the surrounding system is oxygen and the inputs are carbon dioxide and water.

Teacher Talk and Actions

Students may need help remembering that glucose is the only molecule in the system before the reaction and that all the atoms from the glucose molecules leave the system after the reaction.

One glucose model really represents all the glucose molecules that are involved in the reaction. We don't have to make models of all the glucose molecules that reacted because they all behave the same and it would take us forever to model a mole of molecules.

15. Record in Table 1.5 the number of each type of atom in the surrounding system before and after the chemical reaction occurs.

16. What happens to the number of atoms that make up the surrounding system during this chemical reaction? Explain.

The number of atoms increases because for every glucose molecule that reacts with oxygen molecules to form carbon dioxide and water, 12 oxygen atoms leave the surrounding system (as O_2) and 6 carbon atoms and 12 oxygen atoms (from CO_2) and 12 hydrogen atoms and 6 oxygen atoms (from H_2O) enter the surrounding system.

17. What does your answer to Step 16 mean in terms of the mass of the surrounding system? Complete Table 1.5 below to help you think about the answer to this question. (The mass of each atom is given in atomic mass units [amu].)

Table 1.5. Atoms in the Surrounding System

		Before the Reaction		After the Reaction	
Atom	**Mass (amu)**	**Number of Atoms**	**Mass of Atoms (amu)**	**Number of Atoms**	**Mass of Atoms (amu)**
Carbon	12	0	0	6	72
Hydrogen	1	0	0	12	12
Oxygen	16	12	192	18	288
		Total Mass Before	192	Total Mass After	372

If you subtract the mass before the reaction from the mass after the reaction you get 180 amu. Therefore, the surrounding system gains 180 amu for every glucose molecule reacted because a net total of 6 carbon atoms, 12 hydrogen atoms, and 6 oxygen atoms enter the surrounding system during the reaction.

18. Air is a mixture of gases, mainly nitrogen (about 80%) and oxygen (about 20%). Why didn't we include nitrogen atoms from nitrogen gas in Table 1.5? Explain.

Because nitrogen (and other gases in the air) is not involved in the reaction, and we only need to include in our model the components that help us make sense of the phenomenon we are studying. Including all the other substances would have made the table of atoms more complicated and wouldn't have increased our understanding.

19. How did keeping track of the inputs and outputs of the marshmallow system help us understand why the mass of the chemical reaction decreased?

O_2 was an input and CO_2 and H_2O were outputs. Because the mass of the outputs (mass of atoms that left the system as CO_2 + H_2O) was greater than the mass of the inputs (mass of atoms that entered the system), there was a net decrease in mass of the system.

Teacher Talk and Actions

Science Ideas

The activities in this lesson were intended to help you understand important ideas about chemical reactions and systems. Read the ideas below. Look back through the lesson. In the space provided after each science idea, give evidence that supports the idea.

Science Idea #5: Systems are defined by placing boundaries around collections of interrelated things to make them easier to study. Regardless of where the boundaries are placed, a system still interacts with its surrounding environment. Therefore, when studying a system, it is important to keep track of what enters (inputs) or leaves (outputs) the system.

Evidence:

To understand what happened when water molecules interact as an electric current from a battery is applied, it was important to keep track of the gases that left the system (hydrogen and oxygen).

To understand what happened when the marshmallow burned, it was important to keep track of the gas that entered the system (oxygen) and the gases that left the system (carbon dioxide and water vapor).

Science Idea #6: Atoms are neither created nor destroyed during chemical reactions, so the total number of each type of atom remains the same. If the mass of a system changes, it is because atoms have either left the system (and are now in another system) or entered the system (from another system).

Evidence:

When we compared the sum of the masses of the atoms of the reactant molecules (glucose + 6 oxygen molecules) to the sum of the masses of the product molecules (6 carbon dioxide + 6 water molecules), we saw that the sums were the same. We showed that when the glucose reacted with oxygen in an open system, the mass of the marshmallow system decreased. We could explain our observation by showing that the decrease in mass could be accounted for by summing the masses of all the atoms that left the system.

Teacher Talk and Actions

Science Idea #7: An important kind of chemical reaction involves the interaction of oxygen with a fuel, such as hydrogen or glucose. When glucose molecules react with oxygen molecules, carbon dioxide and water molecules are produced.

Evidence:

In Lesson 1.2, we saw that when a candle burns, the candle wax reacts with oxygen to form carbon dioxide and water. This reaction could be considered important because burning a candle provides light and heat.

In Lesson 1.3, we saw that when a marshmallow burns, the glucose in the marshmallow reacts with oxygen to form carbon dioxide and water. This reaction also gave off light and likely heated the surrounding air.

Teacher Talk and Actions

This science idea (and Science Idea #8) are the foundation for two science ideas in Chapter 2. They provide the matter story for photosynthesis and cellular respiration.

Activity 3: A Chemical Reaction in a Living System

Materials

For each team of students
Ball-and-stick models

We are now ready to investigate a chemical reaction (known as **photosynthesis**) that takes place inside a living system. Your goal in this activity is to explain what happens to the mass of a plant as it grows. Just as you did with the reactions you observed in previous activities, you will identify substances entering the plant system (inputs) and substances leaving the plant system (outputs). Then you will examine data that provide evidence of what happens to atoms and molecules of the inputs and outputs, use a chemical equation to model how the inputs could be converted to the outputs, and use information about the masses of the atoms to understand changes in the mass of the system. Finally, you will pull all of this information together to write an explanation of what happens to the mass of a plant as it grows.

Procedures and Questions

1. Watch the video of the plant growing. Record your observations below. Focus on observations that provide evidence for whether or not the mass of the plant system is changing.

As the plant grows, it gets bigger in size and develops new body structures, such as leaves. The plant gets bigger as it adds atoms to its body, so its mass will increase.

Inputs and outputs in the plant system

2. Experiment 1: Watch the video showing an experiment with an aquatic plant. Record your observations below. Focus on observations that provide evidence for the identity of any substances involved in a chemical reaction.

When the bubbles were collected and tested with a glowing splint, the splint ignited. This indicated that oxygen gas was produced.

3. Experiment 2: Watch the video showing an experiment with another plant. Record your observations below. Focus on observations that provide evidence for the identity of any substances involved in a chemical reaction.

The limewater in the flask with the plant stayed clear while the limewater in the flask without a plant turned cloudy. This indicates that the flask with the plant does not contain carbon dioxide, but the flask without the plant does. This is evidence that the plant took in carbon dioxide, thereby removing it from the flask.

Teacher Talk and Actions

For this activity, each team needs 6 pre-built molecules of carbon dioxide (6 red and 2 white for each carbon dioxide) and 6 pre-built molecules of water (1 red and 2 white for each water).

Safety Note
Wear safety goggles or glasses with side shields during the setup, hands-on, and take down segments of the activity.

For Question 1, use the video *Plant Growing.*

For Question 2, use the video *Aquatic Plant in Light and Dark.*

For Question 3, use the video *Plant Experiment With Limewater.*

Students may need help in reasoning from evidence from the limewater test to the conclusion that CO_2 is a reactant. As students watch the video, make sure that they understand that both test tubes started with carbon dioxide because the scientist blew exhaled air into both test tubes. At the end of the experiment, the test tube without the plant still had carbon dioxide (based on a positive limewater test), whereas the test tube with the plant no longer had carbon dioxide (based on a negative limewater test). Since the only difference was the presence of the plant, the experiment provides evidence that the plant used carbon dioxide (i.e., that CO_2 was a reactant).

Keeping track of atoms within a system

In Experiments 3 and 4 described below, scientists used a technique called isotopic labeling to track atoms from reactants to products. You will learn more about this technique in high school chemistry, but two things are important for now. First, scientists can prepare reactant molecules that have some of the labeled atoms. Second, labeled atoms undergo the same chemical reactions as the regular versions of the atoms. This allows scientists to see which product molecules have the labeled atoms at the end of the chemical reaction. In this way, scientists can figure out what the reactants and products are for chemical reactions.

Let's look at two experiments that scientists carried out. Table 1.6 shows the findings of the two experiments.

Experiment 3: Scientists exposed a plant to carbon dioxide made with labeled carbon atoms and then looked to see where the carbon atoms ended up after the carbon dioxide entered the plant.

Experiment 4: Scientists thought that water could be a reactant, so they gave water made with labeled oxygen atoms to a plant, and it absorbed the water through its roots. They looked to see where the oxygen atoms ended up after the water entered the plant.

Table 1.6. Summary of Experiments 3 and 4

Experiment	Molecule With Labeled Atom Highlighted	
	Before Reaction	After Reaction
Experiment 3	CO_2	$C_6H_{12}O_6$ (glucose inside the plant system)
Experiment 4	H_2O	O_2 (oxygen in the surrounding air system)

4. Summarize the findings from Experiment 3. Be sure to include any evidence that indicates which substances are involved in the chemical reaction.

The carbon atoms that make up the carbon dioxide enter the plant and are found in glucose molecules inside of the plant. This suggests that carbon dioxide is a reactant in the chemical reaction and glucose is a product.

5. Summarize the findings from Experiment 4. Be sure to include any evidence that indicates which substances are involved in the chemical reaction.

The oxygen atoms that make up the oxygen gas that is formed are from water molecules that enter the plant. This provides evidence that water is a reactant and oxygen is a product of the chemical reaction.

6. Which data can be used as evidence to support a claim about what substances are the reactants of photosynthesis?

Data from Experiment 2 provide evidence that carbon dioxide is a reactant. Data from Experiment 4 provide evidence that water is a reactant.

Teacher Talk and Actions

7. Which data can be used as evidence to support a claim about what substances are the products of photosynthesis?

Data from Experiments 3 and 4 provide evidence that oxygen and glucose are the products.

8. Using the evidence that you collected, write the chemical reaction equation for the photosynthesis reaction that occurs inside the plant.

$6\ CO_2 + 6\ H_2O \rightarrow C_6H_{12}O_6 + 6\ O_2$

9. If we think of the plant as a system, describe the inputs and outputs of the plant system during photosynthesis.

The inputs are water and carbon dioxide. The output is oxygen. Glucose may remain in the plant, where it can be used to produce cellulose polymers that make up plant body structures.

Understanding changes in mass

10. You will now use the equation for photosynthesis and data about the mass of each atom to explain changes in mass in the plant system and in its surroundings. In the table, use the equation or models to fill in the number of each type of atom before and after the reaction. Use the data about the mass of each atom to calculate the mass that each type of atom contributes to the total mass of each system.

Teacher Talk and Actions

We can refer to it as "isotopic" labeling but not radiolabeling. ^{14}C is an unstable isotope that decays (and hence is radioactive) whereas ^{18}O is not. ^{14}C can be detected by its emission products whereas ^{18}O is detected by its mass.

Students learned this in *THSB* as well.

Tell the students that for every six carbon dioxide and six water molecules that react, one glucose and six oxygen molecules are produced. Do not take time to balance the equation.

Plant System					
		Before the Reaction		**After the Reaction**	
Atom	**Mass (amu)**	**Number of Atoms in the Plant System**	**Mass of Atoms (amu)**	**Number of Atoms in the Plant System**	**Mass of Atoms (amu)**
Carbon	12	0	0	6	72
Hydrogen	1	0	0	12	12
Oxygen	16	0	0	6	96
		Total Mass Before	0	Total Mass After	180

Surrounding System					
		Before the Reaction		**After the Reaction**	
Atom	**Mass (amu)**	**Number of Atoms in the Surrounding System**	**Mass of Atoms (amu)**	**Number of Atoms in the Surrounding System**	**Mass of Atoms (amu)**
Carbon	12	6	72	0	0
Hydrogen	1	12	12	0	0
Oxygen	16	18	288	12	192
		Total Mass Before	372	Total Mass After	192

a. How does the increase in the mass of the plant system compare to the decrease in the mass of the surrounding system?

The changes in mass of the plant system and the surrounding system are the same. When the mass of the plant system increases by 180 amu (180 amu minus 0 amu), the mass of the surrounding system decreases by 180 amu (372 amu minus 192 amu).

b. Predict the change in the mass of the surrounding system if the plant system increased by 200 pounds.

The mass of the surrounding system would decrease by 200 pounds.

c. Explain why you think so using ideas about atoms.

Because atoms aren't created or destroyed, the atoms that are now part of the plant system must have come from the surrounding system.

Teacher Talk and Actions

It could be pointed out to the students that this reaction is the reverse of the glucose/oxygen reaction they investigated in the last activity. This information will help them know how many CO_2, H_2O, and O_2 molecules are involved.

Remind the students that we are only including the atoms that are involved in the reaction because the other atoms were in the plant system before and after the reaction and therefore do not help us make sense of mass changes.

Writing Explanations

11. You will now use the information about atoms and mass to explain what happens to the mass of a plant as it grows. Review the elements of a scientific explanation listed in the table below and use it to guide your explanation. The table describes what each element is and provides criteria for judging its quality.

Elements of an Explanation and Criteria for Evaluating Its Quality		
Element	**What It Is**	**Quality Criteria**
Question	The question to be answered	
Claim	A statement or conclusion that is intended to respond to the question	• The claim responds to the question.
Science Ideas	Widely accepted scientific ideas, concepts, or principles that can be used to show why the evidence supports the claim	•The listed science ideas are relevant to the claim. • The science ideas are used to show why the evidence supports the claim. (This involves applying the science ideas to the specific case in the question.)
Evidence	Data that are relevant to and support the claim and can be confirmed by others. Data are based on observations about the world that are either made with our senses or measured with instruments	•The evidence supports the claim. (All the data should be consistent with the science idea, not just the data cited as evidence.)
Models	Tools for thinking about the world that can give you ideas about how something might work	• The models are consistent with the science ideas listed. •The models show that the claim is reasonable.

12. Write an explanation for what happens to the mass of a plant as it grows. Use the data from the experiments, models of the reactant and products, science ideas, and the tables in Step 10 to help you. On the next page is a table that you can use to organize your thinking before writing the explanation.

Teacher Talk and Actions

Additional support for constructing scientific explanations can be found in the Teacher Edition of *THSB* on pages 48–51, 72–74, 118–127, 218–221, 318–324, and 400–407.

Question	What happens to the mass of a plant as it grows?
Claim	The mass of a plant increases as it grows.
Science Ideas	• During chemical reactions, atoms that make up molecules of the starting substances (called reactants) disconnect from one another and connect in different ways to form the molecules of the ending substances (called products). • Atoms are neither created nor destroyed during chemical reactions, so the total number of each type of atom remains the same. If the mass of a system changes, it is because atoms have either left the system (and are now in another system) or entered the system (from another system).
Evidence	• Data from Experiments 1 and 2 showed that oxygen leaves the plant. • Data from Experiment 3 showed that carbon dioxide enters the plant. • Data from Experiment 4 showed that carbon atoms that make up carbon dioxide end up as part of glucose molecules inside the plant. • Data from Experiment 5 showed that hydrogen atoms that make up water end up as part of oxygen molecules that leave the plant. • Data on the mass of each atom showed that if the plant has more atoms after the reaction, it increases in mass.
Models	Models show that more atoms enter the plant system than leave the plant system during the reaction.

Teacher Talk and Actions

Explanation:

The mass of a plant increases as it grows because the number of atoms that make up the plant increases as it grows. We know that if the mass of a system increases, it is because atoms have entered the system. So if we have evidence that the atoms have entered the system, this can explain the increase in mass. Experimental data show that carbon dioxide and water enter the plant and that the atoms that make up those molecules rearranged to form glucose molecules and oxygen molecules.

The oxygen molecules leave the plant system, but the glucose molecules stay inside of the plant system. This means that for every glucose molecule that is made, the plant system gains 6 carbon atoms, 12 hydrogen atoms, and 6 oxygen atoms.

These additional atoms add to the mass of the plant system.

Teacher Talk and Actions

13. You may have heard of the law of conservation of mass, which is a science idea that says that mass is neither created nor destroyed. A student thinks that the mass of a plant cannot change as it grows because doing so would violate the conservation of mass. Using the data from the experiments, models of the reactants and products, science ideas, and the tables in Step 10, write an explanation for why the growth of a plant does not contradict the law of conservation of mass.

Question	Does the increase in the mass of a plant when it grows contradict conservation of mass?
Claim	*No, the increase in the mass of a plant does not contradict conservation of mass.*
Science Ideas	*• During chemical reactions, atoms that make up molecules of the starting substances (called reactants) disconnect from one another and connect in different ways to form the molecules of the ending substances (called products).* *• Atoms are neither created nor destroyed during chemical reactions, so the total number of each type of atom remains the same. If the mass of a system changes, it is because atoms have either left the system (and are now in another system) or entered the system (from another system).*
Evidence	*• Data from Experiments 1 and 2 showed that oxygen leaves the plant.* *• Data from Experiment 3 showed that carbon dioxide enters the plant.* *• Data from Experiment 4 showed that carbon atoms that make up carbon dioxide end up as part of glucose molecules inside the plant.* *• Data from Experiment 5 showed that hydrogen atoms that make up water end up as part of oxygen molecules that leave the plant.* *• The data tables show that the sum of the mass of the plant and surrounding systems stays the same.*
Models	*Models show that the number of each type of atom stays the same.*

Teacher Talk and Actions

Explanation:

The increase in the mass of a plant does not contradict the law of conservation of mass because the number of each type of atom stays the same, but the atoms are now located in different systems. We know that during chemical reactions, atoms that make up molecules of the reactants rearrange to form the molecules of the products. In this case, data from the labeling experiments showed that the atoms that make up carbon dioxide and water molecules rearranged to form glucose and oxygen molecules. The other experiments showed that carbon dioxide and water molecules entered the plant system and oxygen molecules left the plant system. The glucose molecules stayed inside the plant system. If you consider both the plant and surrounding systems, you see that the number of each type of atom stays the same and therefore the mass must stay the same. If you consider only the plant system, it may seem like mass is being created but that is only because for every glucose molecule that is made the plant gains a net total of 6 carbon atoms, 12 hydrogen atoms, and 6 oxygen atoms. These atoms were part of the surrounding system before the reaction and have entered the plant system, where they were involved in a chemical reaction and became components of the plant system.

Teacher Talk and Actions

Science Ideas

The explanation you wrote uses physical science ideas about atom rearrangement and conservation to explain a biological phenomenon that is essential to human life on earth. The process of photosynthesis is so important for understanding the growth of all plants that scientists have given it its own science idea. You can use this idea to explain the growth of plants ranging from single-celled plants in the ocean to the tallest sequoia tree on land.

Science Idea #8: Plants take in carbon dioxide and water molecules from their surroundings and use them to make glucose and oxygen molecules. Glucose molecules remain inside the plant and oxygen molecules are released to the air surrounding the plant. The increase in mass that accompanies plant growth is due to more atoms entering the plant system than leaving the plant system.

Evidence:

Data from Experiment 2 showed that carbon dioxide enters the plant and data from Experiment 3 showed that carbon atoms that make up carbon dioxide end up as part of glucose molecules inside the plant. Data from Experiment 1 showed that oxygen leaves the plant.

Teacher Talk and Actions

Pulling It Together

Work on your own to answer these questions. Be prepared for a class discussion.

1. Now that you have had more experiences with chemical reactions, how would you answer the Key Question: **How can attending to a system's inputs and outputs help us make sense of what is happening during a chemical reaction?**

Paying attention to the system helps us focus on the atoms that are involved in the reaction and ignore all other atoms that aren't involved in the reaction, even though they may be present. For example, in the water reactions we only needed to focus on the water molecules that reacted but not on the water molecules that were left in the Petri dish after the reaction stopped.

2. For the water reaction inside the Petri dish, can you think of a different way of defining the system such that mass and atoms would be conserved within the system?

Perhaps we could combine the Petri dish system and the surrounding system into a single system. That way all the atoms involved in the chemical reaction would be included in the defined system.

3. A student measures the mass of a candle and then lights the candle. After burning the candle for two hours, she blows the candle out and measures the mass of the candle again. She observes that the mass has decreased. This surprised her because she expected the mass to stay the same because she knows that mass is conserved during a chemical reaction. Write an explanation for why the mass changed. Use systems in your response.

Question	Why does the mass of the candle decrease after it has been burned?
Claim	*The mass of the candle decreases because the wax it is made of reacted with oxygen in the air, resulting in atoms leaving the candle system.*
Science Ideas	• *During chemical reactions, atoms that make up molecules of the starting substances (called reactants) disconnect from one another and connect to other atoms to form the molecules of the ending substances (called products).* • *Atoms are neither created nor destroyed during chemical reactions, so the total number of each type of atom remains the same. If the mass of a system changes, it is because atoms have either left the system (and are now in another system) or entered the system (from another system).*

Teacher Talk and Actions

Evidence	• In Lesson 1.3, we observed that a gas produced by one burning candle turned limewater cloudy, and another gas condensed to form droplets that turned cobalt chloride paper from blue to pink. These observations provide evidence that carbon dioxide and water are produced during the reaction between candle wax and oxygen.
Models	Models show that the number of C, H, and O atoms stays the same during the reaction.

Explanation:

The mass of the candle decreases because the wax that makes it up reacted with oxygen from the surrounding air system, resulting in atoms leaving the candle system. While the candle was burning, a chemical reaction was occurring between candle wax and oxygen. During this reaction, the atoms that make up the candle wax and oxygen rearrange to form carbon dioxide and water molecules. The carbon atoms that make up the carbon dioxide and the hydrogen atoms that make up the water used to be part of the wax molecules that made up the candle. These atoms left the candle system and entered the surrounding air system. Therefore, the candle is made up of fewer atoms and the air system is made up of more atoms after the reaction. If atoms leave a system, the mass of the system decreases. If you were to consider both the candle system and the surrounding air system, you would find that the number of atoms doesn't change. So, while the mass of the candle system changes, the mass of all the matter involved in the chemical reaction does not.

4. Recall from the last Pulling It Together question in Lesson 1.3 that we hypothesized that chemical reactions were occurring inside our bodies because the molecules that enter the body are different from the molecules that leave the body. This suggests that atoms are rearranged inside the body. In which body systems do you think the reaction(s) occur?

Because the food we eat enters the digestive system, it's possible that chemical reactions occur in the digestive system. It is also possible that chemical reactions occur in the body system that includes the body part that is growing.

Teacher Talk and Actions

As students suggest body systems, ask them to give reasons for why they think they are involved.

If there is time to discuss, ask students to give reasons for the body systems they list. If students list the digestive system, they may have learned the phrase "chemical digestion" in middle school. Ask them what they now think chemical reaction means. If students list the respiratory system, they may be thinking (incorrectly) that the lungs take in oxygen and give off carbon dioxide, so a chemical reaction may occur in the lungs. Students don't need to know the correct answers at this point, they will learn them in Lesson 1.5. This question just gives them a chance to express their ideas and gives you a chance to see what ideas they are bringing to the next lesson.

Closure and link to subsequent lessons:

In this lesson, you investigated examples of chemical reactions occurring in simple systems where it was easy to keep track of what substances entered the system and what substances left the system. We used LEGO bricks and ball-and-stick models to help us visualize what happened to the atoms and molecules during the chemical reaction that caused the mass changes and to think about why changes in mass do not violate ideas about atom conservation.

Now we are ready to use what we know about atom rearrangement during chemical reactions, along with ideas about system inputs and outputs to make sense of chemical reactions that occur in complex systems that make up our bodies. In the next lesson, you will use these ideas and experimental data to model and explain how our bodies convert polymers that make up our food into different polymers that make up our body structures.

Lesson Guide
Chapter 1, Lesson 1.5
Converting Food to Body Structures

Focus: In this lesson, students apply what they have learned about atom rearrangement and conservation in simple systems to consider how our body systems work together to convert polymers from food into polymers making up our body structures. Because the link between reactants and products of chemical reactions occurring in the body cannot be observed directly, students learn to identify reactants and products by drawing inferences from data. The lesson concludes the basic story of how matter changes contribute to the growth of living organisms.

Key Question: How do our bodies turn molecules from food into molecules that make up our body?

Target Idea(s) Addressed

The large carbon-based molecules that make up an animal's food must be broken down so that cells can use them. These molecules, mainly protein, carbohydrate, and fat polymers, are broken down in the digestive system into smaller carbon-based molecules (monomers) such as amino acids, glucose, and fatty acids, during chemical reactions with water molecules. (Science Idea #9)

Within body cells, the monomers can react with one another to form polymer molecules that become part of the animal's body. Water molecules are also produced during polymer formation. (Science Idea #10)

Multicellular organisms need a way to transport molecules from one body system to another. The circulatory system (including a heart that pumps blood and vessels of various diameters including capillaries) transports the reactants and products of various chemical reactions in response to changes in the body's needs. (Science Idea #11)

Models are useful for representing systems and their interactions—such as inputs, outputs, and processes—and energy and matter flows within and between systems. (Science Idea #12)

Science Practice(s) Addressed

Analyzing and interpreting data
Developing and using models
Constructing explanations
Obtaining, evaluating, and communicating information

Materials

Activity 1: *Per group of students:* 2D-models of piece of food protein, amino acids, and piece of body protein (place in plastic sleeves); body system posters (from Lesson 1.2)

Activity 2: *Per group of students:* 2D-models of piece of starch, glucose, and a piece of glycogen (place in plastic sleeves); body system posters (from Lesson 1.2)

Pulling It Together Question 2: 2D-models of fat molecule from food, fatty acids, fat molecule from body, and membrane phospholipid; video: *From Food to Body Structures*

Advance Preparation

Organize materials in advance so that each group has materials when class begins. Download and test videos. Decide how you will use the *Food Proteins to Body Proteins* slides available at *www.nsta.org/growthandactivity.*

Phenomena, Data, or Models	Intended Observations	Purpose	Rationale or Notes
Activity 1 Data from body system posters Experimental data on changes in protein concentration in the small intestine after eating a protein-rich meal Experimental data on the concentration of free amino acids and amino acids in small peptides in the small intestine before and after a protein-rich meal 2D ball-and-stick models of a peptide consisting of five amino acids linked together Experimental data on the amount of labeled muscle protein formed with and without an amino acid supplement 2D ball-and-stick models of five amino acids	• The digestive system is where proteins from food are taken in. • The muscular system is where actin and myosin proteins are needed to build muscle fibers. • The circulatory system moves substances from one body system to another. • The amount of protein in the small intestine decreases by 5–10% during the first 30 minutes and by 90% from 1–2 hours after it was eaten. • The amount of free amino acids in the small intestine increases after a protein-rich meal and the amount of amino acids present in the small intestine as small peptides increases after a protein-rich meal. • The amount of amino acids in the small intestine increases as the amount of protein in the small intestine decreases. • Four water molecules are needed to digest a pentapeptide to five amino acids and (n-1) water molecules are needed to digest a protein made up of n amino acids. • Both experimental and control groups produced labeled protein, but the experimental group produced more protein. • Four water molecules are produced when the peptide is produced from five amino acids and (n-1) water molecules are produced from n amino acids.	Give students an opportunity to develop and revise a model in a complex biological system (the human body). Give students opportunities to analyze experimental data, draw conclusions, and use conclusions to revise a model. Give students an opportunity to use an evidence-based model to construct an explanation for how proteins in food become muscle proteins.	Targets the misconception that food is added to body structures unchanged Targets the misconception that chemical reactions are not involved in the growth of living organisms (or that they are involved only in "chemical digestion")

Phenomena, Data, or Models	Intended Observations	Purpose	Rationale or Notes
Activity 2 Data from body system posters Models of a starch polymer and a glycogen polymer Experimental data on changes in blood glucose concentration following ingestion of a carbohydrate-rich meal Experimental data on glycogen formation in infused rat liver Experimental data comparing the amounts of 14C atoms (from 14C-glucose) that ended up in liver glycogen and muscle glycogen Models of glucose monomers and a glycogen polymer	• The digestive system is where carbohydrates from food are taken in. • Both the liver (in the digestive system) and muscles (in the muscular system) store glycogen. • The circulatory system moves substances from one body system to another. • If starch is digested to glucose monomers + water molecules and the glucose monomers are used to make glycogen, then the body could convert starch to glycogen. • In both males and females, the concentration of blood glucose increased to a maximum about an hour after a carbohydrate-rich meal and then decreased to baseline after six hours. • Glycogen formed in the rat liver during the first hour after glucose infusion. When glucose was not infused, no glycogen formed in the rat liver. • Following ingestion of ^{14}C glucose, ^{14}C atoms ended up in both liver glycogen and muscle glycogen (indicating both liver and muscle convert glucose to glycogen). • When n glucose monomers react to form a glycogen polymer, n-1 molecules of water are also produced.	Give students an opportunity to develop and revise a model in a complex biological system (the human body). Give students opportunities to analyze experimental data, draw conclusions, and use conclusions to revise a model. Give students an opportunity to use an evidence-based model to construct an explanation for how carbohydrate polymers in food become glycogen in liver and muscle.	Targets the misconceptions that • cells are "in" the body rather than actually "make up" the body, • all cells are pretty much the same shape, and • atoms/molecules are "in" the body rather than actually "make up" the body

Phenomena, Data, or Models	Intended Observations	Purpose	Rationale or Notes
Pulling It Together Q2: Models of triglyceride, three fatty acids, and glycerol	A triglyceride can react with 3 H_2O to form 3 fatty acid monomers + 1 glycerol monomer. These monomers could react with one another (or with other monomers) to form triglycerides stored in fat tissue or other polymers (such as membrane phospholipids)	Give students an opportunity to use an evidence-based model to explain how fats in food become body fat.	

Lesson 1.5—Converting Food to Body Structures

What do we know and what are we trying to find out?

We saw in Lesson 1.2 that our bodies are organized into body systems that work together to keep us alive and functioning properly. Each body system is composed of organs that work together to carry out the functions of the body system. And each organ is composed of cells that work together to carry out the functions of the organ.

When we zoomed down to the molecular level, we saw that both our body structures and our food are made up of complex molecules, including protein, carbohydrate, and fat polymers. And although the molecules making up our bodies and our food are similar in type, they differ in several ways. For example, the protein ovalbumin from egg whites is composed of 386 amino acid monomers whereas the muscle protein myosin is composed of 1,937 amino acid monomers. In addition to having different numbers of amino acid monomers, these two proteins have very different sequences of amino acids. These amino acid differences give the proteins different properties. It was not clear how our bodies could use polymers from food, such as egg-white ovalbumin, to produce different molecules that make up our body, such as myosin in our muscles.

In Lessons 1.3 and 1.4, we saw that converting reactant molecules to product molecules required chemical reactions in which atoms of reactants rearranged to form products. We learned that by carefully defining the systems we were studying and paying attention to the inputs and outputs of the systems, we could identify the reactants and products of chemical reactions.

In this lesson, we are finally ready to use what we have learned to help us figure out how our bodies use molecules from food to produce molecules that make up the body. In each activity, we will try to identify the relevant components of each system, pay careful attention to inputs and outputs, and create and use models to help us think about what is going on.

Answer the Key Question to the best of your knowledge. Be prepared to share your ideas with the class.

> **Key Question: How do our bodies turn molecules from food into molecules that make up our body?**

Teacher Talk and Actions

Activity 1: Converting Food Proteins to Body Proteins

Materials

For each team of students
Ball-and-stick model images (2D models of piece of food protein, amino acids, and piece of body protein)
Body system posters prepared in Lesson 1.2
Eraseable markers

In this activity, we will try to develop a model that represents how our bodies convert food proteins to muscle proteins. We will start by identifying the body systems and their components that play a role in this process. We will pay attention to inputs and outputs of each system as we create and use a model that can help us think about what chemical reactions occur and where they occur.

Once we have created a model, we will see whether it is consistent with data from several experiments. If necessary, we will revise the model to fit with the experimental data. By the end of the activity, we should have a model that is consistent with all of the data and can be used to explain how food proteins are converted to body proteins.

Procedures and Questions

Designing a model

As a first step in our design process, let's examine our body system posters from Lesson 1.2 and consider what components we should include in our model. Answer the following questions.

1. What body systems do you think play a role in converting food proteins to muscle proteins? Explain why you think each system is relevant.

   Digestive system: that's where proteins enter the body (and digestion to amino acids occurs)

   Muscular system: that's where actin and myosin proteins are needed to build muscle fibers

   Circulatory system: that's how substances get from one part of the body to another

2. What molecules do you think could be inputs and outputs? Explain why you think so.

   Protein molecules in our food, such as ovalbumin, casein, soy protein

   Protein molecules that make up our muscles, such as actin and myosin

3. Are chemical reactions relevant? Explain.

   Chemical reactions must be involved because food proteins and muscle proteins are different molecules.

Teacher Talk and Actions

Safety Note
Wear safety goggles or glasses with side shields during the setup, hands-on, and take down segments of the activity.

Lead the class through the construction of a model. A set of slides showing one way to help students construct the model is available at *www.nsta.org/growthandactivity*. Start by repeating the question we are interested in: "How do proteins from our food become proteins that make up our body?" Then ask students to share their answers to the previous questions:

What body systems do you think are relevant? What molecules do you think are relevant?

Are any chemical reactions relevant?

If the students have good reasons for why a body system, molecule, or chemical reactions are relevant, add them to your model. You should diagram the model on the blackboard/whiteboard so students can see how the model is gradually built. Students should also record it in their books.

If the students mention something that isn't relevant, ask them if it is relevant and why they think so.

Creating a model

4. Discuss your answers to Questions 1–3 with your class. Your teacher will then help the class develop the draft model. Follow along with your teacher and draw the model below.

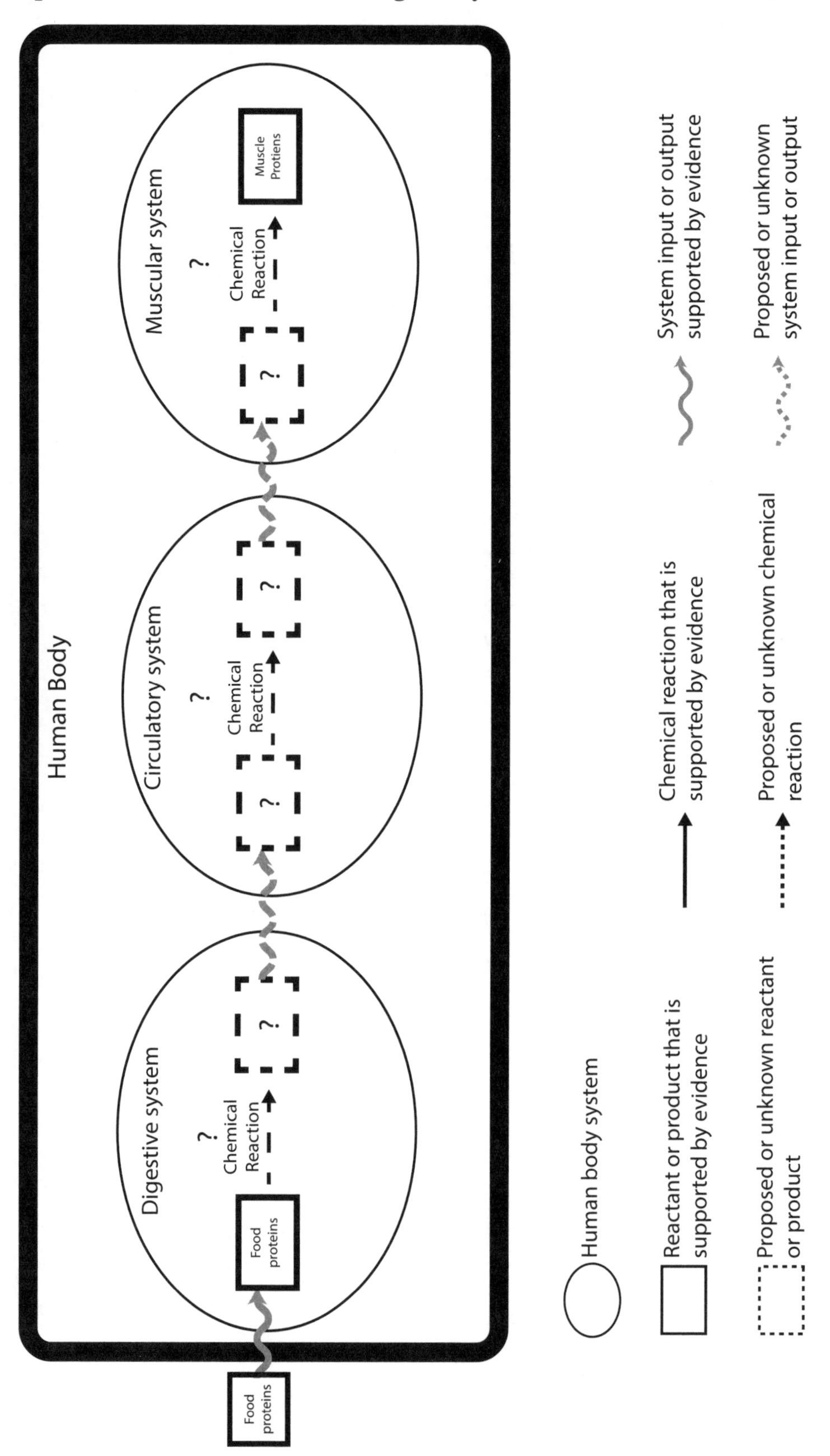

Teacher Talk and Actions

Lead the class through the construction of a model like the one shown, and have students record it in their books as the class reaches consensus on each component. In the model shown, solid lines represent components that are supported by evidence and dashed lines and question marks ("?") represent components we have no evidence to support.

Title the model: "How Our Bodies Convert Food Proteins to Muscle Proteins."

Start by representing the whole body and then ask students what systems we should represent within the body. Students should at least list the digestive and the muscular systems.

Then ask what polymers should be included as components. Based on the table students completed in Lesson 1.2, page 36, students should list one or two specific protein polymers from food and one or two protein polymers making up muscle.

Ask students where we should put the food polymers in our model. If students represent food proteins either outside the body or in the digestive system, ask if food proteins can be considered an input to the digestive system. Hopefully, students will recognize that food polymers start outside the body, are an input to the digestive system, and hence should also be included in the digestive system. (It's fine if students note that water is also an input to the digestive system as long as they give a reason for how they know).

Ask students where we should put the muscle polymers in our model. Students should list them in the muscular system.

If students haven't yet mentioned the circulatory system, ask students to consider how molecules from food could get from the digestive system to the muscles. If no students mention the circulatory system, they can add it to the model later after examining data from Experiment 2 (p. 146).

Ask students whether food proteins, such as ovalbumin, are the same as muscle proteins, such as actin and myosin. Students should know that they are different because they have different arrangements of monomers.

Ask students what that tells us about chemical reactions. Students should know that chemical reactions must be involved in producing actin and myosin from ovalbumin.

Ask students where the chemical reactions could occur. Students should realize that they could occur in any or all of the body systems, but we don't have evidence to support claims. So for now, we can list chemical reaction with a question mark (?) in all three systems.

Now that we have proposed a model, let's step back and consider the inputs and outputs we know and the inputs and outputs we don't know. Students should realize that they don't know (a) the output(s) of the digestive system, (b) the input(s) and output(s) of the circulatory system, (c) the input(s) to the muscular system, and (d) where chemical reactions occur.

Testing our model against evidence

When scientists think that they have identified the relevant components of a model and how the components might interact, they test the model against evidence. The evidence can come from data others have already collected or from newly designed investigations. By checking their models against the data, they can either gain confidence in their model or revise their model to be consistent with the data.

We'll now look at data from several experiments published in a paper in 1973 to see what parts of our model the data support and what parts need to be revised.

5. To find out what happens to proteins in the digestive system, scientists obtained informed consent from human volunteers and then carried out a series of experiments. Read the experiments, examine the data, and respond to the questions.

Experiment 1: What Happens to Food Molecules in the Digestive System?

Introduction
Scientists were interested in what happens to proteins in our food in the digestive system.

Methods
Several volunteers were fed a protein-rich meal (consisting of bovine serum albumin, cornstarch, olive oil, lemon juice, salt, and water). Bovine serum albumin is a cow protein that is easily distinguishable from other proteins and is not normally present in the human digestive system. Scientists then monitored the contents of the volunteers' small intestines over four hours.

Results
The table below shows the amount of bovine serum albumin present in the small intestines for the first four hours after they ate the protein-rich meal.

Table 1. Change in the amount of bovine serum albumin found in the small intestine over four hours after a protein-rich meal

Time	% Protein Remaining
30 minutes	100%
1 hour	90–95%
2 hours	10%
3 hours	6%
4 hours	5%

Source: Adibi S. A., and D. W. Mercer. 1973. Protein digestion in human intestine as reflected in luminal, mucosal, and plasma amino acid concentrations after meals. *Journal of Clinical Investigation* 52 (7): 1586–1594.

Teacher Talk and Actions

By this point, students should understand what is included in their model and why, some of its limitations, and that they need to look at some data to test whether the model makes sense or can be improved.

Students' answers here will vary. For example, students may say that we don't know the output of the digestive system, the input to the circulatory system, the output of the circulatory system, the input to the muscle system, or where the chemical reaction(s) occur in the body.

In addition to measuring how much cow protein was present, scientists also measured the amount of amino acids present in the small intestine. The amino acids were found either as "free" amino acids or as small peptides, which are proteins made up of 2-3 amino acids. The table below summarizes their data before the meal and three hours after the meal.

Table 2. Change in the amount of amino acids and small peptides found in the small intestine before and after protein-rich meal

Time	Free Amino Acids (μmole/μl)	Amino Acids (in small peptides)
Before protien-rich meal	3.21	15.94
After protein-rich meal	29.29	117.97

6. Based on the data in Table 1, what happens to the amount of protein in the small intestine over four hours?

The amount of protein in the small intestine decreases a little during the first hour after a protein-rich meal and then decreases a lot between 1-2 hours until very little remains in the small intestine after 3 hours.

7. Based on the data in Table 2, what happens to the amount of amino acids in the small intestine before the meal and three hours after the meal?

Three hours after a protein-rich meal, the concentration of free amino acids increased from 3 to nearly 30 μmole/μl, and the concentration of amino acids in small peptides increased from about 15 to about 118 μmole/μl.

8. What do you think is happening to the proteins in the small intestine? Explain.

Chemical reactions in the small intestine are probably converting proteins to amino acids and small peptides. My evidence is that as the amount of protein decreases, the amount of amino acids increases. Perhaps proteins are first digested to small peptides, and the peptides are then digested to free amino acids.

9. We will use ball-and-stick models to help us write a reaction equation for a chemical reaction that occurs in the small intestine as shown in Experiment 1. The models below show a small piece of protein consisting of five amino acids bonded together and five "free" amino acids after digestion.

small piece of protein:

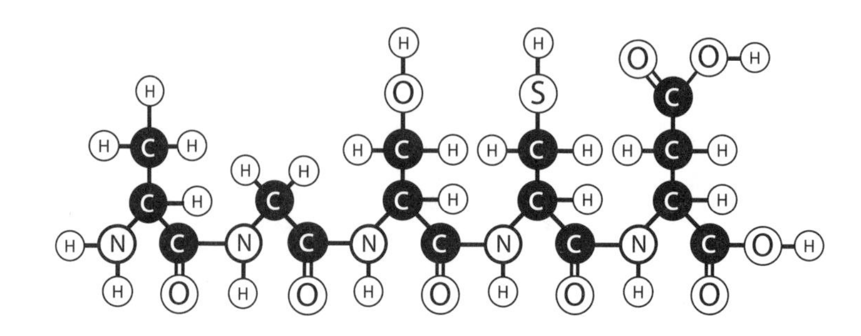

Teacher Talk and Actions

For Steps 9–11, using the 2D models in plastic sleeves and erasable markers, students should identify atoms making up amino acids that are not in the piece of protein and use the information to identify water as a reactant of the chemical reaction.

five "free" amino acids:

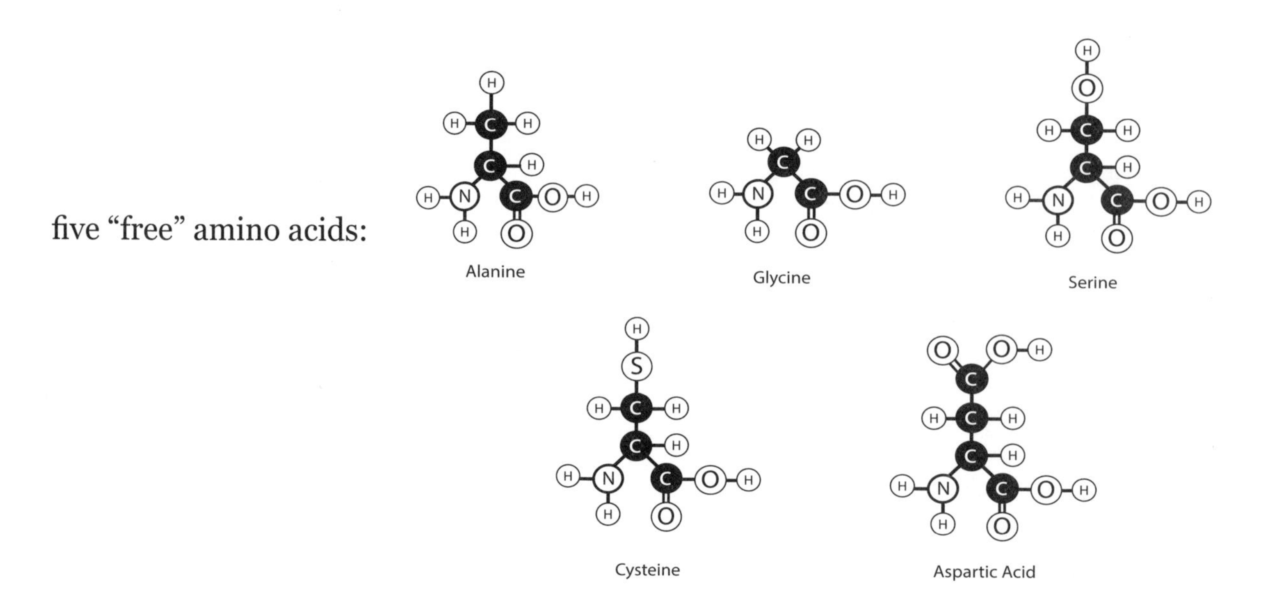

10. Because Experiment 1 shows that the molecules in the small intestine changed, there must have been a chemical reaction. What are the reactants and products of the chemical reaction? You may use the models above and/or ball-and-stick models to justify your answer.

Reactants: $protein_5$ and 4 H_2O

Products: 5 amino acids

11. Write a word equation for the chemical reaction that you just modeled.

$protein_5$ + 4 H_2O → 5 amino acids

12. Incorporate the information from Experiment 1 and the word equation you have just written into the model you created in Step 4.

Teacher Talk and Actions

Have students "digest" the 2D model of protein polymer to five amino acids. They should find that they need H and O atoms from four water molecules. Once they have figured that out, they can write a reaction equation: $protein_5 + 4\ H_2O \rightarrow 5$ amino acids.

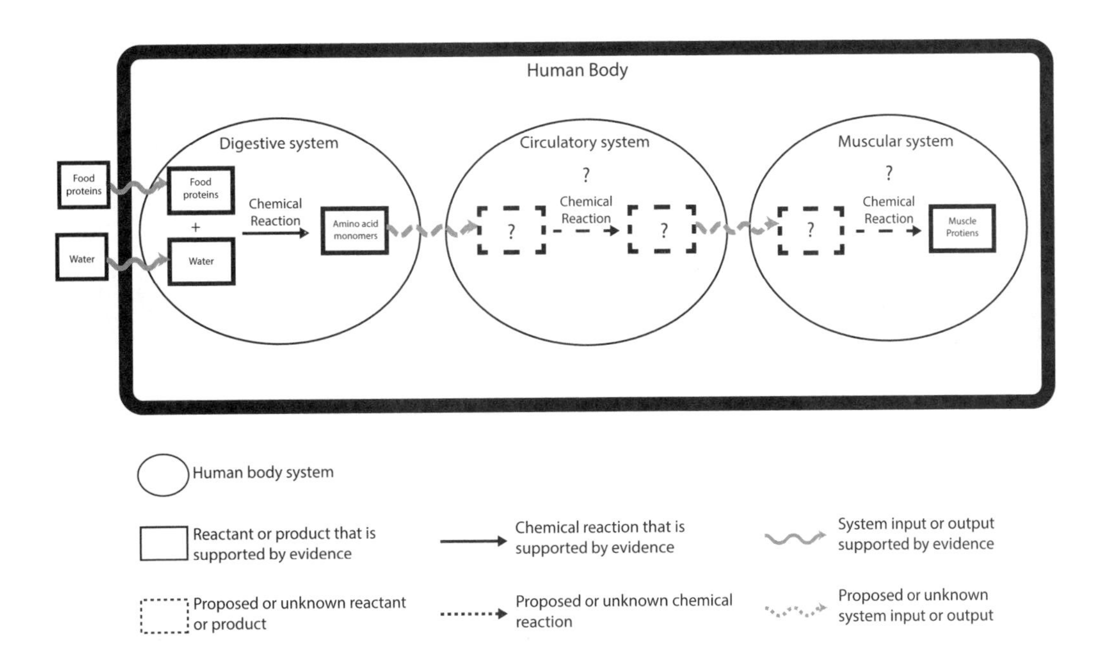

Experiment 2: What Are the Outputs of the Digestive System?

Introduction

The scientists from the first investigation were also interested in what happens to the amino acids once they are in the digestive system.

Methods

They conducted the same experiment. However, now after volunteers were fed the protein-rich meal, the scientists monitored the amount of amino acids in the veins carrying blood away from the small intestine.

Results

The concentration of free amino acids in the blood for the first four hours after the meal is shown in the figure below.

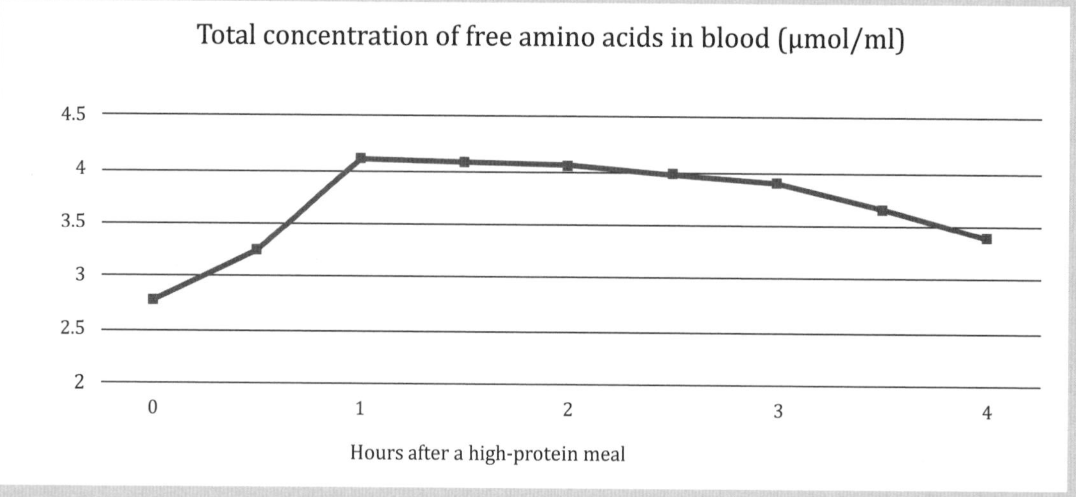

Source: Frame, E.G. 1958. The levels of individual free amino acids in the plasma of normal man at various intervals after a high-protein meal. *Journal of Clinical Investigation* 37 (12): 1710–1723.

13. Summarize what the data show.

The concentration of amino acids in veins carrying blood away from the small intestine starts out low, increases to a maximum at 1 hour, decreases a little from 1–3 hours, and then decreases more rapidly from 3–4 hours. After 4 hours, the concentration is still not down to the concentration measured at time 0.

14. Why do you think the concentration of amino acids in the blood increases?

Amino acids are passing through the lining of the small intestine and into capillaries. Blood in the capillaries flows into veins.

15. Use the evidence from the experiment to update your model.

Teacher Talk and Actions

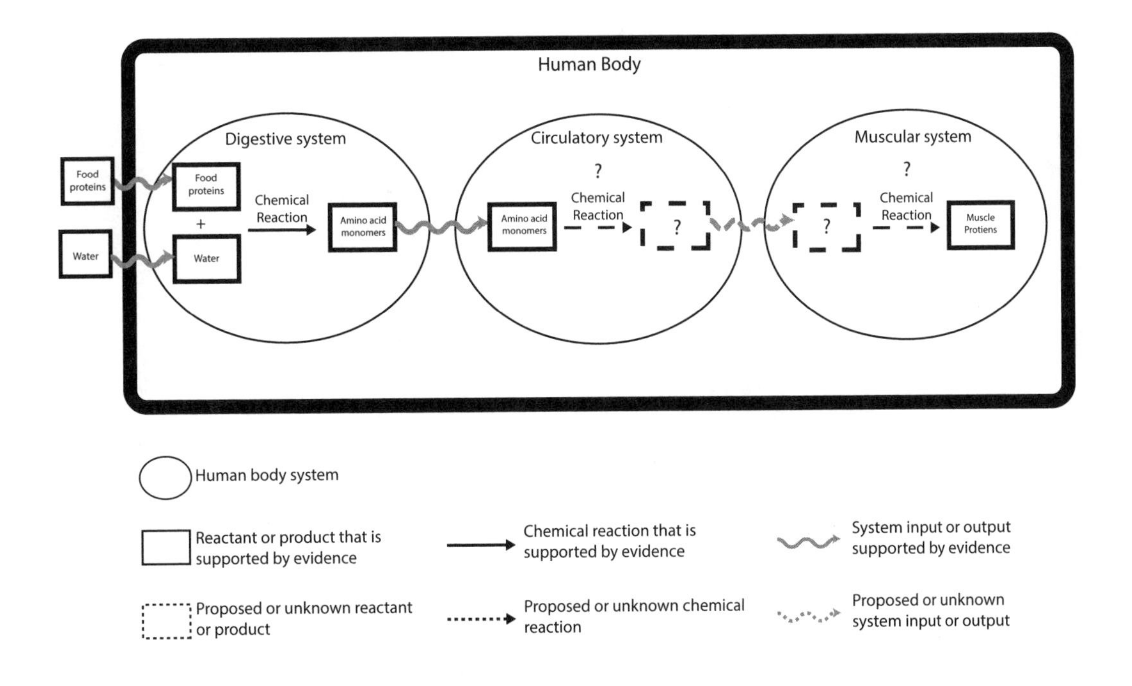

Experiment 3: How Are Muscle Proteins Made?

Introduction

During aging, a gradual decrease in skeletal muscle mass occurs in both rodents (like mice and rats) and humans. This decrease in muscle mass often makes elderly people weaker and affects their ability to move around and their health. Prior studies have shown that when elderly rats were fed high amounts of an amino acid supplement, the rats had more muscle mass and produced more muscle protein (myosin) than similar rats that did not get the amino acid supplement. Scientists wanted to know if the amino acid supplement would also help elderly men produce more muscle protein and if that could stop or slow down the decrease in muscle mass.

Methods

A group of 70-year-old male volunteers were randomly assigned to either an **experimental** or a **control** group. The volunteer men were approximately the same size and weight and in good health. They were informed of potential risks and provided their written consent to participate in the study. The methods were approved by the local committee responsible for overseeing the ethics of research studies of this sort.

The experimental group was fed complete meals (containing proteins, carbohydrates, and fats) plus an amino acid supplement. The control group was fed the complete meals without the amino acid supplement.

To measure the amount of myosin, a muscle protein, produced, researchers injected the volunteers with radiolabeled phenylalanine, an amino acid.

Results

Scientists found that the radiolabeled amino acid they injected had become part of the myosin proteins in the volunteers' muscles. They then compared the amount of radiolabeled amino acid that had been incorporated into myosin proteins for the control and experimental group. They found higher amounts of labeled myosin in the experimental group when compared to the control group. The data are plotted below.

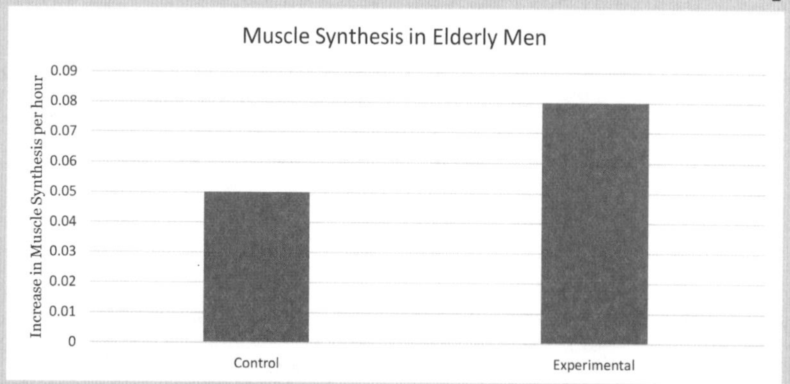

Source: Paddon-Jones, D., M. Sheffield-Moore, X. J. Zhang, E. Volpi, S. E. Wolf, A. Aarsland, A. A. Ferrando, and R.R. Wolfe. 2004. Amino acid ingestion improves muscle protein synthesis in the young and elderly. *American Journal of Physiology-Endocrinology and Metabolism* 286 (3): E321–328.

Teacher Talk and Actions

You can use ball-and-stick models to help students figure out the chemical reaction.

Teacher-led discussion, as suggested below, guides students' interpretation of Experiment 3:

Let's start with the **control** group.

- What was the diet of the control group?

Complete meals (without the amino acid supplement).

- How was protein synthesis measured in the Control group?

Scientists injected members of the control group with a radiolabeled amino acid (phenylalanine). Any protein made after the injection by the control group would have the radioactively labeled amino acid in it. So, the amount of radioactively labeled protein was a measure of how much protein the control group had made.

Now let's look at the **experimental** group. Remember that scientists were testing whether an amino acid supplement caused members of the experimental group to produce more protein than the control group.

- How was protein synthesis measured in the experimental group?

The same way it was measured in the control group: Scientists injected the experimental group with a radiolabeled amino acid and measured how much radioactively labeled protein the treatment group had made.

16. What can scientists conclude from Experiment 3 about the effects of feeding elderly men the amino acid supplement?

Scientists can conclude that volunteers taking the amino acid supplement produced more protein than volunteers not taking the amino acid supplement. From this data, the scientists can conclude that the amino acid supplement caused an increase in muscle protein production.

17. Use 2D ball-and-stick models to help you write the reaction that starts with amino acids and ends with protein.

The ball-and-stick models show that 5 amino acids can react to form a small piece of protein + 4 H_2O.

We know that myosin is made up of 1,937 amino acids, so the reaction that produces myosin could be similar: 1,937 amino acids → myosin + 1936 H_2O.

18. What have we learned from Experiment 3 that can help us revise our model of how our bodies use proteins from food to build muscle proteins?

- *The observation that a radiolabeled amino acid (phenylalanine) injected into the bloodstream ends up in muscle provides evidence that amino acids can get from the circulatory system to the muscular system.*
- *The observation that the radiolabeled amino acid (phenylalanine) becomes part of the myosin protein provides evidence that protein synthesis occurs in the muscle.*

19. Revise your model of how our bodies use proteins from food to build muscle proteins.

- *Student models should show that the chemical reaction that produces protein (and water molecules) from amino acids occurs in the muscular system (see ideal response on TE p. 151).*
- *Students should reason that since amino acids can pass from the digestive system to the circulatory system (Experiment 2) and from the circulatory system to the muscular system (Experiment 3), there is no need for a chemical reaction to occur in the bloodstream of the circulatory system (see ideal response on TE p. 151).*

Teacher Talk and Actions

Optional: If time permits, "react" the amino acids to form a piece of muscle protein. (Ideally students would be able to build a small piece of either actin or myosin from the amino acids they have. If not, they may need to build more.)

At this point, students should have a complete model (example below).

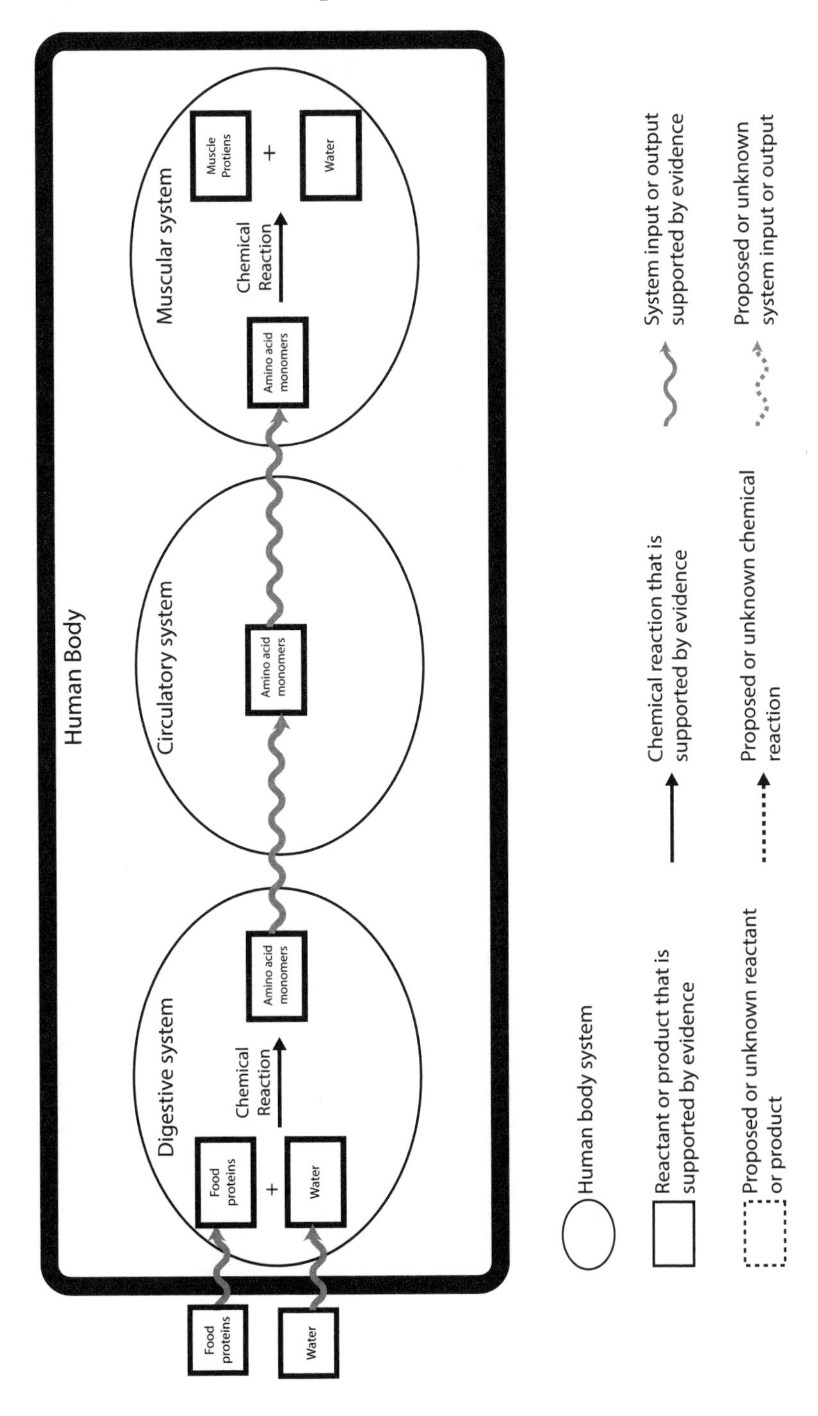

20. Now that you have evidence for all components of your model, use your model to write an explanation for how protein molecules in food become protein molecules that make up muscles.

Protein molecules from food become protein molecules that make up muscles through a series of steps involving chemical reactions and the transportation of matter through body systems. First, protein polymers from food (e.g., ovalbumin, casein, soy protein, wheat gluten) and water molecules (either in food or liquids) are taken into the body and undergo chemical reactions in the digestive system that convert them to amino acids. The amino acids then leave the digestive system, enter into capillaries surrounding the small intestine, and are then carried through vessels of the circulatory system to muscles. From the capillaries surrounding muscle cells, amino acids move across cell membranes to enter the muscle cells.

Once amino acids enter muscle cells, they react to form muscle proteins (e.g., actin, myosin) + water molecules. Evidence for this process was obtained by a variety of experiments that used techniques that allowed scientists to track atoms as they move within and between the body systems.

Teacher Talk and Actions

Activity 2: Converting Carbohydrate Polymers in Food (starch) to Carbohydrate Polymers in the Human Body (glycogen)

Materials

For each team of students
2D models of piece of starch, glucose, and piece of glycogen in plastic sleeves
Body system posters prepared in Lesson 1.2

In the previous activity, we created a model showing how proteins from our food become proteins that make up our body. We revised the model after testing it against data from several scientific studies and used the final model to write an explanation.

In this activity, you will investigate how carbohydrate molecules from your food become carbohydrate molecules that make up your body and create a model that is consistent with your observations. Then your group will test your model by looking at several experiments and revise the model if necessary.

Procedures and Questions

One example of a carbohydrate polymer from our food is **amylose.** In Lesson 1.2 you learned that amylose is found in foods like bread and pasta and is a single chain made from **glucose monomers.**

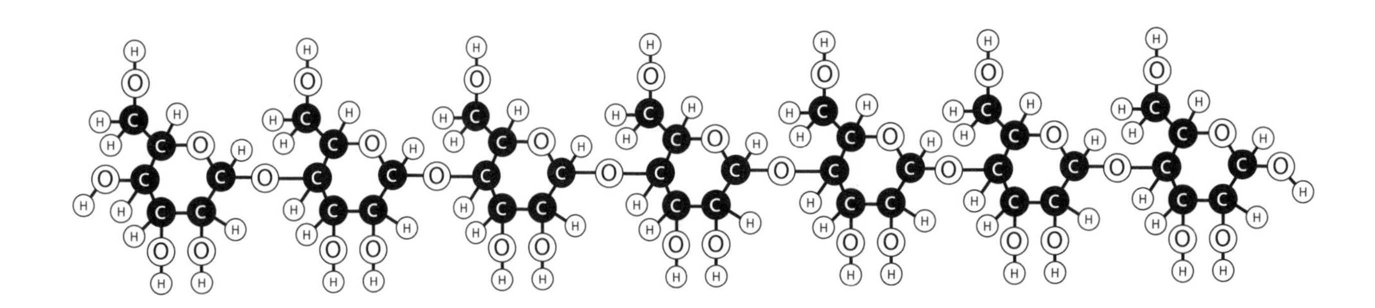

Teacher Talk and Actions

Safety Note
Wear safety goggles or glasses with side shields during the setup, hands-on, and take down segments of the activity.

Point out to students the glucose monomers and that amylose is a single chain of glucose monomers while glycogen has a branch.

You also learned in Lesson 1.2 that another carbohydrate polymer, called **glycogen,** is stored in the human body in liver and muscles cells. Like amylose, glycogen is composed of **glucose monomers.** However, instead of being a single long chain, glycogen is a branched chain. The difference in branching pattern gives glycogen different properties from amylose.

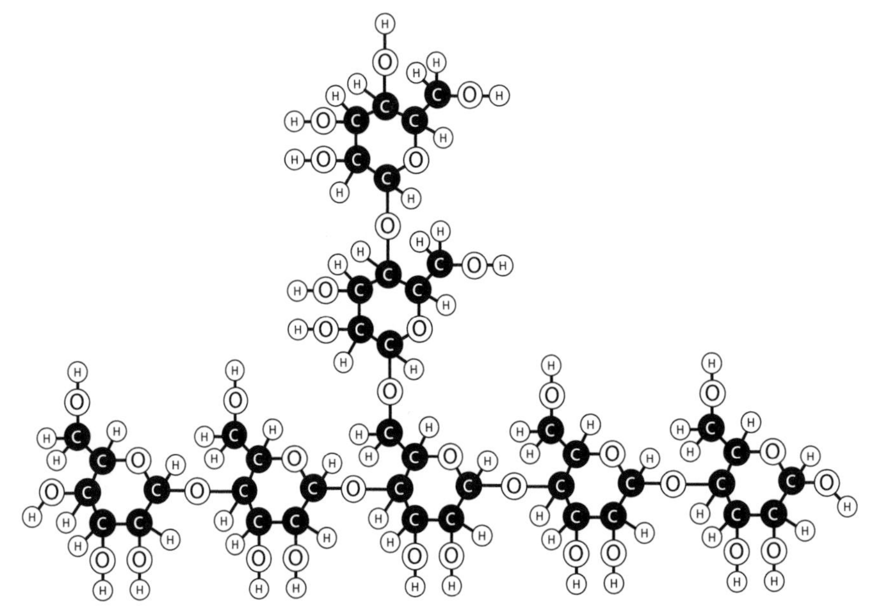

Work through the steps below with your group to create a model that represents how the body converts amylose in food to glycogen that is in human liver and muscle cells.

Designing a model

1. What body systems do you think play a role in converting amylose to glycogen? Explain why you think each system is relevant.

   Digestive system: small intestine and cells lining it and amylose, liver and liver cells and glycogen

   Circulatory system: blood vessels, including capillaries, blood

   Muscular system: muscles and muscle cells, glycogen

2. What molecules do you think could be inputs or outputs? Explain why you think each molecule is relevant.

   Carbohydrate molecules in our food, such as amylose

   Carbohydrate molecules in our body, such as glycogen

   [Some students may say water because water is a reactant of polymer digestion to monomers.]

3. Are any chemical reactions relevant? Explain.

   Chemical reactions: [Some students might reason from the protein example to list amylose + water → glucose in the digestive system and glucose → glycogen + water in the liver and muscle.]

Teacher Talk and Actions

Creating a model

4. Create your model. Remember to clearly label the systems, inputs and outputs of each system, each system's components, and chemical reactions you have evidence for. Make a note on anything you are unsure of or do not have evidence to support. For example, you may want to write a question mark or "no evidence."

Teacher Talk and Actions

Give student groups 10–15 minutes to create their model. Circulate around the class, asking guiding questions like those in Activity 1 if groups seem stuck.

If specific groups have difficulty creating a model, remind them of the model they created as a class in the previous activity. Also remind them of the body system hierarchy maps they created in Lesson 1.2 and the tables they created to summarize the information. Students should notice that the carbohydrate polymer glycogen is listed in both the muscular and digestive systems, suggesting that both of these systems should be included in their models. Students might reason that since the circulatory system carries monomers from protein digestion to the cells, it might also carry monomers from the digestion of carbohydrate polymers, e.g., glucose. Their hierarchy maps show that glucose is a monomer that makes up carbohydrate polymers like glycogen. You should also ask groups questions about their model if they forget specific elements or have model components that aren't relevant to the question of carbohydrate digestion and synthesis. Example questions include the following:

- "Why did you include that system in your model?"
- "What are the inputs and outputs to the system?"
- "Do you have evidence for adding that to the model?"
- "Should you put a question mark next to it to indicate that you don't have evidence?"

A groups' initial model will likely have missing elements and have many question marks. If their reasoning for their model is valid, you may let them proceed through the "testing your model" and "revising your model" stages and help them revise their model later.

After the groups have created, tested, and revised their models you will come together as a class to create a class model. At that time groups should have most of the elements, even if they don't have all of them.

Testing our model against evidence

5. For each of the following experiments, examine the data and decide what evidence you can use to revise your model.

Experiment 1: Monitoring Glucose Appearance and Disposal Following Consumption of a Starch-Rich Meal: Comparison of Male and Female Subjects

Introduction
This study was designed to investigate the initial effects of consumption of a starchy meal on the concentration of blood glucose in a group of healthy men and women.

Methods
Twelve subjects (six men and six women, matched for age and body mass index) participated in this study. All volunteers were healthy and none had recently taken medication likely to affect what was being monitored in the study. All subjects gave written informed consent to the study.

Subjects ate a special high-starch test meal consisting of cooked peas. The peas contained a large amount of the carbohydrate amylose that contained isotopically labeled carbon atoms. The scientists measured the amount of isotopically labeled glucose in the bloodstream at 30-minute intervals over eight hours after an overnight fast. After two hours of collecting baseline measurements, subjects consumed the test meal.

Results
The scientists found the labeled carbon atoms were in glucose molecules in the blood stream. They then quantified the changes in glucose concentrations in the blood throughout the study. Their results are shown in Figure 1.

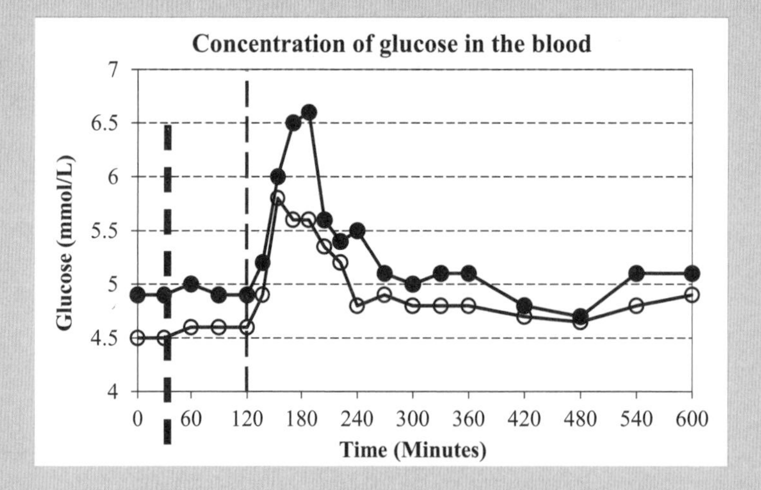

Figure 1. Plasma glucose response following a starch-rich test meal in groups of male (•) and female (o) subjects. (Vertical dashed line at 120 minutes indicates when test meal was consumed.)

Source: Robertson, M.D., G. Livesey, and J. C. Mathers. 2002.Quantitative kinetics of glucose appearance and disposal following a 13 C-labelled starch-rich meal: Comparison of male and female subjects. *British Journal of Nutrition* 87 (6): 569–77.

Teacher Talk and Actions

Circulate around the class, asking guiding questions like those in Activity 1 if groups aren't sure how to answer questions or revise their model.

Note that it is OK if not all the groups get all the components of their model. There is time at the end for the class to come together and create a final model.

6. What evidence do we have that a chemical reaction occurred? Explain your answer.

The participants were given a diet of amylose with labeled carbon atoms, and those carbons atoms were found in glucose molecules in the blood. This indicates a chemical reaction occurred.

7. Write a word equation for the chemical reaction that occurs in the small intestine. You may use the diagrams above or ball-and-stick models to help figure out the reaction.

$\text{Amylose}_6 + 5\ H_2O \rightarrow 6\ C_6H_{12}O_6$

8. What can scientists conclude about the concentration of glucose in the blood after eating the amylose-rich meal?

In both males and females, the concentration of blood glucose increased to a maximum about 1 hour (180 minus 120 minutes) after a meal and then decreased to baseline after 6 hours (480 minus 120 minutes). The initial increase was greater for females than for males.

9. Is your model consistent with the experimental data? If not, make any revisions your group thinks are necessary.

- Add the chemical reaction to the digestive system.
- Add water as an input to the digestive system.
- Add glucose as an output of digestive system and an input to the circulatory system.

Teacher Talk and Actions

Circulate around the class to identify groups that are struggling and provide hints. The guiding questions in Activity 1 may be helpful here as well.

For Steps 6–7, using the 2D models in plastic sleeves and erasable markers, students should identify atoms making up glucose monomers that are not in the piece of starch and use that information to identify water as a reactant of the chemical reaction.

Students could write "amylose → glucose" on their models and reason by analogy from protein digestion that water molecules are probably reactants. However, students will need to look at models (as instructed to do in Step 6) and write a reaction equation to be sure.

Students should look at the first experiment and think to

- add water as an input to the small intestine of the digestive system in the model,
- add the chemical reaction to the small intestine box, and
- add glucose as an output of the small intestine and an input to the circulatory system.

Note that it is OK if not all the groups get all these components in their model. There will be time at the end for the class to come together and create a final model with all these elements.

Experiment 2: Effect of Glucose on Glycogen Deposition in Rat Liver

Introduction

The liver plays a central role in controlling the concentration of glucose in the blood. Immediately after a meal, substances produced by the liver stimulate the pancreas to produce more insulin. Insulin increases the absorption of glucose by body cells, thereby reducing the level of blood glucose somewhat. However, the amount of glucose removed from the blood by body cells does not account for the large decrease in blood glucose observed after its initial spike about an hour after a meal. The scientists hypothesized that the liver may remove glucose from the blood and use it to make glycogen.

Methods

Twelve male Sprague-Dawley rats weighing 100–140 grams were fasted for 18–22 hours, sacrificed, and their livers removed for the study. The livers were bathed in a solution similar to normal body fluids (physiologic medium).

Half the livers were infused with physiologic medium and the other half were infused with physiologic medium with added glucose.

Results

Changes in the amount of glycogen formed in the liver with and without glucose are shown in Figure 2.

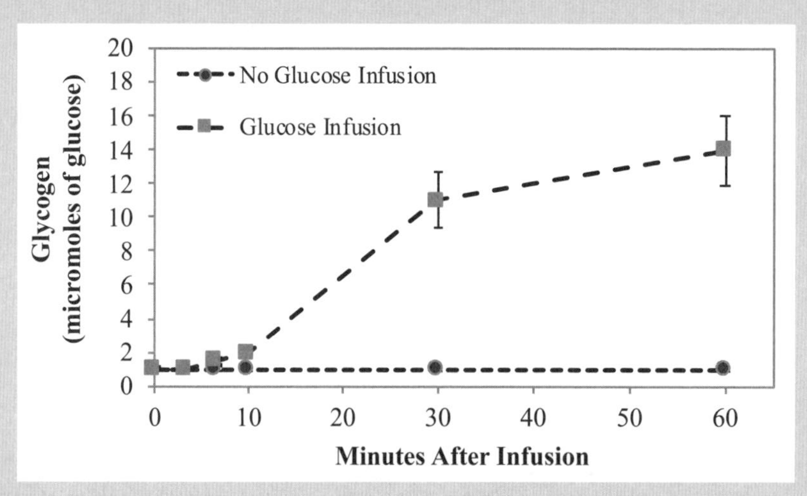

Figure 2. Time course of the effect of glucose infusion on the level of glycogen in perfused livers from fasted rats.

Source: Buschiazzo, H., J. H. Exton, and C. R. Park. 1970. Effects of glucose on glycogen synthetase, phosphorylase, and glycogen deposition in the perfused rat liver. *Proceedings of the National Academy of Sciences* 65 (2): 383–387.

10. What can scientists conclude about glycogen formation from the data?

Glucose infusion into the blood increases the amount of glycogen formed in the liver.

Teacher Talk and Actions

Circulate around the class, asking guiding questions like those in Activity 1 if groups aren't sure how to answer questions or revise their model.

Note that it is OK if not all the groups make the same revisions to their model. There is time at the end for the class to come together and create a final model.

Students should look at the second experiment and think to

- add glucose as an input to the muscle system and
- add glucose as an output to the circulatory system.

11. Write a word equation for the chemical reaction that occurs in the liver and muscle. You can use the diagrams above or ball-and-stick models to help figure out the reaction.

6 glucose ($C_6H_{12}O_6$) → glycogen 6 + 5 H_2O

12. Based on the experimental data, does your group need to revise its model? Explain your answer.

Add glucose as input to muscular system.

Add glucose as output from circulatory system.

Experiment 3: The Conversion of Glucose to Glycogen in the Liver and Muscles of Rats

Introduction
A scientist investigates where isotopically labeled carbon atoms in glucose end up in the liver and muscle.

Methods
Adult male rats were fasted for 24 hours and then glucose that contained isotopically labeled carbon atoms was injected into the rats' circulatory system. After one hour, the rats were sacrificed. The scientist located the isotopically labeled atoms in the body.

Results
An hour after injecting the rats with carbon-labeled glucose, the labeled carbon atoms were in liver glycogen and in muscle glycogen. In addition, the scientist found that the location of the labeled carbon atoms in glucose did not change when they became a part of the glycogen molecule.

Source: Marks, P.A., and P. Feigelson. Pathways of glycogen formation in liver and skeletal muscle in fed and fasted rats. *Journal of Clinical Investigation* 36 (8): 1279–1284.

13. What evidence do we have that a chemical reaction occurred? Explain your answer.

The isotopically labeled carbon atoms in glucose were found in glycogen, indicating a chemical reaction occurred where glucose was a reactant and glycogen was a product.

14. Write a word equation for the chemical reaction that occurs in the small intestine. You may use the diagrams above or ball-and-stick models to help figure out the reaction.

6 glucose → glycogen_6 + 5 H_2O

15. Based on the experimental data, does your group need to revise its model? Explain your answer.

- Add glucose as input to muscular system.
- Add water to the muscular system.
- Add chemical reaction (glucose → glycogen + water) to the muscle system.

Teacher Talk and Actions

For Step 11, using the 2D models in plastic sleeves and erasable markers, students should identify atoms making up glucose monomers that are not in the piece of glycogen and use that information to identify water as a product of the reaction.

As a result of Step 12, students should

- add glucose as an output of the circulatory system and an input to the liver,
- add water as an output of the liver, and
- add the chemical reaction between glucose molecules to form glycogen and water molecules to the liver system.

Circulate around the class, asking guiding questions like those in Activity 1 if groups aren't sure how to answer questions or revise their model.

After the third experiment students should think to

- add the chemical reaction glucose → glycogen + water,
- add glucose as an output to the circulatory system, and
- add water to the muscular system.

Forming a class model

We will review each group's model to see what the models have in common and how they differ. Our goal is to have a model that the whole class agrees with and that is consistent with all of the evidence that was collected in the three experiments. Draw the model below:

Class consensus model of the Carbohydrate Digestion and Synthesis Model that you and your students may come up with:

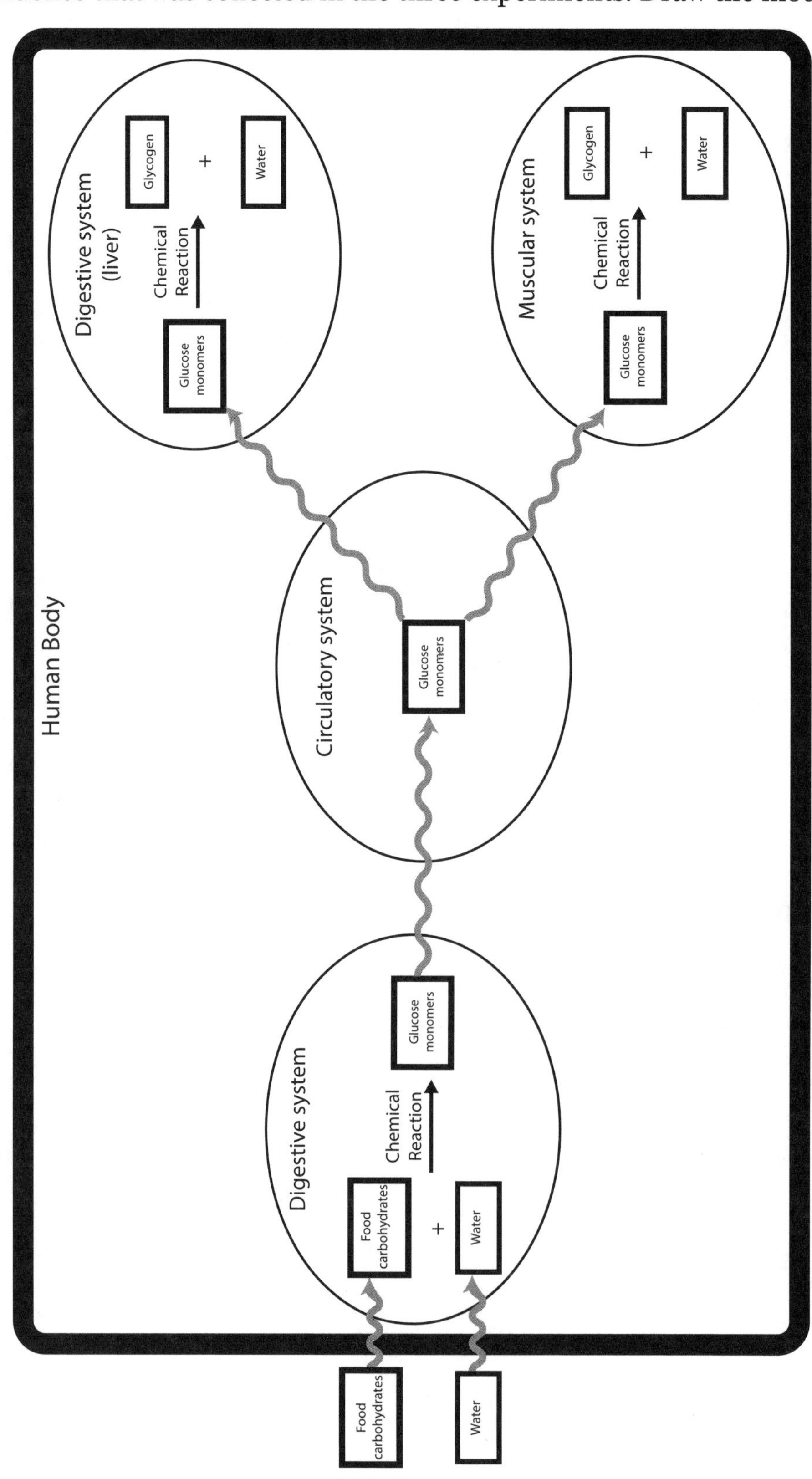

Teacher Talk and Actions

Forming a consensus model:

After groups have completed their models, bring the whole class together for a discussion that leads to the creation of a consensus model that represents how the students think carbohydrates in our food become carbohydrates in our body.

Use this discussion to get students to help you create a model on the blackboard/whiteboard.

Groups could have similar models, so you may want to poll the class to save discussion time. For example, "How many groups had the digestive system in their model?" If all groups raise their hands, you can add that system to your class model and proceed to the next question.

Examples of discussion questions:

•What systems did your group include in its model? (If there are significant differences in the systems groups used, ask them to explain why those systems are in their model.)

•What components did you include for each system and what were the inputs and outputs? (Go system by system and have groups answer. Ask students what their models have in common and how they differ.)

Final model:

By the end of this discussion your class should have created a model that they can use to write their explanation for how carbohydrates in food become carbohydrates in the body. At a minimum, the model should input all the components shown in the example model on the facing page.

Writing an explanation

16. Based on the consensus model, write an explanation for how carbon atoms in amylose become carbon atoms in glycogen. Be sure to include

- chemical reactions that are important;
- body systems involved in the process, including each system's inputs and outputs; and
- evidence from the experiments and model that support your explanation.

Carbon atoms of amylose (starch) polymers from food become carbon atoms of glycogen polymers that make up the liver and muscles through a series of steps involving chemical reactions and the transportation of matter through body systems. First, amylose polymers from food and water molecules (either in food or liquids) are taken into the body and undergo chemical reactions in the digestive system that convert them to glucose monomers. The carbon atoms that were part of amylose are now part of glucose. The glucose then leaves the digestive system, enters into capillaries surrounding the small intestine, and is then carried through vessels of the circulatory system to the liver and muscles. From the capillaries surrounding liver and muscle cells, glucose molecules move across cell membranes to enter the liver and muscle systems. Once inside these systems, they react to form glycogen and water molecules. Now the carbon atoms that are part of glucose become carbon atoms of glycogen. This is how carbon atoms that were initially part of amylose become carbon atoms of glycogen. Evidence for this process was obtained by a variety of experiments that used techniques that allowed scientists to track atoms as they move within and between the body systems.

Teacher Talk and Actions

Optional: The reaction is "enabled/made possible/facilitated" by carbohydrate-digesting enzymes such as amylase.

Students will learn in Lesson 2.7 that between meals, the reverse reaction occurs in liver cells, releasing glucose into the capillaries.

Science Ideas

The activities in this lesson were intended to help you understand important ideas about chemical reactions that take place in various body systems and how models can help us clarify our thinking. Read the ideas below. Look back through the lesson. In the space provided after the science idea, give evidence that supports the idea.

Science Idea #9: The large carbon-based molecules that make up an animal's food must be broken down so that cells can use them. These molecules, mainly protein, carbohydrate, and fat polymers, are broken down in the digestive system into smaller carbon-based molecules (monomers) such as amino acids, glucose, and fatty acids during chemical reactions with water molecules.

Evidence:

Protein polymers in food → monomers:

- *2 hours after eating a protein-rich meal, the amount of protein in the small intestine decreased and the amounts of amino acids in the small intestine increased.*

Carbohydrate polymers in food → monomers:

- *1 hour after eating isotopically labeled cooked peas (rich in amylose), the amount of isotopically labeled glucose in the blood increased. (This is evidence that digestion must have occurred.)*

In both examples, molecular models showed that water molecules were also needed to form the monomers.

We didn't examine evidence of fat digestion.

Science Idea #10: Within body cells, monomers from food can react with one another to form polymer molecules that become part of an animal's body. Water molecules are also produced during polymer formation.

Evidence:

- *Myosin protein formation: after injecting labeled amino acids into blood vessels of human volunteers, the amount of labeled myosin protein increased.*
- *Glycogen formation: after injecting ^{14}C-glucose into blood vessels of rats, the amount of ^{14}C glycogen increased in both livers and diaphragm muscle.*
- *We didn't examine evidence of fat synthesis.*

Teacher Talk and Actions

Science Idea #11: Multicellular organisms need a way to transport molecules from one body system to another. The circulatory system (including a heart that pumps blood and vessels of various diameters, including capillaries) transports the reactants and products of various chemical reactions in response to changes in the body's needs.

Evidence:

- *Labeled amino acids injected into an arm vein are incorporated into muscle protein (providing evidence the circulatory system carried the amino acids from the arm vein to muscles).*
- *Labeled glucose injected into blood vessels of rats was incorporated into glycogen in liver and muscle (providing evidence the circulatory system carried the glucose from the injection site to liver and muscle cells).*

Science Idea #12: Models are useful for representing systems and their interactions—such as inputs, outputs, and processes—and energy and matter flows within and between systems.

Evidence:

Drafting models helped us (a) think about how atoms/monomers from food flow within the digestive system and between digestive, circulatory, and muscular systems and (b) examine and interpret data in terms of inputs and outputs of matter and the rearrangement of atoms during chemical reactions.

Teacher Talk and Actions

Students do not yet have evidence for the last phrase (shown in gray text) of Science Idea #11. After completing Lesson 2.9, students will have evidence for this part of the idea.

Evidence for Science Idea #12 should come from Lesson 1.4 as well as Lesson 1.5. Students do not yet have evidence for energy flows within and between systems. In Lesson 2.2, students will find evidence for the usefulness of models in representing energy flows within and between systems, encounter Science Idea #12 again, and this time they will be asked to list the evidence that supports the energy part of this idea.

Pulling It Together

Work on your own to answer these questions. Be prepared for a class discussion.

1. Now that you have had more experience with chemical reactions, how would you answer the Key Question: **How do our bodies turn molecules from food into molecules that make up our body?**

Students should summarize Science Ideas #10 and #11 in their own words.

Teacher Talk and Actions

2. You've created models for how protein and carbohydrate molecules in food become molecules in the body. What about fats? Can fat molecules in our food become fat molecules in our body?

a. Create a model for how fat molecules in our food could become fat molecules in our body. Represent or describe your model in the space below.

Fats could go through a chemical reaction in the digestive system to produce monomers, such as fatty acids. (We examined models of fatty acids in Lesson 1.2.) Fatty acids could then be transported to various parts of the body via blood vessels making up the circulatory system. They could go through a chemical reaction to form triglycerides and be stored in fat cells or be converted to phospholipids that make up cell membranes.

b. What kind of experiments would be helpful to test your model?

Experiments where fats with some of the atoms labeled are eaten would be helpful, since we would be able to track where the atoms associated with fats end up in the body.

Teacher Talk and Actions

For Question 2, students can use the 2D models in plastic sleeves of fat molecule from food, fatty acids, fat molecule from body, and membrane phospholipid to identify water as a reactant in the digestion of fat molecule from food to fatty acids and water as a product of the synthesis of both fat molecule from body and membrane phospholipid.

The video can be shown before or after the discussion of what students have learned in Chapter 1. For each bullet below, students could be asked to give an example.

Show the video *From Food to Body Structures* (5 minutes)

Lead a brief discussion about what students have learned in Chapter 1:

- The importance of chemical reactions in converting large molecules (polymers) from food into the large molecules (protein, fat, and carbohydrate polymers) that make up our body structures
- The contribution of components of various body systems to this process
- The usefulness of thinking about the body as a system that includes sub-systems and considering matter inputs, outputs, and processes (chemical reactions, transport processes) to figure out what is going on
- The interplay of models and data in explaining growth phenomena

And about what we still need to learn:

- We haven't figured out where the energy comes from for competing in races and just staying alive.

Tell students that Chapter 2 will focus on energy.

3. You know from the food cards in Lesson 1.2 and from Activity 2 of this lesson that plant-based foods are made up of amylose polymers. In Lesson 1.4 you explained how the mass of a plant could increase by thinking about systems, inputs, outputs, and processes. However, the only chemical reaction you considered was photosynthesis, which forms glucose not amylose. Use this information to create a model that represents how plants grow by making amylose.

Your model should include relevant systems, system components, system inputs and outputs, and system processes.

Minimally, students should have a plant system, inputs consisting of CO_2 + H_2O, the photosynthesis chemical reaction, and the polymerization chemical reaction of glucose monomers to amylose + water.

Teacher Talk and Actions

Students who used *THSB* may be familiar with the reaction that converts glucose monomers to cellulose polymers and water molecules. (See *THSB* Lesson 3.4, pp. 294–309 in the Teacher Edition).

Chapter 2

Making Sense of Energy Changes Involved in Human Growth and Activity

Teacher Edition

Chapter 2 Overview

Chapter 2 tells the basic story of how our bodies use energy-releasing chemical reactions—mainly the oxidation of monomers from food polymers—to drive energy-requiring chemical reactions and processes involved in growth and motion. Much of the reasoning from evidence that students are expected to do rests on the idea that energy is conserved during chemical reactions. The unit offers no quantitative proof for this idea but does provide evidence for the more intuitive idea that an energy increase (or decrease) in one system is always accompanied by an energy decrease (or increase) in another system that interacts with it.

In contrast to the matter story in Chapter 1, where concrete physical models are used to help students account for what happens to all the atoms involved in a chemical reaction, concrete objects are not used to represent energy. Concrete objects could give students the mistaken impression that energy is a material substance and exacerbate the difficulties students have in distinguishing changes in matter from changes in energy (though of course they are related). To avoid reinforcing this confusion, the chapter engages students in using bar graphs and system boxes to represent changes in energy within systems and energy transfer models to represent transfers of energy between systems. To help students become comfortable with these abstract representations, the chapter first engages them in representing energy changes in simple physical systems where indicators of energy changes can be readily observed and/or measured (e.g., changes in motion when billiard balls collide, changes in temperature when a flask containing hot water is immersed in a beaker of cooler water). Then, students are asked to use the representations to make sense of energy changes involving one or more chemical reaction systems and transfers of energy between systems.

Energy-releasing reactions, such as the reaction between hydrogen and oxygen to form water and the combustion of glucose to produce carbon dioxide and water, are introduced in physical systems where indicators of energy release are readily observed. Then, building on their knowledge of the reactants and products of glucose combustion and that monomers such as glucose result from digestion, they are ready to (a) interpret data that provides evidence that the same reactants and products are involved in the energy-releasing oxidation of glucose in cells and (b) make predictions about how it might be coupled to muscle contraction.

Similarly, energy-requiring reactions and processes are also introduced in simple physical systems: a toy solar car moves farther under a higher-intensity light than under a lower-intensity light; more oxygen and hydrogen gas are produced when water is connected in a complete circuit to a new battery versus a used battery; and plants produce more oxygen gas under a higher-intensity light than under a lower-intensity light. This foundation enables students to interpret data that provide evidence for an energy-transfer model from the Sun system to the photosynthesis system in plants (TE p. 314). Other evidence enables students to construct a complex energy-transfer model that couples cellular respiration to ATP synthesis and ATP breakdown to the sliding of muscle filaments (shown on the next page) and to protein synthesis (TE p. 426).

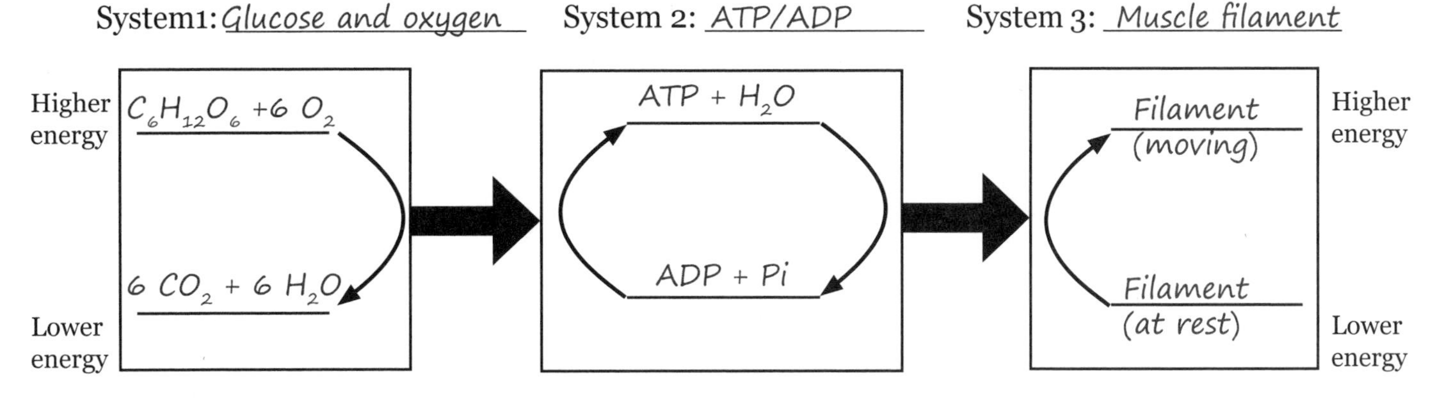

Chapter 2 also tackles a misconception that is prevalent among high school and college students—the incorrect idea that energy is stored in the bonds of "energy-rich molecules" such as glucose and is released when the bonds are broken, rather than the correct idea that a large amount of energy is released when glucose and oxygen react to form carbon dioxide and water because the energy required for breaking bonds of glucose and oxygen is considerably less than the amount of energy released when bonds of carbon dioxide and water form. Introducing the ideas that (a) energy is always required to break bonds between atoms of reactant molecules and (b) energy is always released when bonds form between atoms to make product molecules allows students to develop a coherent picture of energy-releasing and energy-requiring reactions and to understand why some energy must be added to get an energy-releasing reaction started (e.g., a donut must be ignited in order to burn and hydrogen gas and oxygen gas must be sparked to react to form water).

By the end of the chapter, students are ready to use ideas and models about matter and energy changes associated with chemical reactions within systems and energy transfer between systems to show how a hamster running on a wheel makes a light glow and how a corn plant uses energy from the Sun to build its body structures.

The unit's final project engages students in considering how the body maintains a nearly constant environment in its muscles during intense exercise. This connects what students learned about the role of body systems in storing digestion products for later use, matter and energy changes and energy transfers during exercise, and students' experiences with increases in breathing and heart rate during exercise.

Lesson Overviews

Lesson 2.1 To begin to make sense of phenomena related to energy changes, students must first be able to detect them. To answer the question "How can we detect energy changes?" students observe several phenomena involving energy changes and compile a list of indicators of energy changes. In the first phenomenon, students observe and compare changes in motion when athletes run and play tennis, changes in motion when two billiard balls collide, changes in motion of a battery-powered toy car with a new battery versus a used battery, changes in motion of a solar-powered toy car under a 100 w versus a 40 w light bulb, and changes in motion in coffee grounds in response to sound. Students then observe temperature changes that result when a flask of hot water is placed in a beaker of cooler water. Finally, students observe a video of a hamster running on a wheel that is connected to a light that flashes when the wheel spins fast. Students discuss the observations they made that provide evidence of changes in energy and see that increases in one object's energy tend to accompany decreases in another object's energy when the two objects interact. After generalizing ideas about energy changes and energy transfers, students use them to describe energy changes and transfers when a spring shoots a ball in a pinball machine and a cell phone is used.

Lesson 2.2 The lesson highlights the role of models in representing systems and their interactions in simple physical systems in preparation for using similar models to represent energy changes and transfers in chemical reactions in complex living systems. For each of the physical systems observed in the previous lesson, students use bar graphs and energy-transfer models to represent energy changes within a system and energy transfers between two systems.

Lesson 2.3 To answer the question "How can we tell if chemical reactions involve changes in energy?" students observe three systems in which chemical reactions release energy as the reactants are converted to products—forming water from hydrogen and oxygen, a candle reacting with oxygen in the air to produce carbon dioxide in water, and a donut reacting with oxygen in the air and with pure oxygen to produce carbon dioxide and water—and list indicators that energy was released in each reaction. Students use bar graphs and energy-transfer models to represent the energy changes they observe in each of the three reactions and the direction of energy transfer from each system to its surroundings. Additionally, students use a table of enthalpy data (this term is not used in the unit) to compare the amounts of energy released from various combustion reactions in living and nonliving systems. The donut-burning-in-air example is also used to address the misconception that the energy released during combustion comes solely from the fuel. Students observe the same donut being burned in air (~20% oxygen) and in liquid oxygen (~100% oxygen) and see that more energy is released during the reaction with 100% oxygen in the same amount of time.

Lesson 2.4 To answer the question "Do all chemical reactions release energy?" students observe three systems in which a net input of energy is required to convert the reactants to products—breaking water into hydrogen and oxygen gas, the reaction of ammonium thiocyanate with barium hydroxide, and the synthesis of glucose and oxygen by an aquatic plant. Students compare the role of a battery in increasing the motion of a battery-powered toy car to its role in producing bubbles during the breakdown of water. They also compare the role of light in increasing the motion of a solar-powered toy car and in increasing the production of oxygen and glucose from carbon dioxide and water. Students use bar graphs to represent the energy changes during these reactions and energy-transfer models to represent the direction of energy transfer between systems. Finally, students use data to compare the energy required to carry out various biological processes (e.g., synthesis of starch, synthesis of fat, synthesis of protein).

Lesson 2.5 To answer the question "Why do energy-releasing chemical reactions require an initial input of energy?" students observe that while the number of each type of atom doesn't change during a chemical reaction, the types of bonds between atoms do change, which leads to the question of whether changes in energy result from changes in the arrangement of atoms. Students are given the rules that bond breaking requires an input of energy and bond forming releases energy and use them to represent hypothetical intermediates in the water splitting, glucose combustion, and photosynthesis reactions on bar graphs. Using data on bond energies, students calculate the energy associated with breaking and making bonds in these three reactions and observe that, in each case, their bond energy calculations are consistent with their observations about whether these reactions release energy or require a net input of energy. Moreover, representing the energy associated with a hypothetical intermediate (consisting of separate atoms) with a higher-energy bar than that of either reactants or products allows students to make sense of their observations that both the oxidation of hydrogen and oxidation of glucose reactions require an initial input of energy even though the overall reactions release energy.

Lesson 2.6 To answer the question "Can energy released by an energy-releasing chemical reaction provide the energy for an energy-requiring chemical reaction?" students use the analogy of two circus performers on a seesaw, in which the seesaw transfers energy from the downward-moving performer to lift the other performer upward, to think about the coupling of an energy-releasing chemical reaction to an energy-requiring reaction. Students then consider how a chemical reaction occurring inside the battery transfers energy to the water in the water-breaking reaction from Lesson 2.4 and represent the transfer using an energy-transfer model. In the third activity, students consider how energy released from the combustion of octane is transferred to move a car. Students then use what they have learned about energy transfer in simple systems to model and explain why an aquatic plant produces more oxygen gas when grown under a 100 w light bulb than under a 40 w light bulb.

Lesson 2.7 To answer the question "Does our body use the energy from chemical reactions to move and build body structures?" students collect and analyze data that provide evidence that the oxidation of glucose to carbon dioxide and water provides energy for motion during exercise. Students (a) use a limewater test to compare the amount of carbon dioxide exhaled before and after exercise, (b) analyze and interpret data from a published study on the effect of exercise duration on oxygen uptake and on the effect of exercise intensity on glucose uptake, (c) use models to suggest that H_2O is a likely product, (d) use all the data to write an equation for an energy-releasing chemical reaction that could provide the energy for exercise, and finally (e) represent the transfer of energy from the glucose + oxygen reaction to an athlete's motion on an energy-transfer model. A video zooms from a tennis player serving a ball down to the sliding of fibers of actin and myosin past one another that leads to muscle shortening. Students predict what will happen when a glucose solution is dripped on a preparation of isolated muscle fibers and represent their predictions with an energy-transfer model. When shown that, contrary to predictions, dripping glucose on muscle fibers does not lead to shortening but that dripping an ATP solution does lead to shortening, students revise their energy transfer model to take account of the finding.

In the second activity, students investigate whether the same chemical reaction can provide energy needed for building body structures. The lesson ends by focusing students on the observation that muscles can begin exercising before sufficient oxygen is available, which leads to the question of whether chemical reactions that don't require oxygen could supply energy for motion.

Lesson 2.8 To answer the question "What other chemical reactions can organisms use to release energy?" students examine data on bacterial growth and human exercise intensity. Students observe that *Lactoccocus lactis,* the bacteria used in making yogurt, grow without oxygen (though more slowly than with oxygen) and that the bacteria use glucose and produce lactic acid as they grow. Students then examine data that provide evidence that humans also produce lactic acid from glucose and that lactic acid production increases with exercise intensity. The lesson concludes by having students examine the contributions of anaerobic and aerobic systems to running events from 100 meters to a marathon.

Lesson 2.9 To answer the question, "How do various body systems respond to the matter and energy demands of exercise?" students (a) use their knowledge of cellular respiration to predict matter and energy changes in muscle cells during exercise and consider how to test their predictions; (b) examine data on the body temperature of a cross-country team before and after a race that challenge their predictions; and (c) examine data showing the acceptable range of oxygen, carbon dioxide, and glucose levels in muscle cells that challenge their predictions and use information from a homeostasis handout to find out how various body systems maintain the levels of those substances within the acceptable range.

Lesson Guide
Chapter 2, Lesson 2.1
Observing Energy Changes All Around Us

Focus: This lesson begins the story of how energy changes in chemical reactions contribute to body motion by starting with energy changes in simple systems. Students learn that changes in motion or temperature or the emission or absorption of light or sound all indicate that energy is changing in a system.

Key Question: How can we detect energy changes?

Target Idea(s) Addressed

Changes in temperature or motion and the production or absorption of light or sound are indicators of energy changes. (Science Idea #13)

When the energy of one system increases, the energy of another system or systems decreases. Likewise, when the energy of one system decreases, the energy of another system or systems increases. (Science Idea #14)

When two systems interact, and one system decreases in energy while the other system increases in energy, we say that energy was transferred from the first system to the second. (Science Idea #15)

Science Practice(s) Addressed

Analyzing and interpreting data
Carrying out investigations
Obtaining, evaluating, and communicating information

Materials

Introduction: *Per class:* Videos: *Olympic Athletes* and *Tennis Player*

Activity 1: *Per group of students:* new, unused 9-volt battery; partially used 9-volt battery; battery-powered toy car; 40 w light bulb with reflector attached to ring stand with solar-car track taped in position underneath; 100 w light bulb with reflector attached to ring stand with solar-car track taped in position underneath; solar-powered toy car; *per class:* battery testers; videos: *Balls Colliding, Sound-Powered Coffee Grounds, Wheel and Battery,* and *Solar Car in Sunlight and Shade* (optional)

Activity 2: *Per group of students:* 400 ml beaker; 125 ml flask; 2 thermometers; *per class:* hot water, cold water (see Advance Preparation)

Activity 3: Video: *Hamster*

Advance Preparation

Organize materials in advance so that each group has materials when class begins.

Activity 1: Teachers should set up this experiment as stations—four solar-car stations and four battery-car stations. For the solar-car station, see photo in online resources for setup that will enable you to control variables such as height of bulb and starting position of car under bulb.

Activity 2: *Per class:* Heat 1,000 ml water (hot plate on setting 1 recommended to heat water to 60–70°C); Chill 1,000 ml water to between 2–5°C. (It is OK to cool the water in the refrigerator but be sure that there is no ice in the water during the investigation! Use insulated containers to maintain temperature.)

Phenomena, Data, or Models	Intended Observations	Purpose	Rationale or Notes
Activity 1 Two balls collide. Sound is used to move coffee grounds. A battery is used to power a toy car. Sunlight is used to power a toy car.	Balls colliding • The rolling ball hits a stationary ball. The rolling ball slows down and the stationary ball starts to roll. A sound is heard when the balls collide. Sound-powered coffee grounds • Coffee grounds move more when the frying pan makes a louder sound than when it makes a quieter sound. Battery-powered toy car • The wheels of the car spin and the car moves when it is connected in a closed circuit to a new 9-volt battery. • The wheels of the car stop spinning when the battery is disconnected. • The wheels of the car spin slower and hum at a lower pitch when the car is connected to a partially used 9-volt battery. Solar-powered toy car • The wheels of the car spin and the car moves when the light shines on it. • The wheels of the car stop spinning when the light is not shining on it. • The car moves farther when a 100 w light bulb is shone on it than when a 40 w light bulb is shone on it.	Introduce students to change in motion as an indicator that a change in energy has occurred: • An increase in the motion of an object indicates that its energy has increased. • A decrease in the motion of an object indicates that its energy has decreased. Introduce students to production of sound as an indicator that energy has been transferred from the systems. Show that an increase in motion can result from a force, an electric current, the absorption of light, and sound.	Students should recognize that going from not moving to moving is a change in motion and that going from moving to not moving is a change in motion. Students should recognize that there is a relationship between the intensity of the light bulb and the distance the solar car travels, which suggests that the higher-intensity light transfers more energy.

Continued

Phenomena, Data, or Models	Intended Observations	Purpose	Rationale or Notes
Activity 2 A system containing hot water interacts with a system containing cold water.	• When a flask containing hot water is placed inside a beaker containing cold water, the temperature of the cold water increases and the temperature of the hot water decreases.	Introduce students to temperature change as an indicator that a change in energy has occurred: • An increase in the temperature of an object indicates that its energy has increased. • A decrease in the temperature of an object indicates that its energy has decreased.	This is a simple example to start with because students can directly measure changes in temperature.
Activity 3 A hamster runs on an exercise wheel.	• As the hamster runs on a wheel and turns it fast enough, a light bulb emits light. • As the hamster runs slower and the wheel slows down, the bulb stops emitting light.	Introduces a more complex phenomenon in a life science context, gives students a chance to look for indicators of energy change, and gives teachers a chance to find out what they know about energy changes in the hamster.	Students may struggle to apply an analytical approach to a more complex system.
Activity 4	• There are no new intended observations during this activity.	Students develop a list of indicators of energy changes. These include a change in temperature or motion and the production or absorption of light and sound. Students should also start to appreciate that when one system increases in energy, another system decreases in energy.	This activity helps students practice talking about energy changes and energy transfers.

Continued

Phenomena, Data, or Models	Intended Observations	Purpose	Rationale or Notes
Pulling It Together Q2: Describing energy changes in a simple physical system (compressed spring moves ball)	• When a compressed spring is used to push a stationary ball, the spring goes from a compressed state to a not-compressed state and the ball's motion increases.	Gives students a chance to apply science ideas about energy changes within systems Gives students a chance to correctly use language about energy changes and transfers and to distinguish evidence of energy changes (e.g., observations of changes in motion or temperature) from their descriptions of energy changes and transfers (which they can only infer from the evidence)	
Pulling It Together Q3: Describing energy changes in a cell phone	• As a cell phone rings, vibrates, and gives off light, its battery charge decreases.	Gives students a chance to apply science ideas about energy changes within systems and energy transfer between systems to explain the transfer of energy in a more complex context Gives students a chance to correctly use language about energy changes and transfers and to distinguish evidence of energy changes (e.g., observations of changes in motion or temperature) from their descriptions of energy changes and transfers (which they can only infer from the evidence)	Students should realize that the battery is transferring energy to the cell phone because its motion increases and that the cell phone is transferring energy to its surroundings as light and sound.

Lesson 2.1—Observing Energy Changes All Around Us

What do we know and what are we trying to find out?

In the previous chapter, you learned that our bodies convert polymers from our food into polymers that make up our body structures by carrying out chemical reactions in various body systems. You concluded from data and modeling activities in Chapter 1 that chemical reactions involve rearranging atoms of molecules of the starting substances (reactants) to make new substances (products) with different characteristic properties. For example, during digestion a few atoms of food polymers and water molecules disconnect and then connect to different atoms to form monomers. In body cells, the monomers react to form water molecules and other polymers that are used to build body structures like muscle fibers and cell membranes. Food provides the atoms and molecules that form the building blocks that our bodies need for growth and repair.

If you are thinking that building new membranes and muscle fibers isn't sufficient to win an athletic competition, you are correct! Watch the videos of the Olympic runner and tennis player. Instead of focusing on changes in matter like we did in Chapter 1, focus on observations that provide us with evidence of changes in energy. Where do the athletes get the energy to run or play tennis?

We often answer this question by saying that the athlete gets the energy from food. But what exactly does that mean? To understand how molecules from food can help us run, play tennis, do a flip on parallel bars, and just stay alive while we're sleeping, we need to consider the second essential role of food and chemical reactions—namely, to provide the energy our body needs to carry out all its functions. In this chapter, we will investigate how chemical reactions are related to energy, first in simple systems, where we can observe firsthand what is happening, and then in complex biological systems, where we will need to draw inferences from data.

In Chapter 1, we used firsthand observations of increased size and mass measurements to detect mass changes. In this lesson, you will observe some changes in living and nonliving things and begin thinking about the Key Question (there is no need to respond in writing now).

Key Question: How can we detect energy changes?

Teacher Talk and Actions

Show the *Olympic Athletes* and *Tennis Player* videos here. Ask students what observations in the videos provide evidence that energy is changing.

It would be a good idea to have a class brainstorming session at this point to find out whether students have any ideas about indicators of energy changes.

Activity 1: Observing Changes in Motion

Materials	For each class
For each team of students	1–2 battery testers
New 9-volt battery	
Partially used 9-volt battery	
Battery-powered toy car	
40 w light bulb with reflector	
100 w light bulb with reflector	
Solar-powered toy car	

In this activity, we will observe changes in motion in simple physical systems and use our observations to support claims about changes in energy.

Procedures and Questions

Balls colliding

1. Watch the video of two balls colliding and record your observations below. Focus on observations that provide evidence of changes in energy.

The first ball stopped moving once it hit the second ball. The second ball started moving after the first ball collided with it. I heard a sound when the balls collided.

2. Does either ball end up with *more* energy than it had in the beginning? If so, what evidence do you have that the energy of the ball increased?

The second ball had more energy. It started moving, and an increase in motion is evidence of an increase in energy.

3. Does either ball end up with *less* energy than it had in the beginning? If so, what evidence do you have that the energy of the ball decreased?

The first ball had less energy. It stopped moving after the collision. A decrease in motion is evidence of a decrease in energy. Also the sound is evidence that energy was transferred from the ball systems to the surrounding environment system.

Sound-powered coffee grounds

4. Watch the video of the coffee grounds and record your observations below. Focus on observations that provide evidence of changes in energy.

When a frying pan is struck and makes a sound, the coffee grounds move.

5. How does the amount of energy of the moving coffee grounds change as the volume of the sound is increased? Cite evidence for your answer.

The energy of the coffee grounds is greater with the louder sound because they move more.

Teacher Talk and Actions

Safety Notes

1. Wear safety goggles or glasses with side shields during the setup, hands-on, and take down segments of the activity.

2. Use caution when using sharp tools and materials.

3. Use caution when working with glass bulbs, which can shatter if dropped and cut skin.

4. Use caution when working with hot bulbs, which can cause skin burns or electric shock.

5. Use only GFI-protected circuits when handling electrical equipment, and keep the equipment away from water sources to prevent shock.

6. Properly clean up and dispose of waste materials.

7. Wash your hands with soap and water immediately after completing this activity.

Show the video *Balls Colliding* here.

Show the video *Sound-Powered Coffee Grounds* here.

Battery-powered toy car

6. Use the battery tester to test the new 9-volt battery. Record the results below:

The needle on the tester points to the green "good" region.

7. Use the battery tester to test the used 9-volt battery. Record the results below:

The needle on the tester points to the yellow "low" region.

8. Place the new 9-volt battery into the toy car and turn it on. Then put the car on the floor and let it go. Catch the car and turn it off.

9. What did you observe when the car was switched on?

The wheel attached to the motor spins very fast when the switch is turned on. The car rolls across the floor when let go.

10. What did you observe when the car was switched off?

The wheel slowed down and stopped spinning when the switch was turned off.

11. Now repeat Step 6 using the used 9-volt battery. What observations do you make?

The wheel attached to the motor spins slowly when the switch is turned on. The car slowly rolls across the floor when let go. The wheel stops spinning when the car is switched off.

12. How are your observations different from the ones you made when using the new 9-volt battery?

The wheel rotated slower with the "used" battery than it did with the "new" battery. Also, the motor made less sound.

13. Watch the *Wheel and Battery* video that shows what would happen if we left the car switched on for a long period of time. Record your observations below.

The wheel spun faster at the beginning when the battery was new. Then the wheel spun more slowly in the middle when the battery tester showed that the battery was "low." At the end the wheel did not spin and the battery tester showed that the battery was "dead."

14. How does the amount of energy of the car moving with the used battery compare to the amount of energy of the car moving with the new battery? Cite evidence to support your answer.

The wheel/car connected to either battery has the least energy when the battery is turned off. The car has the least energy when it is not moving.

Teacher Talk and Actions

The battery tester indicates the amount of energy the battery has. The scale has been set to have the line pointing completely to the right for a fully charged battery, with a line pointing to the left of the reference line indicating that the battery has less energy than when it was fully charged. Thus, the battery tester indicates relative amounts of energy but does not provide quantitative information about the amount of energy in a battery.

Show the video *Wheel and Battery* here.

Solar-powered toy car

15. Place the solar car in the starting position centered under the 40 w light bulb on the ruled paper. Switch on the light, observe what happens, and record the distance the car travels in the table below.

16. Turn off the light.

17. Now repeat Step 15 two more times and record the distance the car travels below.

18. Now repeat the above procedure using the 100 w bulb and record the results in the table.

Table 2.1. Observations of a Solar Car Under Different Wattage Bulbs

Wattage	40 W Bulb			100 W Bulb		
Observations	*The car moved when the light was turned on and then stopped when it was no longer under the light.*			*The car moved when the light was turned on and then stopped when it was no longer under the light.*		
Distance traveled in each trial						
Average distance traveled						

19. How are your observations of the solar car different under the different bulbs?

The car under the 100 w bulb traveled farther than it did under the 40 w bulb. If it traveled farther under the 100 w bulb, it must have moved more.

20. How does the amount of energy of the car moving under the 40 w bulb compare to the amount of energy of the car moving under the 100 w bulb? Cite evidence to support your answer.

The car under the 40 w bulb moves less, which means it has less energy.

21. Give two examples from your observations that energy was changing.

When the motion of the car was increasing, its energy was increasing. When the motion of the car was decreasing, its energy was decreasing.

Teacher Talk and Actions

Students' answers will vary depending on the distance traveled. The important observation to make is that the car under the 100 w bulb travels farther.

Show the optional video *Solar Car in Sunlight and Shade* here.

Activity 2: Observing Changes in Temperature

Materials

For each team of students
400 ml beaker
125 ml flask
2 thermometers

For each class
Hot water
Cold water

In this activity, we will observe what happens when something cold is placed in contact with something warm. We will focus on making observations that provide us with evidence of changes in energy.

Procedures and Questions

1. Add approximately 125 ml of cold water to the 400 ml beaker (System 1).

2. Add approximately 125 ml of hot water to the 125 ml flask (System 2).

3. Use the thermometers to measure the temperatures of the hot and cold water and record them in Table 2.2 in the "Initial" row.

4. Carefully place the flask inside the beaker of cold water.

5. Place one thermometer in the beaker and one thermometer in the flask and hold them so that they are not touching the sides of the beaker or flask.

6. After two minutes, record the temperature of the water in the beaker and flask. Continue recording the temperature every two minutes until six minutes have passed.

Table 2.2. Temperature

Time	System 1 Temperature of Water in Beaker	System 2 Temperature of Water in Flask
Initial	10°C	43°C
2 min.	19°C	32°C
4 min.	23°C	28°C
6 min.	25°C	25°C

Teacher Talk and Actions

The amount of energy an object has cannot be directly measured. The thermometer is not measuring how much energy the water has. It is measuring the temperature of the water which is an indicator that can be used to calculate how much energy the water has.

Safety Notes

1. Wear safety goggles, nonlatex aprons, and thermal gloves during the setup, hands-on, and take down segments of this activity.

2. Use caution when working with glassware and the thermometer, which can shatter if dropped and cut skin.

3. Immediately wipe up any spilled water on the floor to avoid a slip-and-fall hazard.

4. Use caution when working with hot water, which can burn skin.

5. Properly clean up and dispose of waste materials.

6. Wash your hands with soap and water immediately after completing this activity.

7. What pattern did you observe in the temperature of the water in the beaker (System 1)?

The temperature of the cold water in the beaker increased as time passed.

8. What pattern did you observe in the temperature of the water in the flask (System 2)?

The temperature of the hot water in the flask decreased as time passed.

9. Does either system end up with *more* energy than it had in the beginning? If so, what evidence do you have that the energy of the system increased?

Yes, the beaker system (System 1) has more energy at the end because the water got warmer. The thermometer showed us that the temperature increased, and an increase in temperature is evidence that the amount of energy in the water has increased.

10. Does either system end up with *less* energy than it had in the beginning? If so, what evidence do you have that the energy of the system decreased?

Yes, the flask system (System 2) has less energy at the end because the water got cooler. The thermometer showed us that the temperature decreased, and a decrease in temperature is evidence that the amount of energy the water has decreased.

11. You may have learned in middle school that the temperature of a substance is related to the average speed of the molecules that make up that substance. This means that the average speed of the water molecules of the hot water is greater than the average speed of the water molecules of the cooler water. How does this support your answers to Steps 9 and 10?

An increase in speed is an indicator that energy is increasing. As the cool water warms, its molecules speed up and have more energy. The opposite is true for the warmer water. As the warm water cools, its molecules slow down and have less energy.

Teacher Talk and Actions

Some students may be familiar with the idea that temperature is related to the average speed of the molecules that make up an object. If they know this, they may understand that the molecules have more motion energy (kinetic energy) when they are hotter than when they are cooler. This may be helpful to students who know that moving faster means more energy but are struggling to understand why a hotter substance has more energy than the same substance when it is cooler.

Activity 3: Observing Changes in Living and Nonliving Things

In this activity, we will look at a more complex phenomenon involving both living and non-living things. Again, we will focus on making observations that provide us with evidence of changes in energy.

As we observed in the videos we saw at the beginning of Chapter 1 and earlier in this lesson, changes in energy may be difficult to detect in the video. You can use your personal experiences if they help you generate ideas.

Procedures and Questions

1. Watch the video of a hamster running on a wheel and record your observations below. Focus on observations that provide evidence of changes in energy.

As the hamster runs on the wheel, the wheel spins faster. When the hamster slows down and stops running, the wheel slows down and stops spinning.

When the wheel spins fast enough, the light bulb shines. When the wheel slows down, the light bulb stops shining.

2. Does anything have *more* energy at some point during this video? Explain.

The wheel has more energy when it's spinning/moving than when it is not because more motion means more energy.

3. Does anything have less energy at some point during this activity? Explain.

The wheel has less energy when it's not spinning/moving than when it is.

4. Describe two examples where the energy is changing—one where the energy of something is increasing and another where the energy of something is decreasing.

The energy of the wheel increases as the wheel's motion increases and decreases as the wheel's motion decreases.

Teacher Talk and Actions

Show *Hamster* video here.

Students may also say that the hamster gets tired as it runs and stops to rest. If so, ask what they think might be happening when the hamster is resting so that it can run again. While students are not expected to know the answer, they should have more ideas after Lessons 2.7 and 2.8.

Students should be able to describe energy changes in the wheel but may have difficulty describing energy changes in the hamster. For example, some students may realize that the hamster has more motion energy when it is running, but other students may think that the hamster has less energy because it is getting tired. Students are not expected to understand what is happening with the hamster at this point; they will revisit the example in Lesson 2.7. For now, ask students to give reasons for their answers.

Activity 4: Describing Energy Changes and Energy Transfers

In this activity, we will use the observations we made during this lesson to generalize about how we can detect energy changes and think about how energy changes are related to the transfer of energy.

Procedures and Questions

Energy changes

1. Looking across our observations in this lesson, what are some indicators that the energy of an object has changed?

The object's temperature increases (or decreases), indicating the amount of energy the object has increased (or decreased).

The object's motion increases (or decreases), indicating the amount of energy the object has increased (or decreased).

The object gives off (or absorbs) light, which means the amount of energy it has is decreasing (or increasing).

The object gives off sound, which means the amount of energy it has is decreasing.

2. Consider the observations you made during this lesson. Are there any situations where an increase in energy of one object or system is associated with a decrease in energy of another object or system? Give two examples.

Hot/cold water: When the flask with hot water was put in the beaker with cold water, the energy of the cold water increased (because its temperature increased) and the energy of the hot water decreased (because its temperature decreased).

Balls colliding: When the fast-moving ball collided with the ball at rest, the energy of the fast-moving ball decreased (because its motion decreased) and the energy of the ball at rest increased (because its motion increased).

Energy transfers

Think about the example of the balls colliding. The white ball slowed down after it hit the black ball and the black ball started to move. In this case, the white ball decreased in energy and the black ball increased in energy. A scientist would say that in this case the white ball transferred energy to the black ball.

3. Pick two situations we observed during this lesson and describe what happened in terms of energy transfer.

When the flask of hot water was placed in the beaker of cold water, energy was transferred from the hot water to the cold water, causing it to warm up.

When the balls collided, the white ball transferred energy to the black ball, causing it to move.

Teacher Talk and Actions

The purpose of this activity is to get students to start talking about energy scientifically so that they can communicate their ideas with one another and start to link increases in energy in one system to decreases in energy in another system.

Other examples that students could give include the following:

Battery-powered toy car: When the energy of the toy car increased (because its motion increased), the energy of the battery decreased (based on evidence from battery tester).

Tennis player/ball: The ball increases in speed and, therefore, in energy while the tennis player decreases in energy.

More advanced students may list the solar-powered toy car. The car increases in energy when it absorbs the light from the light bulb because its motion increases. We can't directly observe a decrease in energy in the bulb, but the pattern we observe in the other cases suggest that some other system must be decreasing in energy. Students could be encouraged to work their way back to the wires of the light bulb, to the outlet, and to the power station and could look up how a power station works. They will get to a fuel source and will learn in later lessons about the chemical reaction between fuel and oxygen.

Other examples include the following:

When the hamster runs on the wheel, the hamster transfers energy to the wheel and the wheel begins to move.

The tennis player transfers energy to the ball by exerting a force on the ball.

4. Think about the new and used batteries that you used with the toy cars in Activity 1. Describe your observations in terms of the amount of energy transferred in each case.

Each battery transferred energy to the toy car, and the toy car started moving. The new battery transferred more energy to the toy car than the used battery did because the car moved faster with the new battery.

5. Think about the light bulbs that you used with the solar-powered cars in Activity 1. What can you conclude about the brightness of the light and the amount of energy transferred?

The brighter the light, the more energy was transferred because the car traveled farther (moved more) under the brighter light.

6. Think about the sound from the frying pan that was used to make the coffee grounds move in Activity 1. Describe your observations of the amount of energy transferred with the moderate sound versus the louder sound.

The sound transferred energy to the coffee grounds. The louder the sound was, the more the coffee grounds moved. This means that the louder the sound, the more energy was transferred to the coffee grounds.

Teacher Talk and Actions

Science Ideas

Science ideas are accepted principles or generalizations about how the world works based on a wide range of observations and data collected and confirmed by scientists. Because these science ideas are consistent with the available evidence, you are justified in applying science ideas to other relevant observations and data.

The activities in this lesson were intended to help you understand important ideas about energy changes and energy transfers. Read the ideas below. Look back through the lesson. In the space provided after the science idea, give examples of evidence that supports the idea.

Science Idea #13: Changes in temperature or motion and the production or absorption of light or sound are indicators of energy changes.

Evidence:

When the balls collided, they both changed in motion. The black one went from still to moving, and the white one went from moving to still. The collision also produced sound. These are all indicators of energy changes.

Coffee grounds near a speaker move more when the sound is louder, which probably means they absorb more sound.

Science Idea #14: When the energy of one system increases, the energy of another system or systems decreases. Likewise, when the energy of one system decreases, the energy of another system or systems increases.

Evidence:

In each of the examples we saw during this lesson, an increase in energy of something was accompanied by a decrease in energy of something else. For example, the hot water got cooler, which is a decrease in energy, and the cold water got warmer, which is an increase in energy.

Science Idea #15: When two systems interact, and one system decreases in energy while the other system increases in energy, we say that energy is transferred from the first system to the second.

Evidence:

The hot-water flask (System 1) transferred energy to the cold-water beaker (System 2). The hot water decreased in energy while the cold water increased in energy.

Teacher Talk and Actions

Students could also cite as evidence the increase in water temperature that occurs in a swimming pool when sunlight shines on it. That suggests that the water is absorbing energy from the Sun.

In Science Idea #15, the word *interact* is key. The systems must interact in order for energy to be transferred between them.

Pulling It Together

Work on your own to answer these questions. Be prepared for a class discussion.

1. Now that you have learned more about the indicators of energy changes, how would you answer the key question: **How can we detect energy changes?**

We can detect energy changes by observing indicators such as a change in motion or temperature or the production or absorption of light or sound.

2. In a pinball machine, a spring is used to shoot the ball forward. A person pulls the lever, which compresses a spring. When they let go of the lever the spring pushes the ball and the balls speed increases. Describe the energy changes and energy transfers that occur as the ball speeds up and the spring returns to its original length. Be sure to include the indicators of the energy changes.

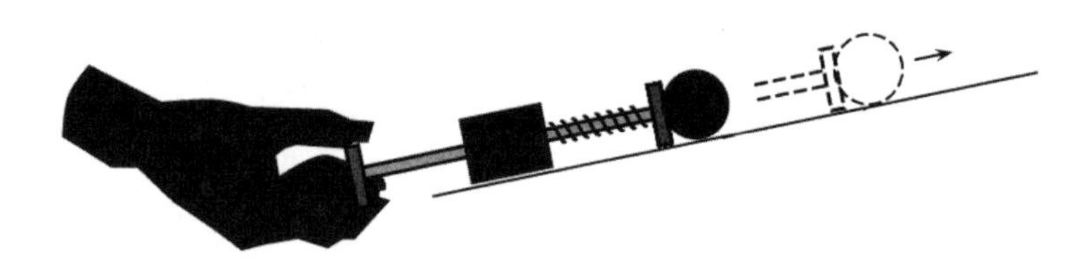

At first the spring is compressed and the ball is at rest. Then the spring returns to its uncompressed length and the ball starts to move. During this change, the ball increases in energy and the spring decreases in energy. The increase in speed is evidence that the ball increased in energy. Because increases in energy are accompanied by decreases in energy, the spring has less energy. (Some students may know that the amount of energy a spring has decreases as the amount it is compressed decreases.)

Teacher Talk and Actions

Students can write this using either increase/decrease terminology or changes terminology. For example, increases or decreases in motion or changes in motion both work as answers to this question.

3. Imagine a cell phone that is ringing. The battery of the phone is 100% charged. You observe that the phone is vibrating and making a sound, and the screen is flashing. After the phone stops ringing, you notice that the battery is now 99% charged. Describe the energy changes and energy transfers that occur while the phone is ringing. Be sure to include the indicators of the energy changes.

While the phone is ringing, the battery is transferring energy to the phone, and the phone is transferring energy to the surrounding air. The increased motion of the phone is evidence that the phone is increasing in energy (that is, the battery is transferring energy to it), and the sound and light being emitted from the phone is evidence that the phone is transferring energy to the surrounding air. The decrease in charge of the battery is evidence that the battery has less energy after the phone rings.

4. Describe another example of an energy change and an energy transfer from your own experience and list the indicators associated with that energy change. Try to include an example that involves living organisms.

Students responses will vary. When a student provides an example, be sure to ask them what indicators helped them decide that energy was transferred.

Teacher Talk and Actions

Closure and Link to subsequent lessons:

In this lesson, you observed examples of athletes exercising, balls colliding, water changing temperature, using a battery or light to make a car move, using sound to move coffee grounds, and a hamster running on a wheel. We talked about some indicators of changes in energy that we observed, but sometimes it was difficult to find clear evidence that the amount of energy was changing. Because the human body is so complex, it can be challenging to determine how energy is changing. This is why we looked at simpler systems like the balls colliding and the hot and cold water interacting. In these cases, we were able to clearly see changes in motion or temperature that indicated changes in energy. When we looked across the examples, we noticed a correlation between increases of energy of one system and decreases of energy of another system and learned to describe our observations in terms of energy transfers between interacting systems. In the next lesson, we will learn how to use models to represent these energy transfers.

Lesson Guide
Chapter 2, Lesson 2.2
Representing Energy Changes and Transfers

Focus: In this lesson, students learn to use bar graphs and system boxes with upward and downward arrows to represent energy changes within systems and energy-transfer models to represent energy transfers between systems. Phenomena move from simple contexts involving colliding balls to more complex contexts involving battery-powered cars, solar-powered cars, and a hamster running on a wheel.

Key Question: How can models help us make sense of energy changes within systems and energy transfers between systems?

Target Idea(s) Addressed

Models are useful for representing systems and their interactions—such as inputs, outputs, and processes—and energy and matter flows within and between systems. (Science Idea #12)

Science Practice(s) Addressed

Developing and using models

Materials

None

Advance Preparation

None

Phenomena, Data, or Models	Intended Observations	Purpose	Rationale or Notes
Activity 1 Bar graphs as models of energy changes within systems	Hot and cold water • In the flask system, the decrease in energy of the water in the flask is represented with two bars, using a higher bar to represent the higher energy of the water before the change and a lower bar to represent the lower energy of the water after the change. • In the beaker system, the increase in energy of the water in the beaker is represented with two bars, using a lower bar to represent the lower energy before the change and a higher bar to represent the higher energy after the change.	Introduce students to pairs of bars on a graph as a model of energy changes within a system inferred from changes in temperature. Get students to start noticing that when one system increases in energy, we should look for a system that is decreasing in energy.	This is a simple example to start with because students can relate changes in the height of the bars to their temperature measurements. Be sure to note that the bars are in energy units, not temperature units.
	Balls colliding • For the ball that is rolling and hits the other ball, the decrease in energy of the ball is represented by a higher bar before the ball collides than after it collides and stops moving. • For the ball that is stationary and then starts moving, the increase in energy of the ball is represented by a lower bar before it starts to move than after it is moving.	Use bars on a graph to represent energy changes inferred from changes in motion. Encourage students to notice that when one system increases in energy, we should look for a system that is decreasing in energy.	This is another simple example where the change in motion of both systems can be directly observed.
	Battery-powered toy car • In the car system, the increase in energy of the car is represented by a lower bar when the car is still and a higher bar when the car is moving. • In the battery system, the decrease in energy of the battery is represented by a higher bar when the battery is new and a lower bar after it has been used.		This is a more complex example that requires inferences that the battery system is decreasing in energy because (a) the car increases in energy and (b) the car moves less with the used versus the new battery.

Phenomena, Data, or Models	Intended Observations	Purpose	Rationale or Notes
Activity 1 (continued) Bar graphs as models of energy changes within systems	Solar-powered toy car • In the car system, the increase in energy of the car is represented by a lower bar when the car is still and a higher bar when the car is moving. • In the electrical system, the decrease in energy of the electrical system is represented by a higher bar before the light is switched on and a lower bar after light is switched on.		
	Hamster on a wheel • In the wheel system, the increase in energy of the wheel is represented by a lower bar when the wheel is still and a higher bar when the wheel is moving. • In the hamster system, the hamster goes from being rested to being tired, which is represented by a higher bar at the beginning and a lower bar at the end.	Give students a more complex system as a context for thinking about how the components and boundaries of a system must be clearly defined to understand what is happening in that system. Encourage students to notice that in all four examples a decrease in energy in one system is always accompanied by an increase in energy in another system. Encourage students to notice that in the four examples the two systems are always interacting in some way.	This is an even more complex example, requiring inferences that the hamster system is decreasing in energy because the wheel is increasing in energy. It's OK if students infer that the hamster is tiring because it stops for a bit. If students say the hamster is decreasing in energy because it is burning fuel, ask how they know.

Phenomena, Data, or Models	Intended Observations	Purpose	Rationale or Notes
Activity 2 System boxes with curved arrows as models of energy changes within systems Thick horizontal arrows representing energy being transferred link the system boxes as models of energy transfers between systems	Balls colliding • The decrease in energy of a moving ball can also be represented by a curved arrow pointing downward in a system box; the increase in energy of a still ball can also be represented by a curved arrow pointing upward in a system box; the transfer of energy between the two colliding (interacting) balls can be represented by a thick horizontal arrow pointing from the moving ball system to the still ball system. Battery-powered toy car • The decrease in energy of the battery can also be represented by a curved arrow pointing downward in a system box; the increase in energy of the toy car can also be represented by a curved arrow pointing upward in a system box; the transfer of energy between the battery and the toy car (when the two systems are connected so they interact) can also be represented by a thick horizontal arrow pointing from the battery system to the toy car system. Hot and cold water • The decrease in energy of the hot water in a flask can also be represented by a curved arrow pointing downward in a system box; the increase in energy of the cold water in the beaker can also be represented by a curved arrow pointing upward in a system box; the transfer of energy between the hot water and the cold water can be represented by a thick arrow pointing from the flask system to the beaker system. Hamster on a wheel • The decrease in energy of the hamster can also be represented by a curved arrow pointing downward in a system box; the increase in energy of the wheel can also be represented by a curved arrow pointing upward in a system box; the transfer of energy between the hamster system and the wheel system can be represented by a horizontal arrow pointing from the hamster system to the wheel system. Solar-powered toy car • The decrease in the electrical system (which includes the light bulb and power plant) can also be represented by a curved arrow pointing downward in a system box; the increase in energy of the wheel can also be represented by a curved arrow pointing upward in a system box; the transfer of energy between the electrical system and the car can be represented by a horizontal arrow pointing from the electrical system to the car system.	Introduce students to another model for representing energy change within a system and a model for representing energy transfer between systems. Encourage students to generalize and use the idea that for energy to be transferred between two systems the two systems must interact.	For systems where the decrease in energy in System 1 is not directly observable, have students complete the System 2 box first and then reason from the science idea that when the energy of one system increases, the energy of another system must have decreased to complete the System 1 box. The need for the two systems to interact can be brought home with the question about what would happen if the flask were made of an insulating material rather than of glass.

Phenomena, Data, or Models	Intended Observations	Purpose	Rationale or Notes
Pulling It Together Q3: Representing energy changes within systems and energy transfer between systems in a simple physical system (balls colliding)	• When a compressed spring is used to push a stationary ball, the spring goes from a compressed state to a not-compressed state and the ball's motion increases.	Give students a chance to apply science ideas about energy changes within systems and energy transfer between systems to explain the transfer of energy from the spring to the ball.	
Pulling It Together Q4: Describing energy changes and transfers in a more complex system (batter hitting a ball)	• When a baseball player swings a bat, the bat goes from not moving to moving.	Give students a chance to apply science ideas about energy changes within systems and energy transfer between systems to explain the transfer of energy in a biological context.	Gives teachers a chance to see what students think about energy changes and transfers involving a human

Lesson 2.2—Representing Energy Changes and Transfers

What do we know and what are we trying to find out?

In the last lesson, we observed several examples in which the amount of energy in one system increased as the amount of energy in another system decreased and we explained the relationship between these two changes in terms of energy transfer between the two systems. For example, we observed a decrease in temperature of the hot water in the flask, which indicated that the water's energy had decreased energy. At the same time, we observed an increase in the temperature of the cold water in the beaker, which indicated that the water's energy had increased. We explained the link between these two observations in terms of the transfer of energy from the hot water in the flask to the cold water in the beaker.

In Chapter 1, we used models to represent the matter changes during chemical reactions. These models helped us make sense of the different chemical reaction systems we investigated. We can also use models to represent energy changes. Because energy is not a physical object we won't use balls or LEGOs like we did for atoms. We will use more abstract models, like bar graphs, to represent energy changes.

In this lesson, you will revisit the examples you examined in Lesson 2.1 and begin thinking about the Key Question (there is no need to respond in writing now).

Key Question: How can models help us make sense of energy changes within systems and energy transfers between systems?

Teacher Talk and Actions

Ask students how they would represent energy changes and energy transfers to see what kinds of models they come up with.

Possible options include bar graphs, pie charts, and so on.

Activity 1: Representing Energy Changes

In this activity, we will use models to represent some of the energy changes we observed during Lesson 2.1. Our models will help us think about and make sense of what is happening to energy during these phenomena.

The models we will use are bar graphs. We will use the height of the bar to represent the amount of energy in the system, with higher bars representing higher energy and lower bars representing lower energy. In the following example, our system consists of an object before a change and the same object after the change. The higher bar on the right indicates that the object had more energy after the change than it did before the change.

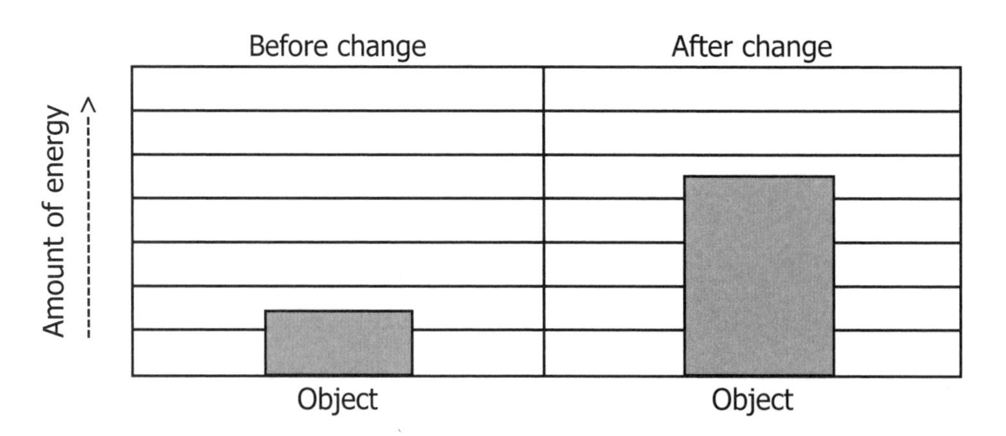

Procedures and Questions

1. For each of the examples from Lesson 2.1, draw a bar graph to represent the amount of energy each system had before and after the change. Be ready to justify why you drew the bars as you did.

Hot and cold water

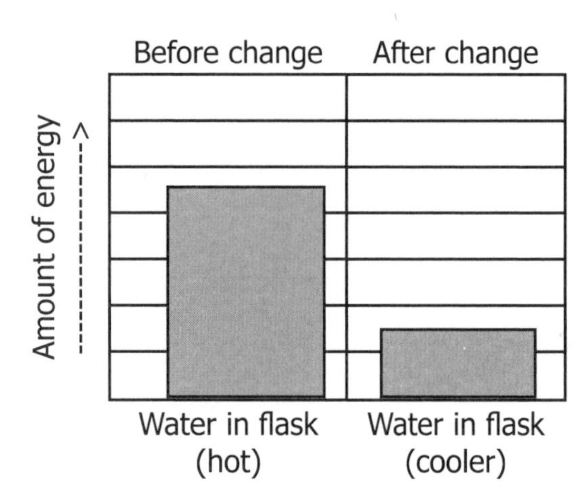

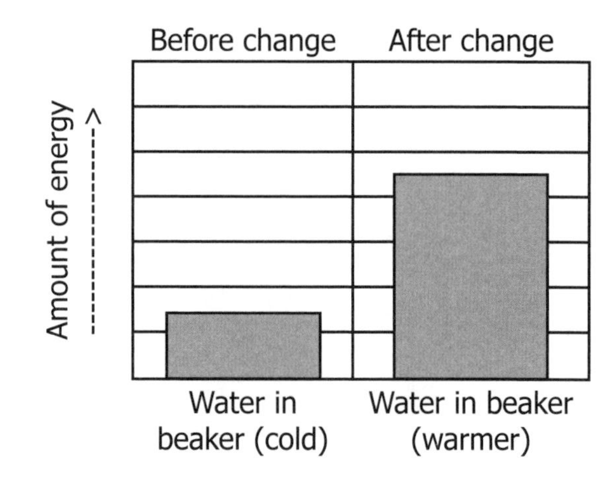

Teacher Talk and Actions

Because everything has some amount of energy at all times, we will never draw a bar with no height.

The absolute height of the bar is not important. What is important is the relative heights of the two bars. We are looking at changes in energy, not the exact quantity of energy.

Balls colliding

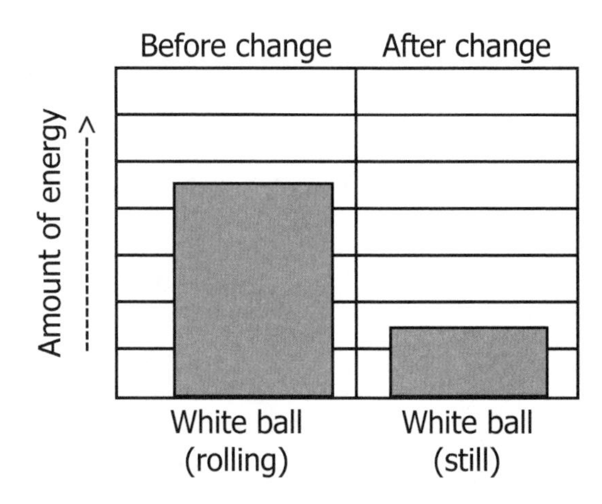

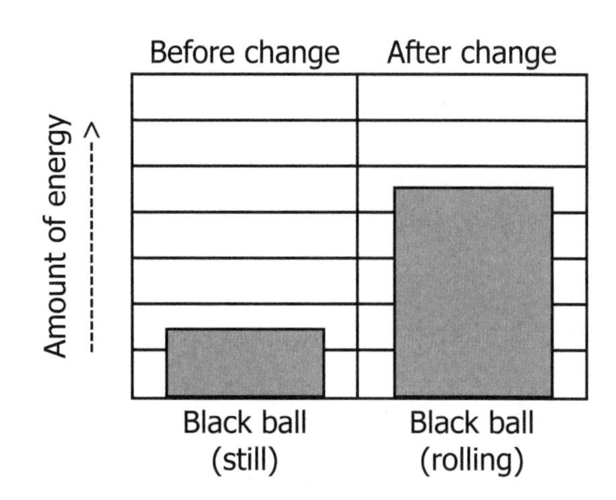

Battery-powered toy car

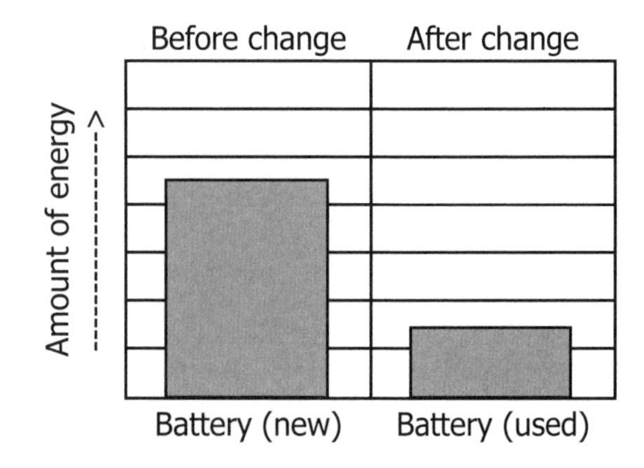

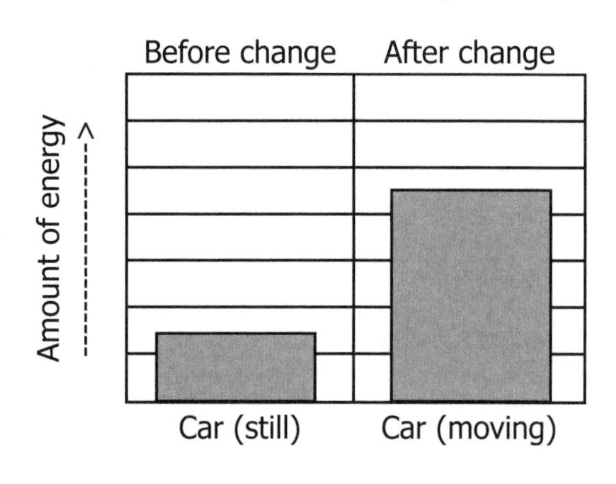

Hamster on a wheel

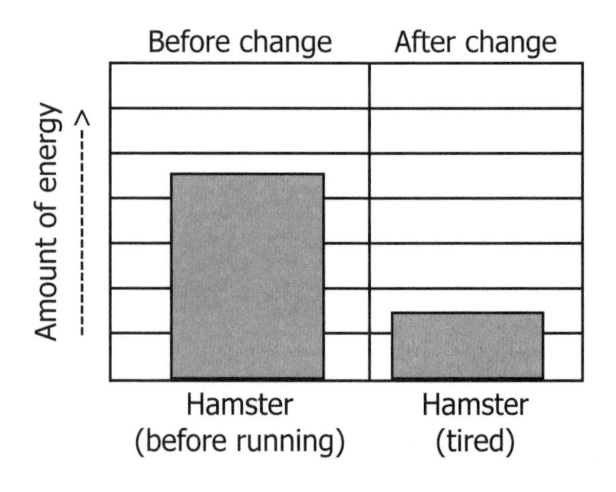

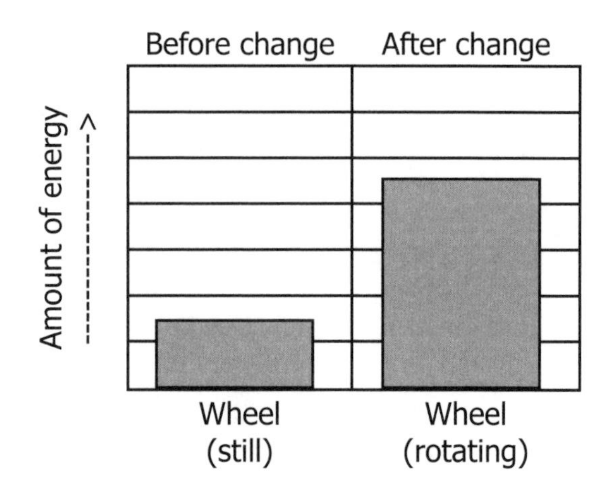

Teacher Talk and Actions

Activity 2: Representing Energy Transfers

In this activity, we will use models to represent some of the energy transfers we observed during Lesson 2.1. In each of the examples you graphed in Activity 1, the energy of one system decreased while the energy of another system increased and we said that energy was transferred from the first system to the second. We can represent the transfer of energy from one system to another using a model like the one shown below.

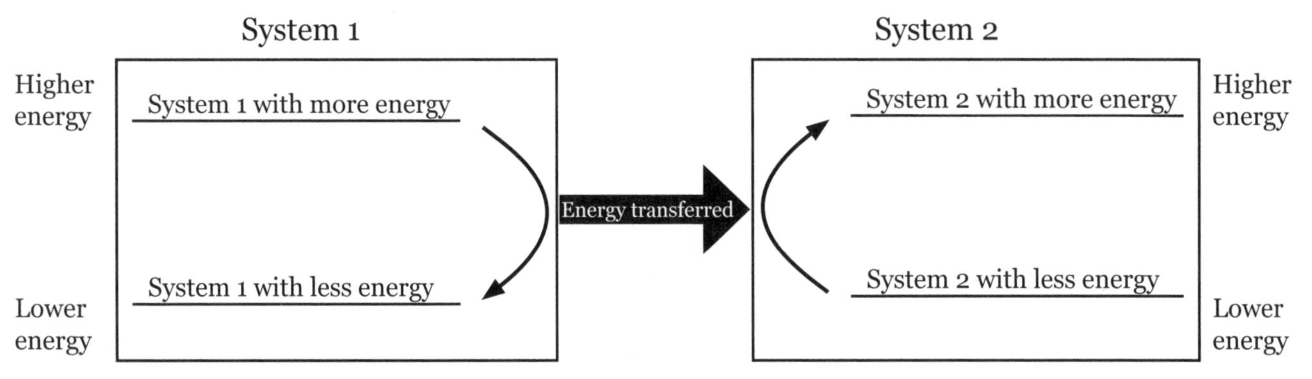

An energy-transfer model like the one above shows the direction of energy transfer between two interacting systems. Each box in the diagram represents a system, and the thick arrow between the boxes represents the direction of energy transfer. Each system box displays a process that either increases or decreases the energy of the system. Processes that involve a decrease in energy are represented with a curved arrow pointing down and processes that involve an increase in energy are represented by a curved arrow pointing up.

In this activity, you will use the bar graphs you created in Activity 1 to build energy-transfer models to describe the energy transfers that took place during the examples you observed in Lesson 2.1.

Procedures and Questions

Balls colliding

1. Let's consider the example of the balls colliding. We can think of this example as being made up of two systems. The first system is the white ball that is moving at the start. The second system is the black ball that is stationary at the start.

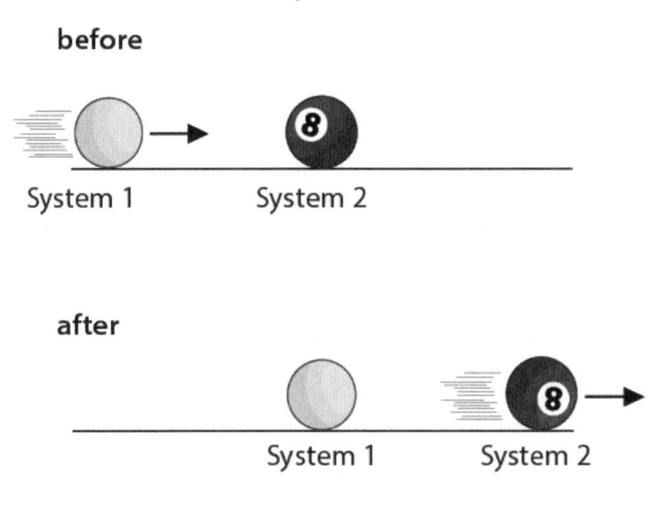

Teacher Talk and Actions

System 1 = White ball

As a result of the collision, the white ball slows down, indicating that its energy has decreased. The bar graph below represents this change in energy. We could also represent this change using a system box, shown below on the right. The downward curved arrow represents the decrease in energy of the system as the white ball goes from rolling (higher energy) to still (lower energy).

System = White ball

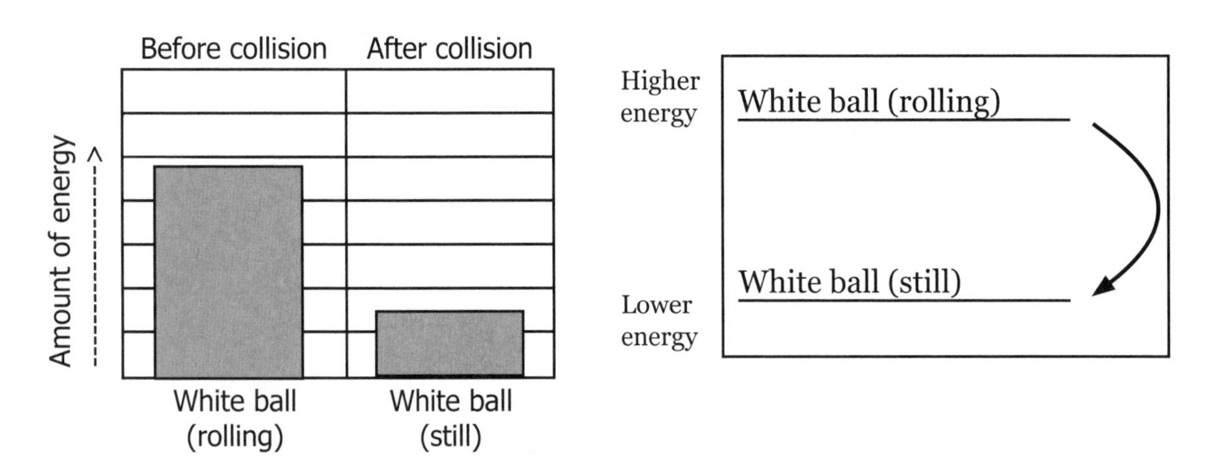

System 2 = Black ball

Now let's look at the black ball. This ball started out not moving and then started moving after the collision. So this system has more energy after the collision than before the collision. The bar graph and box below represent this change in energy. The stationary ball is represented at the bottom of the box and the rolling ball is represented at the top. An arrow is drawn from the black ball (still) to the black ball (rolling), indicating that this system increased in energy during the collision.

System = Black ball

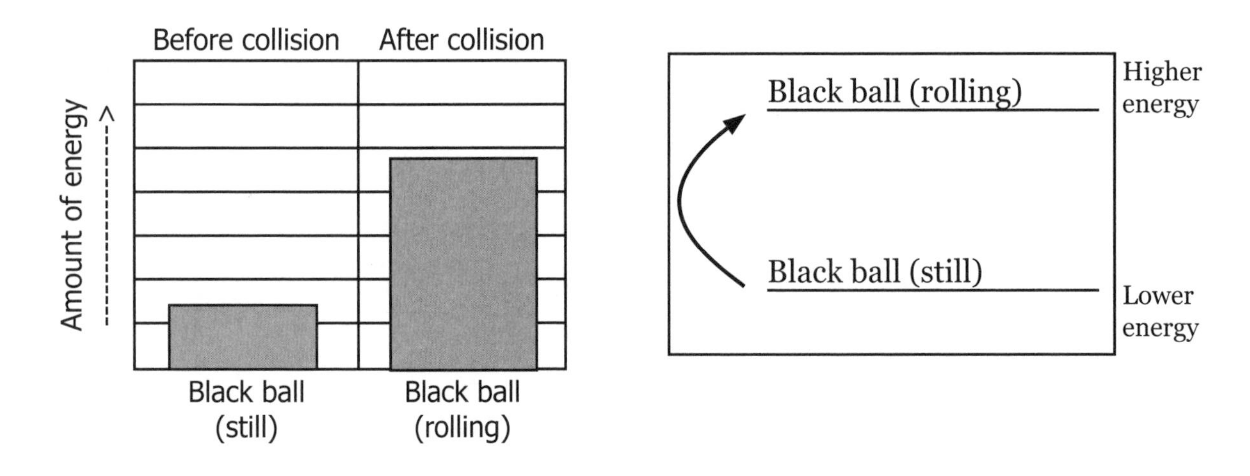

Teacher Talk and Actions

Because energy was transferred from the white ball to the black ball, we can complete our energy-transfer model by connecting the two boxes with a large arrow pointing to the right to represent the direction of energy transfer from one ball to the other ball.

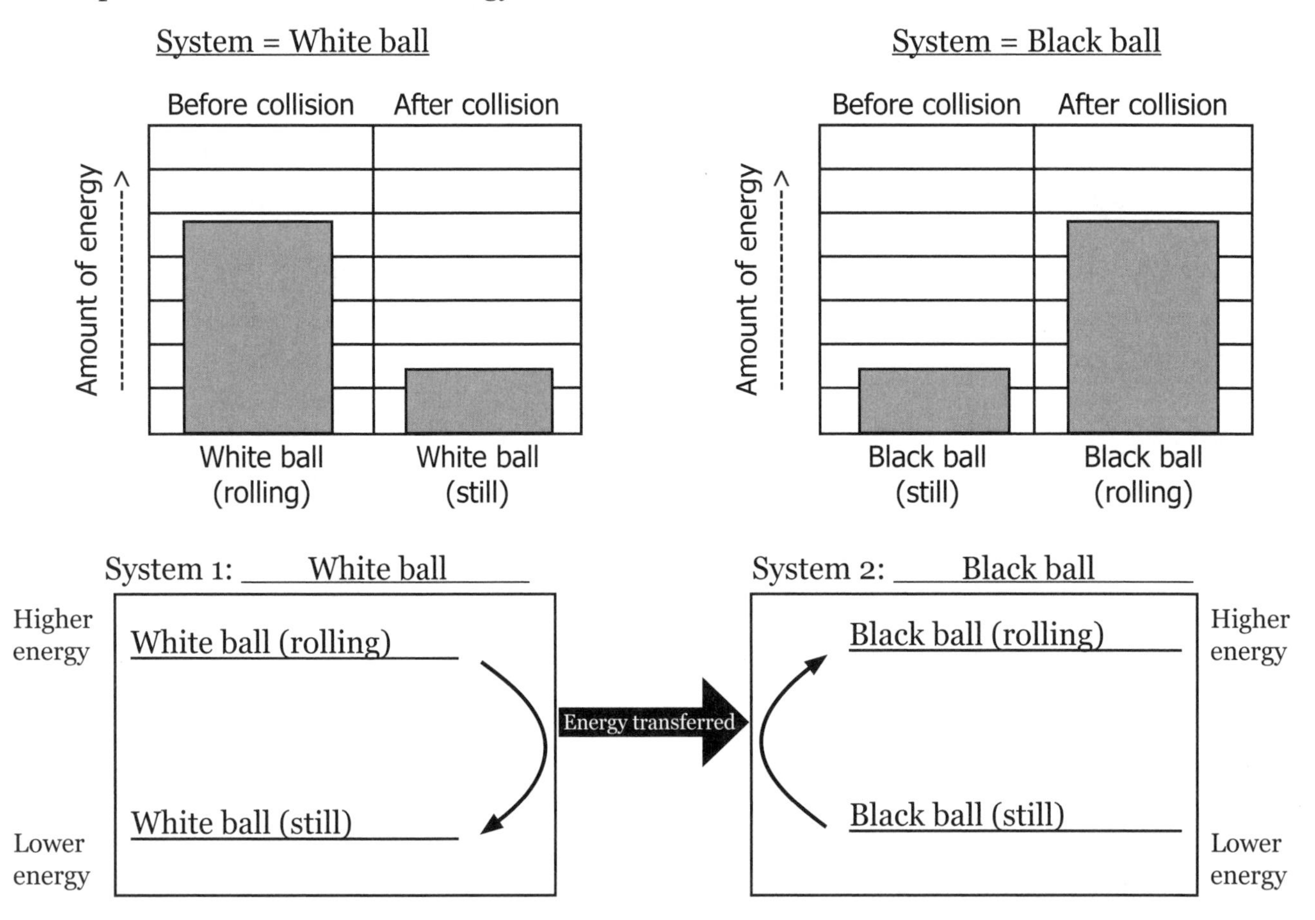

Battery-powered toy car

2. Next, you will create an energy-transfer model that represents the transfer of energy that occurred when the batteries were used to power the toy car. Draw a bar graph to illustrate the amount of energy in the battery system before and after it was used to power the toy car.

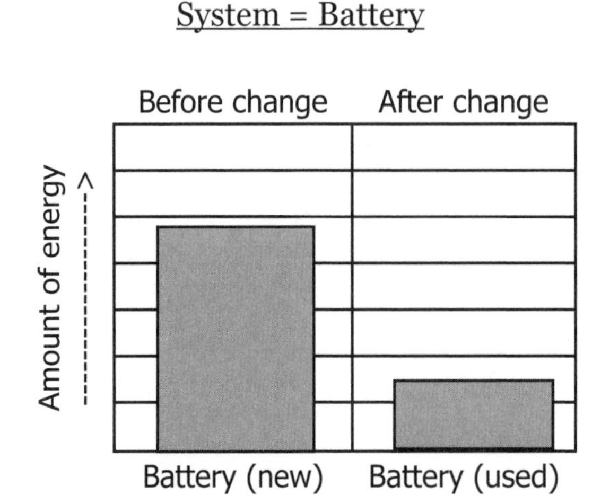

Teacher Talk and Actions

3. Using the bar graph in Step 2, select the system box from those shown on page 99 that accurately represents energy changes in the battery, cut it out, and paste it on the left side of the space below Step 6.

4. Cut out the appropriate energy-transfer arrow shown on page 99 and paste it next to the system box. If energy is transferred from the system, the arrow should point away from the box, and if energy is transferred to the system, the arrow should point toward the box.

5. Draw a bar graph to illustrate the change in energy of the toy car when the battery was connected to it in a complete circuit.

System = Car

Before change | After change

Amount of energy ------->

Car (still) | Car (moving)

6. Using the bar graph in Step 2, select the system box from those shown on page 99 that accurately represents energy changes in the toy car, cut it out, and paste it on the right side of the space below.

Construct the energy-transfer model here:

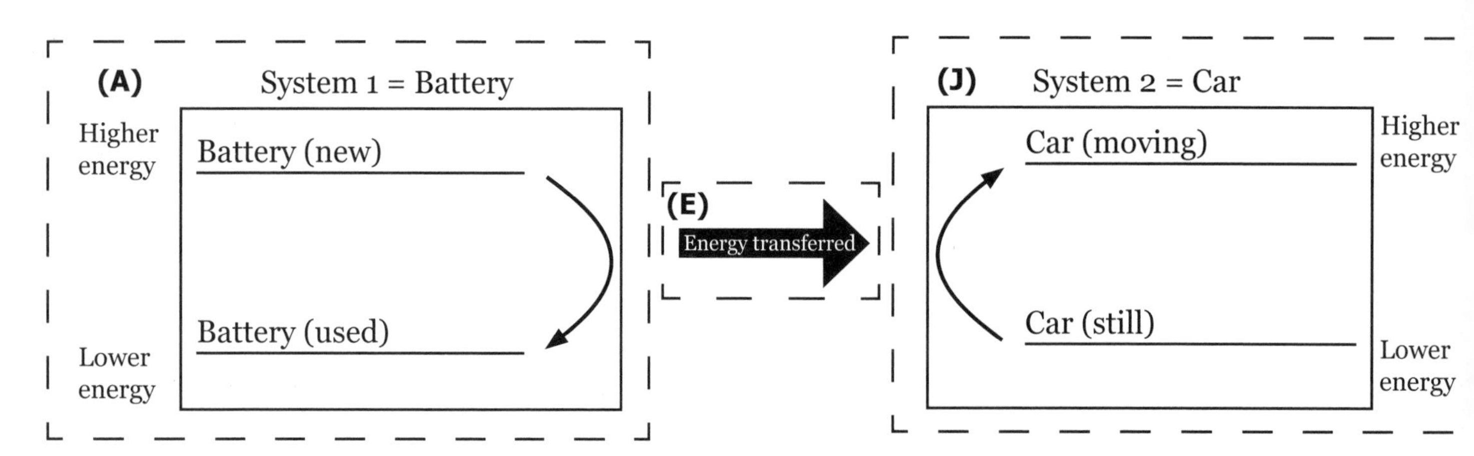

Teacher Talk and Actions

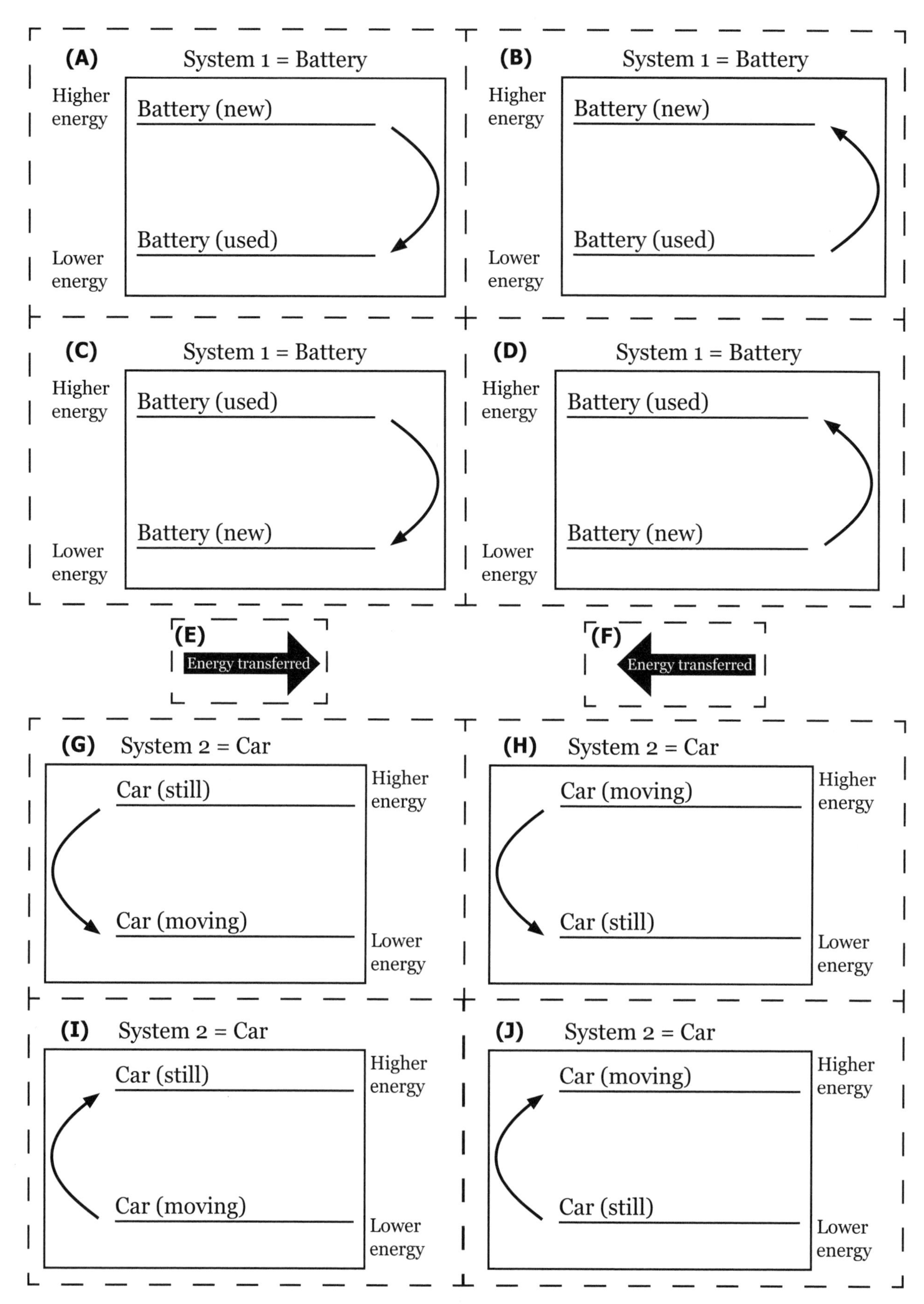
(A)
System 1 = Battery
Higher energy
Battery (new)
Battery (used)
Lower energy
(B)
System 1 = Battery
Higher energy
Battery (new)
Battery (used)
Lower energy
(C)
System 1 = Battery
Higher energy
Battery (used)
Battery (new)
Lower energy
(D)
System 1 = Battery
Higher energy
Battery (used)
Battery (new)
Lower energy
(E)
Energy transferred
(F)
Energy transferred
(G)
System 2 = Car
Car (still)
Car (moving)
Higher energy
Lower energy
(H)
System 2 = Car
Car (moving)
Car (still)
Higher energy
Lower energy
(I)
System 2 = Car
Car (still)
Car (moving)
Higher energy
Lower energy
(J)
System 2 = Car
Car (moving)
Car (still)
Higher energy
Lower energy

Teacher Talk and Actions

Teacher Talk and Actions

7. Complete an energy-transfer model for each of the other phenomena you observed in Activity 1. Refer to the bar graphs you drew in Activity 1 to help you fill in the missing information in the energy-transfer models below.

Hot and cold water

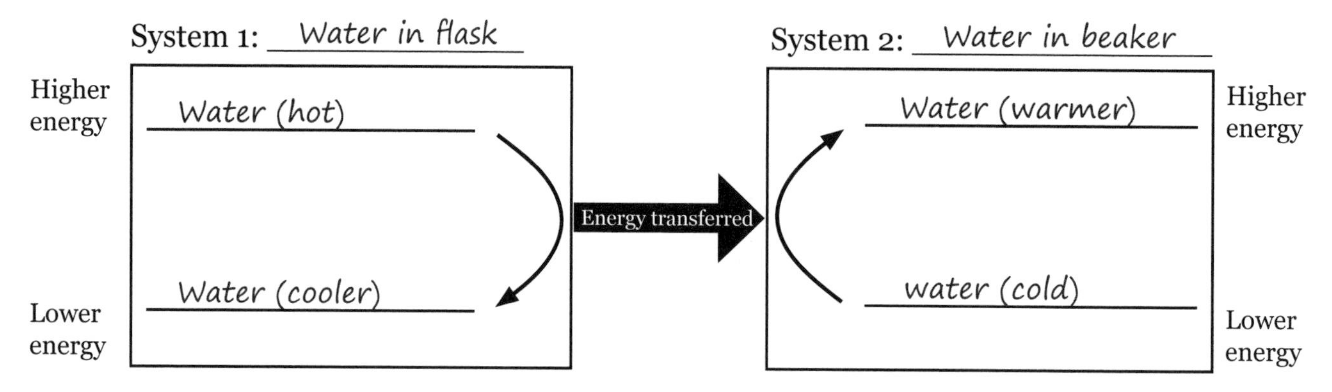

Solar-powered toy car

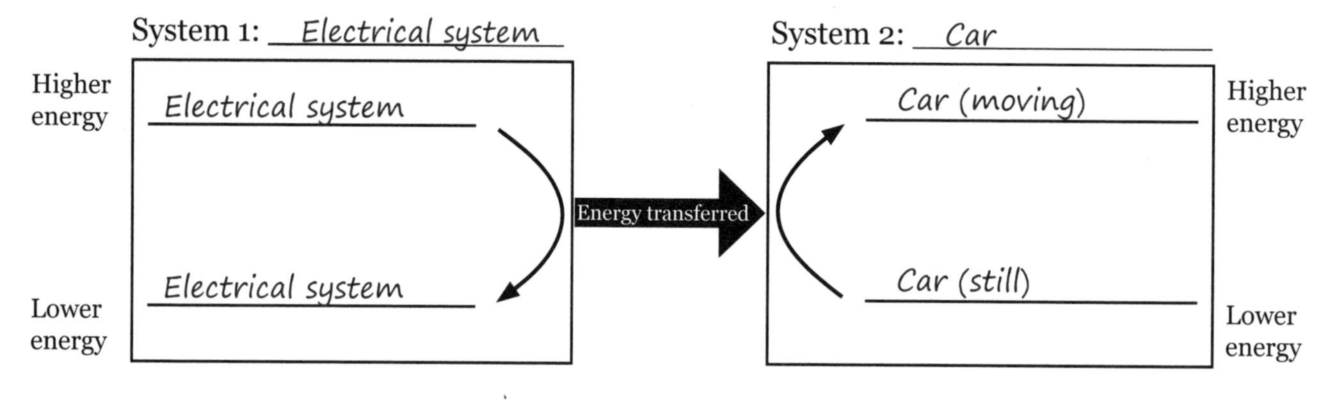

Hamster on a wheel

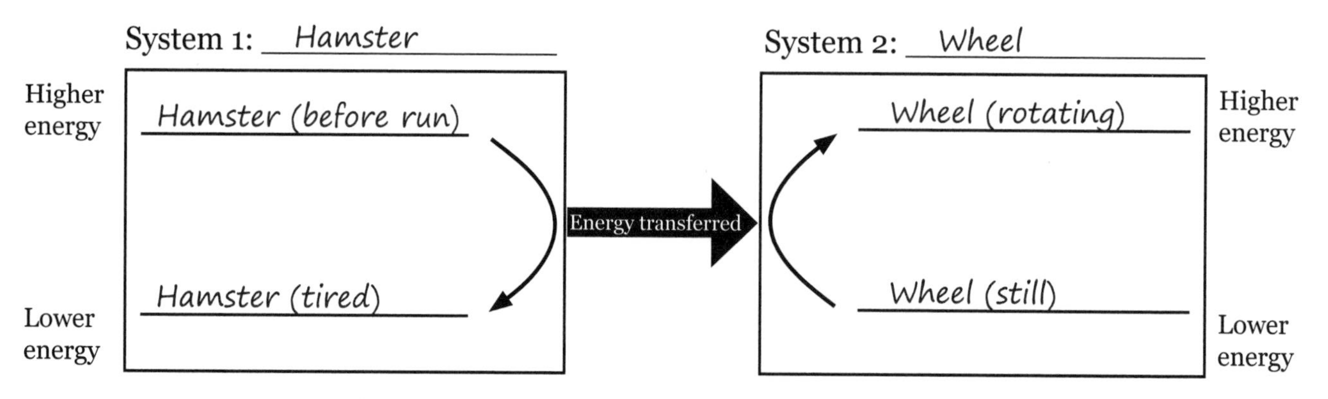

Teacher Talk and Actions

The hot-and-cold water example is straightforward, whereas the others are not. Guide students to reflect on the hot-and-cold water example so they can apply their thinking to the other examples by asking the following questions:

a. In the hot-and-cold water example, which system is increasing in energy? Explain.

The beaker with cold water increases in energy because its temperature increases.

b. Which system is decreasing in energy?

The beaker with hot water

c. Which way is energy transferred? Explain.

Energy is transferred from the flask with hot water to the beaker with cold water. We know that System 2 increases in energy and System 1 decreases in energy. Since energy is not created or destroyed, the energy to warm System 2 must have come from somewhere. Since the two systems are in contact, System 1 could have transferred energy to System 2. Our energy transfer model represents the energy transfer, helping us to visualize what we can't see.

Encourage students to use the same questions to guide their thinking about this example:

a. In the solar-powered-toy-car example, which system is increasing in energy? Explain.

The car increases in energy because its motion increases.

b. Which way is energy transferred? Explain.

Energy is transferred from the solar panel to the car. Since the energy of the car is increasing, another system must be transferring energy to it (because energy is not created or destroyed during chemical reactions).

c. Which system is decreasing in energy? Explain.

The solar panel must be decreasing in energy because that system is transferring energy to the car.

There are two reasons for this that can be discussed if appropriate:

- There is no longer an input of energy once the solar panel is no longer under the light bulb. The optional *Solar Car* video (available at *www.nsta.org/growthandactivity*) provides evidence that the car's motion depends on light, allowing students to draw an arrow from another system (sun) to the solar panel system.
- There is an output of energy from the car system. The moving car transfers energy to its surroundings as heat. The smoother the surface the car rides on, the longer it takes for the car to slow down. This is why the *Solar Car* video was filmed on a marble bench rather than on the sidewalk.

(Teacher Talk and Actions continued on p. 245)

Science Ideas

The activities in this lesson were intended to help you understand an important idea about using models to represent energy changes. Read the idea below. Look back through the lesson. In the space provided after the science idea, give examples of evidence that support the idea.

Science Idea #12: Models are useful for representing systems and their interactions—such as inputs, outputs, and processes—and energy and matter flows within and between systems.

Evidence:

Bar graphs and system boxes were useful for representing the energy changes within a system, and energy-transfer models were useful for representing energy transfers between two systems. When we observed a system increase in energy, we knew (because energy is conserved) that the energy must have come from a system that decreased in energy. Using the models helped us think about where the energy might be coming from in the case of a battery-powered Tesla and a hamster running on a wheel.

Models were also useful for representing matter changes during chemical reactions in the human body. Whether we were constructing a model for how our bodies convert food proteins like ovalbumin and casein to proteins like keratin (to build skin and hair) or actin and myosin (to build muscle), the models helped us keep track of inputs and outputs of matter. Underlying our models of matter changes was the idea that atoms aren't created or destroyed during chemical reactions. So, as we did with energy, we used the idea of matter conservation to make sense of data that showed correlations between increases in matter somewhere (e.g., amino acids increased in blood vessels surrounding the small intestine) and decreases in matter somewhere else (e.g., amino acids decreased in the small intestine).

Teacher Talk and Actions

(Teacher Talk and Actions continued from p. 243)

This would be a good opportunity to point out that our energy-transfer model represents the energy transfers, helping us to visualize what we can't see.

d. Extension questions: If the car were a battery-powered Tesla, what would we want to add to our model?

With the toy car, we could just replace the used battery with a fresh one. With the Tesla, that would be quite costly. So, we would need to "recharge" the battery before it is totally uncharged or dead, which would require an input of energy.

What could we add to our model to represent this?

We would add an upward arrow from the dead battery to the charged battery in the System 1 box. That would tell us that we need a source of energy to transfer to System 1.

Where could it come from?

(Students are not expected to come up with an answer at this point [e.g., from burning coal in a power plant], but they should at least realize that the energy must come from a system that is decreasing in energy and that using energy-transfer models helps them focus on where to look for that system.)

Encourage students to use the same questions to guide their thinking about this example:

a. In the hamster-on-a-wheel example, which system is increasing in energy? Explain.

The wheel increases in energy because its motion increases.

(Encourage students to focus on the wheel, rather than on the light at this point.)

b. Which system is decreasing in energy? Explain.

The hamster system must be decreasing in energy because the wheel only turns when the hamster runs on it, so we know the hamster must be transferring energy to the wheel.

c. What system in the hamster could be decreasing in energy and transferring energy to make the hamster's legs move?

(Students aren't expected to come up with an answer at this point [e.g., from burning glucose in its muscle cells], but they should realize that the hamster can't keep running, because it would get tired. Some students might wonder if the hamster has a kind of battery that runs down like the Tesla battery. However, whereas the owner of the Tesla can recharge its battery by plugging it into an electrical outlet, it isn't obvious how a hamster could recharge its battery. Some students may realize that food has something to do with how a hamster recharges its battery, and by the end of Chapter 2 students should understand the role of food in body motion. The important point is for students to realize that the energy for the hamster's motion must come from a system that is decreasing in energy and that using energy-transfer models helps them focus on what to look for in that system.)

Pulling It Together

Work on your own to answer these questions. Be prepared for a class discussion.

1. Now that you have learned more about representing energy changes, how would you answer the Key Question: **How can models help us make sense of energy changes within systems and energy transfers between systems?**

We can represent energy changes within systems with bar graphs (where the height of each bar shows the amount of energy before and after the interaction) and with system boxes (where the higher-energy version is on the higher line, the lower-energy component is on the lower line, and the arrow shows the direction of the energy change) and we can represent energy transfers between systems with energy-transfer models, where a thick arrow connecting the systems shows the direction of energy transfer between interacting systems.

2. Recall the spring and ball in the pinball machine. Draw bar graphs for the spring and ball that represent the energy changes that occur when the spring pushes the ball.

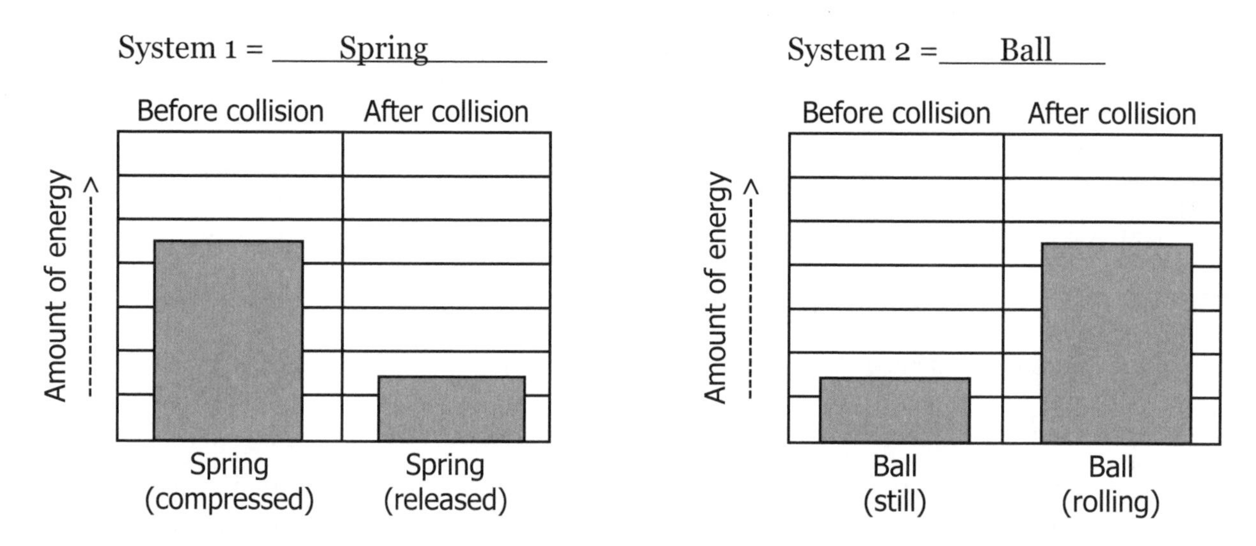

3. Use the bar graphs you drew in Question 2 to draw an energy-transfer model.

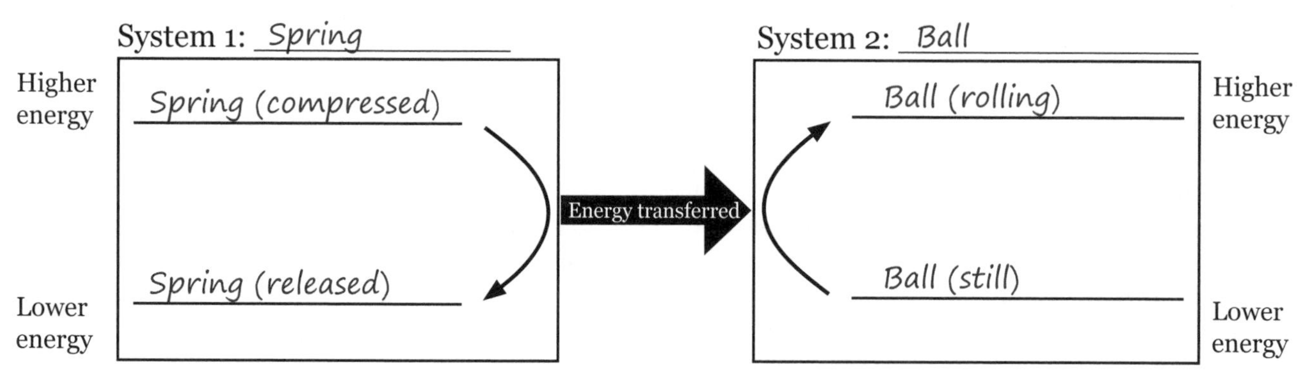

Teacher Talk and Actions

4. A baseball player swings a bat to try to hit a ball that was pitched to him.

Describe any energy changes and transfers you think may be occurring while the player is swinging the bat. Be sure to include the indicators of energy changes.

Option 1: The bat increases in energy because its motion increases (it goes from being stationary to moving).

The player decreases in energy because (a) he is transferring energy to the bat by swinging it, or (b) he will feel tired after practicing.

Option 2: The players arms are increasing in energy because they are going from being stationary to moving, so some system inside the body is transferring energy to the player's arms.

Teacher Talk and Actions

Teacher should look to see if students use an energy-transfer model here.

This is a foreshadowing question, so students may not have all of the answers.

If students take the approach of Option 1, listen for responses that talk about the player getting tired and ask whether "getting tired" is an indicator of an energy change. Follow up by asking students to restate their response in terms of indicators. Students could respond that the player could not continue to swing the bat forever, so his motion would have to stop. Students will have opportunities in Lesson 2.7 to learn which system is decreasing in energy.

Closure and Link to subsequent lessons:

In this lesson, you practiced representing energy changes using bar graphs and energy transfers using energy-transfer diagrams that are made up of system boxes and arrows. These models help us make sense of our observations and communicate our ideas about energy changes and energy transfers to others. In the next lesson, we will see if these models are helpful for making sense of energy changes that we may observe during chemical reactions.

Lesson Guide
Chapter 2, Lesson 2.3
Energy Changes During Chemical Reactions

Focus: This lesson focuses on energy-releasing chemical reactions, on energy transfers between chemical reaction systems and their surroundings, and on representing energy changes within systems and transfer between systems using energy-transfer models.

Key Question: How can we tell if chemical reactions involve changes in energy?

Target Idea(s) Addressed

A chemical reaction system is defined by the atoms that make up the reacting molecules. Defining the system this way keeps the amount of matter constant so that we can focus on inputs and outputs of energy. (Science Idea #16)

Some chemical reactions release energy. An increase in temperature or motion in surrounding systems or the detection of light or sound is evidence that energy was released during the chemical reaction, which means that the chemical reaction system has less energy. (Science Idea #17)

Energy changes within a system and energy transfers between systems usually result in some energy being released to the surrounding environment. (Science Idea #18)

Science Practice(s) Addressed

Analyzing and interpreting data
Developing and using models
Constructing explanations

Materials

Activity 1: Videos: *Reaction Between Hydrogen and Oxygen; Making Water (Animation), Part I*

Activity 2: *Per group of students:* candle; 400 ml glass beaker

Activity 3: Videos: *Reaction Between Glucose (in Donut) and Oxygen (in Air)*; *Reaction Between Glucose (in Donut) and Nearly Pure Oxygen Gas*

Activity 4: Video: *Reaction Between Hydrogen and Oxygen*

Advance Preparation

Organize materials in advance so that each group has materials when class begins.

Phenomena, Data, or Models	Intended Observations	Purpose	Rationale or Notes
Activity 1 Energy-releasing chemical reaction: • $O_2 + 2\ H_2 \rightarrow 2\ H_2O$ Bar graphs and system boxes as models of energy changes within systems and arrows in energy-transfer models to represent energy transfer between systems	• When hydrogen gas and oxygen gas react in a bottle, light and sound are emitted and the bottle's motion increases. • The bottle's increase in motion is represented on a bar graph with a lower bar before the reaction and a higher bar after the reaction and in the System 2 box with the bottle at rest before the reaction on the lower-energy line and the moving bottle after the reaction on the higher-energy line with a curved arrow pointing upward. • The energy decrease in the chemical reaction system (System 1) is represented with a higher bar for reactants ($O_2 + 2\ H_2$) and a lower bar for products ($2\ H_2O$) and in a system box with the reactants on the higher-energy line and the products on the lower-energy line with a curved arrow pointing downward. • An arrow from System 1 to System 2 represents the energy transfer from the chemical reaction system to the surroundings. • Scientists use a technique called calorimetry to determine how much energy is released during a particular chemical reaction. The reaction $O_2 + 2\ H_2 \rightarrow 2\ H_2O$ releases 136 kcal. This amount is the difference in the heights of the bars in the chemical reaction bar graph.	Provide evidence that the chemical reaction between hydrogen and oxygen releases energy and a way to represent energy changes within systems and energy transfer between systems. The change in energy of the surrounding system is represented first because this is what students have evidence for. Students can then apply the science idea to infer that if the surrounding system has increased in energy then the chemical reaction system must have decreased in energy. Introduces students to a unit of the amount of energy, the kilocalorie (kcal), which is the same as Cal on nutrition labels. Introduces students to a unit of the amount of matter, the mole (6.02×10^{23}), a "weighable" number of molecules, which provides a way to compare the amount of energy released in various chemical reactions.	If students are puzzled as to why energy must be added to get the reaction started, tell them they will learn more about this in Lesson 2.5. Students learned the formulas and equation for the reverse reaction in Lesson 1.4.

Phenomena, Data, or Models	Intended Observations	Purpose	Rationale or Notes
Activity 2 Energy-releasing chemical reaction: • $C_{11}H_{24} + 17\ O_2 \rightarrow 11\ CO_2 + 12\ H_2O$ Bar graphs and system boxes as models of energy changes within systems and arrows in energy-transfer models to represent energy transfer between systems	• When candle wax reacts with oxygen in the air, the temperature of the surrounding air/beaker increases. • The air/beaker's increase in temperature is represented on a bar graph with a lower bar before the reaction and a higher bar after the reaction and in the System 2 box with the cooler air before the reaction on the lower-energy line and the warmer air after the reaction on the higher-energy line with an arrow pointing upward. • The energy decrease in the chemical reaction system (System 1) is represented with a higher bar for the reactants ($C_{11}H_{24} + O_2$) and a lower bar for products ($11\ CO_2 + 12\ H_2O$) and in a system box with the reactants on the higher-energy line and the products on the lower-energy line with an arrow pointing downward. • An arrow from System 1 to System 2 represents the energy transfer from the chemical reaction system to the surrounding air/beaker. • The reaction $C_{11}H_{24} + 17\ O_2 \rightarrow 11\ CO_2 + 12\ H_2O$ releases 1,776 kcal. This amount is the difference in the heights of the bars in the chemical reaction bar graph.	Provide evidence that the chemical reaction between candle wax and oxygen releases energy and a way to represent energy changes within systems and energy transfer between systems. The change in energy of the surrounding system is represented first because this is what students have evidence for. Students can then apply the science idea to infer that if the surrounding system has increased in energy then the chemical reaction system must have decreased in energy.	If students are puzzled as to why energy must be added to get the reaction started, tell them they will learn more about this in Lesson 2.5. Students learned the formulas and equation for the reaction in Lesson 1.3.

Phenomena, Data, or Models	Intended Observations	Purpose	Rationale or Notes
Activity 3 Energy releasing chemical reaction: • $C_6H_{12}O_6 + 6\ O_2 \rightarrow 6\ CO_2 + 6\ H_2O$ Bar graphs and system boxes as models of energy changes within systems and arrows in energy-transfer models to represent energy transfer between systems	• A bigger flame and a more vigorous reaction occurs when a piece of donut reacts with a higher concentration of oxygen (nearly 100% from liquid oxygen) than a lower concentration of oxygen (air is 20% oxygen). • The increase in energy of the surroundings (e.g., donut's motion) is represented on a bar graph with a lower bar before the reaction and a higher bar after the reaction and in the System 2 box with the donut at rest before the reaction on the lower-energy line and the donut in motion after the reaction on the higher-energy line with an arrow pointing upward. • The energy decrease in the chemical reaction system (System 1) is represented with a higher bar for the reactants ($C_6H_{12}O_6 + 6\ O_2$) and a lower bar for products ($6\ CO_2 + 6\ H_2O$) and in a system box with the reactants on the higher-energy line and the products on the lower-energy line with an arrow pointing downward. • An arrow from System 1 to System 2 represents the energy transfer from the chemical reaction to the surrounding bottle. • The reaction $C_6H_{12}O_6 + 6\ O_2 \rightarrow 6\ CO_2 + 6\ H_2O$ releases 670 kcal. This amount is the difference in the heights of the bars in the chemical reaction bar graph.	Provide evidence that the chemical reaction between glucose and oxygen releases energy and a way to represent energy changes within systems and energy transfer between systems. Provide evidence that what releases energy is the reaction of the fuel with oxygen, not the fuel alone. The change in energy of the surrounding system is represented first because this is what students have evidence for. Students can then apply the science idea to infer that if the surrounding system has increased in energy then the chemical reaction system must have decreased in energy.	The observation that more energy is released to the surroundings when the donut reacts with more concentrated oxygen is intended to contradict the common misconception that the energy released was stored only in the fuel, not in the fuel + oxygen system. Students learned the formulas and equation for the reverse reaction in Lesson 1.4.

Phenomena, Data, or Models	Intended Observations	Purpose	Rationale or Notes
Activity 4 Revising models to account for the fact that not all of the energy from System 1 is transferred to System 2. Some energy is transferred to the surrounding environment.	• When oxygen and hydrogen react, the increase in motion of the bottle is evidence that some energy is transferred from the chemical reaction system to the bottle and the production of sound and light is evidence that some energy is also transferred to the surrounding environment. The energy-transfer model can be revised to account for this by splitting the thick energy-transfer arrow in two. • When candle wax and oxygen react, the increase in temperature of the air and beaker is evidence that some energy is transferred from the chemical reaction system to the air/beaker and the production of light is evidence that some energy is also transferred to the surrounding environment. The energy-transfer model can be revised to account for this by splitting the thick energy-transfer arrow in two. • When glucose and oxygen react, the increase in motion of the donut is evidence that some energy is transferred from the chemical reaction system to the donut and the production of light and sound is evidence that some energy is also transferred to the surrounding environment. The energy-transfer model can be revised to account for this by splitting the thick energy-transfer arrow in two.	Introduce students to ideas about energy dissipation by identifying evidence that energy is transferred to the surrounding environment in addition to System 2 and revising the models to account for this evidence.	The fact that every energy transfer results in some energy being transferred to the surrounding environment will be important for learning about coupling chemical reactions in Lesson 2.6. When two reactions are coupled, the energy-releasing reaction must release more energy than required by the energy-requiring reaction because some energy will be transferred to the surrounding environment.

Phenomena, Data, or Models	Intended Observations	Purpose	Rationale or Notes
Activity 5 Data table listing the energy released when various substances react with oxygen, including carbohydrate, fat, and protein polymers and their constituent monomers, and other familiar fuels	• The amount of energy released during a chemical reaction depends on the molecules that react. • Chemical reactions between carbohydrates or proteins/ amino acids and oxygen release 3–4 kcal/g, whereas chemical reactions between fats/fatty acids and oxygen release more (7–9.5 kcal). • Chemical reactions between hydrocarbons (carbon-based fuels) and oxygen release 11–12 kcal/g.	Introduce quantitative data on the energy released when various carbon-based substances react with oxygen.	If students are puzzled as to why energy must be added to get the reaction started, tell them they will learn more about this in Lesson 2.5. Students learned the formulas and equation for the reaction in Lesson 1.4.

Lesson 2.3—Energy Changes During Chemical Reactions

What do we know and what are we trying to find out?

In the previous lesson, we examined energy changes in several systems, used our observations to compile a list of indicators of energy changes, and used bar graphs and energy-transfer models to represent the energy changes. Now we will use the same set of indicators to help us detect energy changes in chemical reactions and see if the same models can help us think about the energy changes.

Recall from Chapter 1 that chemical reactions produce new substances with different characteristic properties from the starting substances. The starting substances of a chemical reaction are called reactants and the ending substances are called products. During a chemical reaction, bonds are broken between atoms of reactant molecules, and new bonds form between the atoms to make product molecules. Even though the products and reactants are made up of the same atoms, the products have different properties than the reactants, because some atoms are arranged differently.

We will start by investigating chemical reactions that happen in simple systems, where it is easier to observe when products form. In later lessons, we will apply what we have learned to chemical reactions that occur in complex living systems.

Answer the Key Question to the best of your knowledge. Be prepared to share your ideas with the class.

Key Question: How can we tell if chemical reactions involve changes in energy?

Teacher Talk and Actions

Activity 1: Hydrogen Reacts With Oxygen

In Chapter 1, we examined changes in matter that occur during several chemical reactions. We saw that an electric current can be used to rearrange the atoms that make up water molecules into hydrogen and oxygen molecules. In this activity, we will examine what happens when hydrogen and oxygen molecules react to form water molecules and look to see whether we can detect any changes in energy.

Procedures and Questions

Observing and representing energy changes during the reaction

1. Watch the video that shows what happens when hydrogen and oxygen gas react in a plastic bottle. The chemical equation for the reaction is: $2H_2 + O_2 \rightarrow 2\ H_2O$.

2. Record any observations that provide evidence that energy changes are occurring during the chemical reaction.

I saw a flash of light and heard a sound during the reaction. Also, the bottle moved.

3. During this reaction, does anything appear to end up with more energy or less energy than it had at the beginning? Cite evidence to support your answer.

The bottle has more energy after the reaction. My evidence is that the bottle started moving, which indicates that it has more energy than at the beginning.

4. Let's start by representing energy changes you observed in the surrounding bottle system. Sketch a bar graph to represent your observation that the bottle system increased in energy.

System = Bottle

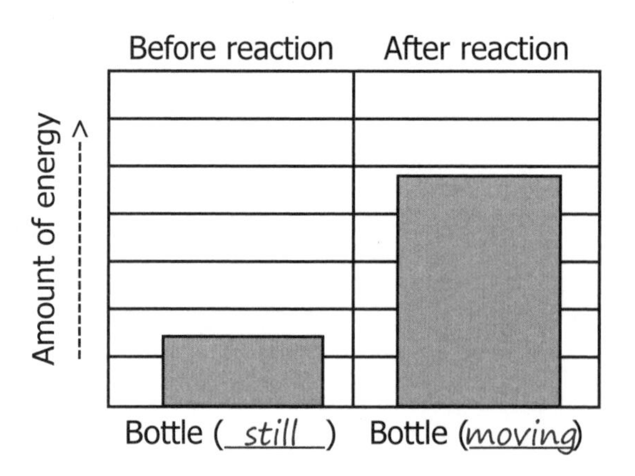

Teacher Talk and Actions

Show the video *Reaction Between Hydrogen and Oxygen* here.

5. According to Science Idea #14, when the energy of one system increases, the energy of another system or systems decreases. What other system could be providing energy to the bottle system? Why do you think so?

Perhaps the energy came from the chemical reaction between hydrogen and oxygen, because (a) the chemical reaction $2H_2 + O_2 \rightarrow 2\ H_2O$ occurred inside the bottle and therefore the chemical reaction was in contact with the bottle and (b) the bottle did not move until the chemical reaction occurred.

6. Do you think the products of the chemical reaction system (water molecules) have *more*, *less*, or *the same* amount of energy as the reactants (hydrogen and oxygen molecules)? Explain your answer.

I think the products have less energy than the reactants. If the bottle system increased in energy, that energy had to have come from somewhere. Since the chemical reaction system was in contact with the bottle system, I think the chemical reaction system provided the energy. And if the chemical reaction system provided the energy, the products must have less energy than the reactants.

7. Sketch bar graphs to represent the amount of energy associated with the reactants and the products. Justify why you drew the bars that way.

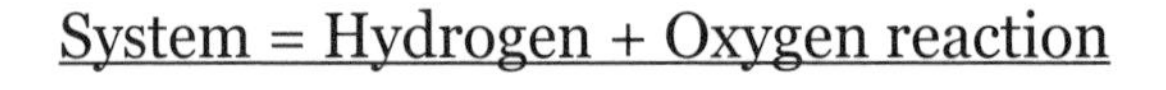

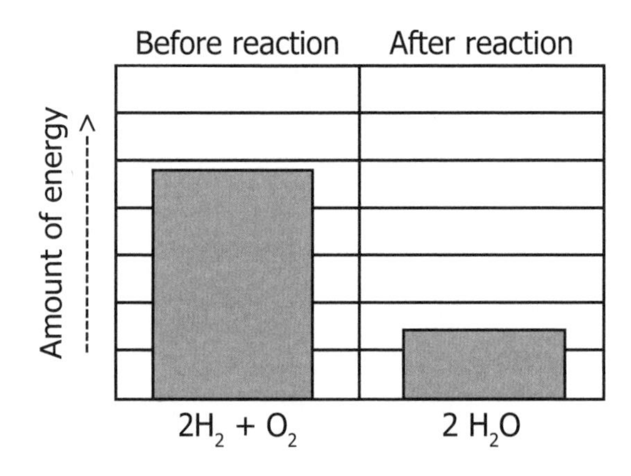

I drew a taller bar for the reactants (hydrogen and oxygen molecules) than the products (water molecules), because the chemical reaction system decreased in energy.

8. According to Science Idea #15, when two systems interact and one system decreases in energy while the other system increases in energy, we say that energy was transferred from the first system to the second system. If the hydrogen + oxygen chemical reaction system interacts with the bottle system, then we can represent the energy transfer between the systems with an arrow. Complete the bar graphs and energy-transfer models on the next page to represent the energy changes within each system and the direction of energy transfer between systems.

Teacher Talk and Actions

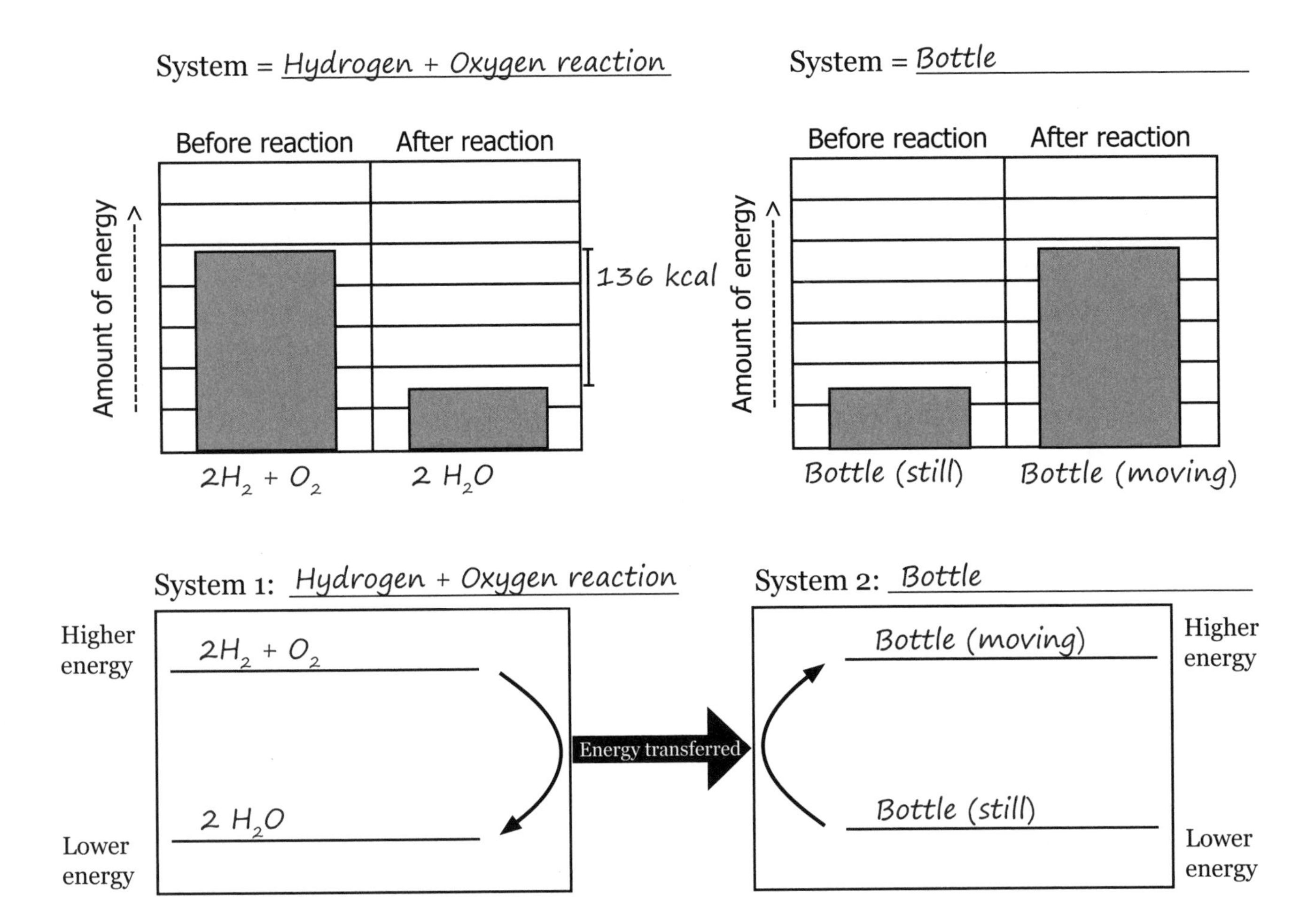

9. The energy-transfer model we just drew shows energy being transferred from the chemical reaction system to the bottle system. What evidence do we have to support this? The answer is that we only have circumstantial evidence. We observed light and sound and the bottle moving when hydrogen and oxygen reacted to form water. But just because things happen in two systems at the same time doesn't necessarily mean that the systems are interacting. Because the bottle system was open to the environment, both matter and energy could enter or leave. Therefore, we cannot be certain that the increase in energy of the bottle system actually came from the chemical reaction system.

If we don't want matter or energy to enter or leave the system, we can carry out the reaction in a calorimeter, like the one shown in Figure 2.1. Watch the video to learn more about using a calorimeter.

Teacher Talk and Actions

Show the video, *Making Water (Animation), Part I,* here.

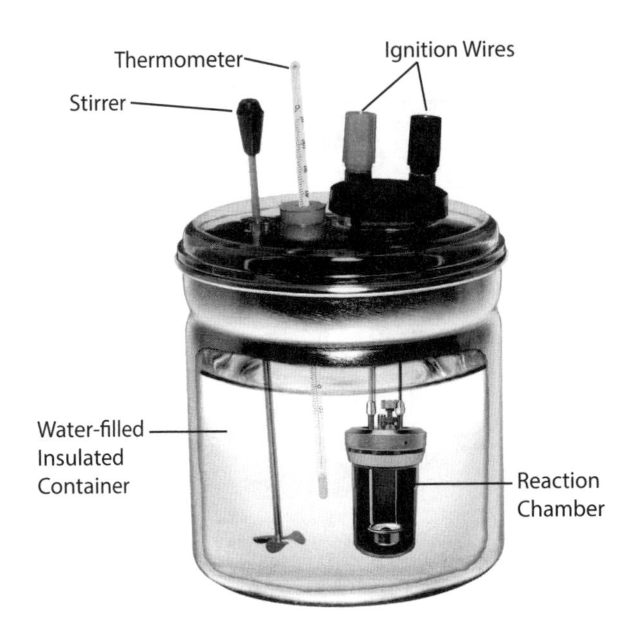

Figure 2.1. Setup for a calorimetry experiment

A calorimeter is an insulated container that does not allow atoms or energy to enter or leave it. Inside the calorimeter, the chemical reaction occurs inside a sealed reaction chamber that does not allow atoms to enter or leave it. The reaction chamber is surrounded by water. If a chemical reaction inside the chamber releases energy, then the energy will be transferred from the reaction chamber to the surrounding water, resulting in an increase the temperature of the water. Because the container of water is insulated, any increase in temperature of the water must have come from the chemical reaction.

10. If hydrogen and oxygen gas are placed in the reaction chamber and ignited with a spark, what do you think will happen to the temperature of the surrounding water? Justify your prediction.

The temperature of the water should increase. The energy released by the reaction between hydrogen and oxygen will warm up the water in the surrounding container.

11. Scientists reacted hydrogen and oxygen in the reaction chamber, measured the temperature of the water before and after the reaction, and used the temperature change to calculate the energy change caused by the reaction. They determined that the reaction $2\ H_2 + O_2 \rightarrow 2\ H_2O$ releases 136 kcal. (Note: A kcal, or kilocalorie, is the amount of energy needed to raise the temperature of 1 liter of water 1°C.) That means that for every 2 moles of H_2 that react with 1 mole of O_2 to form 2 moles of H_2O, 136 kcal are released. Add this data to the bar graphs on the previous page.

12. How many kcal would have been released if the scientists had reacted only 1 mole of H_2 with oxygen? Explain.

68 kcal/mole H_2, because if half as many molecules react with oxygen then only half as much energy would be released.

Teacher Talk and Actions

Students should show that 136 kcal is the difference in the heights of the $2H_2 + O_2$ bars and the $2\ H_2O$ bar.

You may be wondering what the term *mole* means in chemistry. In chemistry, a mole does not refer to a small furry animal. Rather, it is a unit for the amount of a substance. When you buy eggs at a store, you don't buy a single egg, you buy a dozen eggs. A dozen means 12. Similarly, when we weigh atoms or molecules on a balance, we don't weigh one atom or molecule, or even a dozen atoms or molecules, because our balances can't detect such a small amount of weight. For standard laboratory balances to detect the weight of atoms or molecules, we need to weigh an enormous number of them. Scientists define a mole as the number of hydrogen atoms that weigh 1 gram and the number of oxygen atoms that weigh 16 grams. In both cases, the number turns out to be 602,214,154,000,000,000,000,000 or 6.02×10^{23} molecules.

Teacher Talk and Actions

Activity 2: Candle Wax Reacts With Oxygen

Materials

For each team of students
Candle
Glass beaker

In the last activity, we examined changes in energy that occur during the reaction between hydrogen and oxygen and represented the energy changes and transfers with bar graphs and an energy-transfer model. In this activity, we will examine the chemical reaction between candle wax and oxygen to see whether we can detect any changes in energy. Then we will represent the changes we observe. In Chapter 1, you observed that candle wax reacts with oxygen in the air to produce carbon dioxide and water. The chemical equation for the reaction is:

$$C_{11}H_{24} + 17\ O_2 \rightarrow 11\ CO_2 + 12\ H_2O$$

Procedures and Questions

Observing and representing energy changes during the reaction

1. To observe if there are any changes in energy during the reaction, place the candle in the beaker and then ask your teacher to light the candle.

2. Record any observations that provide evidence that energy changes are occurring during the chemical reaction.

The beaker and the air around it get warmer during the reaction, and the candle gives off light.

3. During this reaction, does anything appear to end up with either more energy or less energy than it had at the beginning? Cite evidence to support your answer.

The beaker or the air around the beaker have more energy after the reaction. My evidence is that the beaker or the air around the beaker increased in temperature and an increase in temperature indicates it has more energy than at the beginning.

4. On the next page, sketch bar graphs to represent the change in energy of the chemical reaction system and the beaker system. Below, justify why you drew the bars that way.

I drew a taller bar for the beaker after the reaction than the beaker before the reaction because the beaker system increased in energy. The increase in temperature is evidence of this change.

I drew a taller bar for the reactants (wax and oxygen molecules) than the products (carbon dioxide and water molecules) because the chemical reaction system decreased in energy.

Teacher Talk and Actions

Safety Notes

1. Wear safety glasses with side shields or safety goggles and nonlatex aprons during the setup, hands-on, and take down segments of the activity.

2. Use caution when working with glassware, which can shatter if dropped and cut skin.

3. Use caution when working with a lighted candle and heated beaker, which can burn skin.

4. Properly clean up and dispose of waste materials.

5. Wash your hands with soap and water immediately after completing this activity.

Because we have to use a flame to light the candle, some students may be confused about whether energy is required or released when the candle burns. The answer will become clear in Lesson 2.5. For now, focus students' attention on what happens while the candle is burning (i.e., energy is released, as indicated by an increase in temperature of the surrounding beaker and air and the appearance of light).

If the students have trouble feeling the temperature increase, use a thermometer to measure the temperature of the beaker.

System = *Candle wax + Oxygen reaction*

Before reaction | After reaction

Amount of energy ---->

1,776 kcal

$C_{11}H_{24}$ + 17 O_2 11 CO_2 + 12 H_2O

System = *Glass beaker*

Before reaction | After reaction

Amount of energy ---->

Beaker (*cool*)   Beaker (*warmer*)

5. Complete the energy-transfer models below to represent the energy changes within each system and the direction of energy transfer between systems.

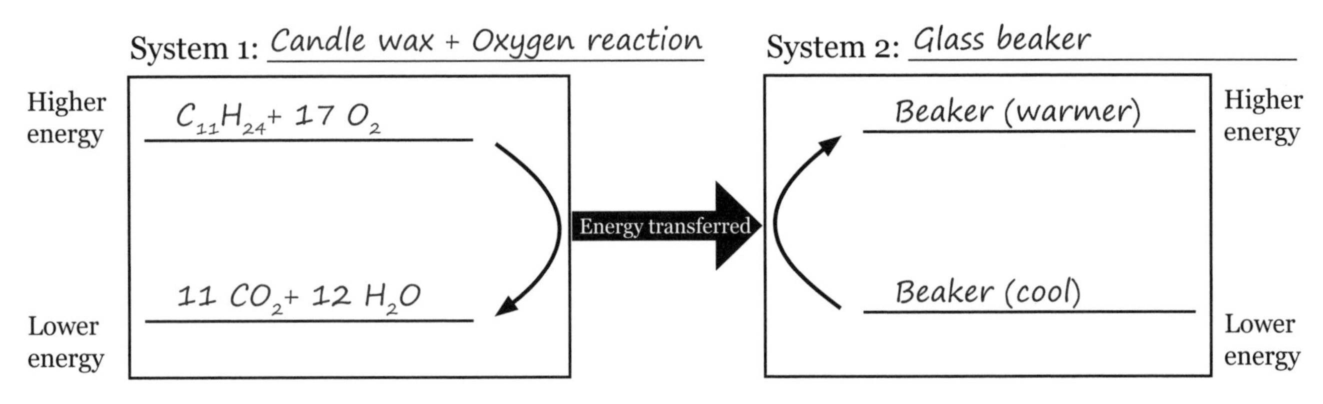

6. When scientists reacted candle wax ($C_{11}H_{24}$) and oxygen (O_2) in the reaction chamber of a calorimeter, they determined that 1,776 kcal were released per mole of candle wax ($C_{11}H_{24}$). Add this data to the bar graphs above.

Teacher Talk and Actions

The 1,776 kcals is the difference in the heights of the bars in the chemical reaction system bar graphs, not the actual height of either bar.

Activity 3: Glucose Reacts With Oxygen

In the last activity, we examined changes in energy that occur during the reaction between candle wax and oxygen and represented the energy changes and transfers with bar graphs and an energy-transfer model. In this activity, we will examine the chemical reaction between glucose and oxygen to see whether we can detect any changes in energy. Then we will represent the changes we observe. As we discovered in Chapter 1, glucose and oxygen molecules react to form carbon dioxide and water. The chemical equation for the reaction is as follows:

$$C_6H_{12}O_6 + 6\ O_2 \rightarrow 6\ H_2O + 6\ CO_2$$

Procedures and Questions

Observing and representing energy changes during the reaction

1. Watch the video of the donut and gaseous oxygen reaction. In the video, a donut reacts with oxygen gas in the air. A donut contains a large amount of glucose and the air is 20% oxygen gas.

2. Record your observations below. Focus on observations that provide evidence that energy is being released by the reaction.

We observed that the flame emitted light. While we can't observe an increase in temperature in the video, we would expect the temperature of the surrounding air to increase (just as the temperature of the beaker increased when the candle burned).

3. During this reaction, does anything appear to end up with either more energy or less energy than it had at the beginning? Cite evidence to support your answer.

Since the chemical reaction emitted light, the energy of the chemical reaction system decreased. Since the temperature of the surrounding air probably increased, its energy should increase.

4. On the next page, sketch bar graphs to represent the amount of energy associated with the reactants and the products. Below, justify why you drew the bars that way.

I drew a taller bar for the reactants (glucose and oxygen molecules) than the products (carbon dioxide and water molecules), because the chemical reaction system decreased in energy.

I drew a taller bar for the air after the reaction than the air before the reaction, because the surrounding air system increased in energy. The increase in temperature is evidence of this change.

Teacher Talk and Actions

Show the video *Reaction Between Glucose (in Donut) and Oxygen* here.

Many students are confused about what a flame indicates. If students say, "We observed a flame," ask them what indicators of an energy change are evident when they observe a flame. Students should realize that the flame emits light. Based on students' experiences with flames, they may also realize that the flame would warm the surrounding air.

Because we have to use a flame to start the reaction, some students may be confused about whether energy is required or released when glucose burns. The answer will become clear in Lesson 2.5. For now, focus students' attention on what happens once the donut is already burning (i.e., energy is released, as indicated by an increase in temperature of the surrounding air and the appearance of light).

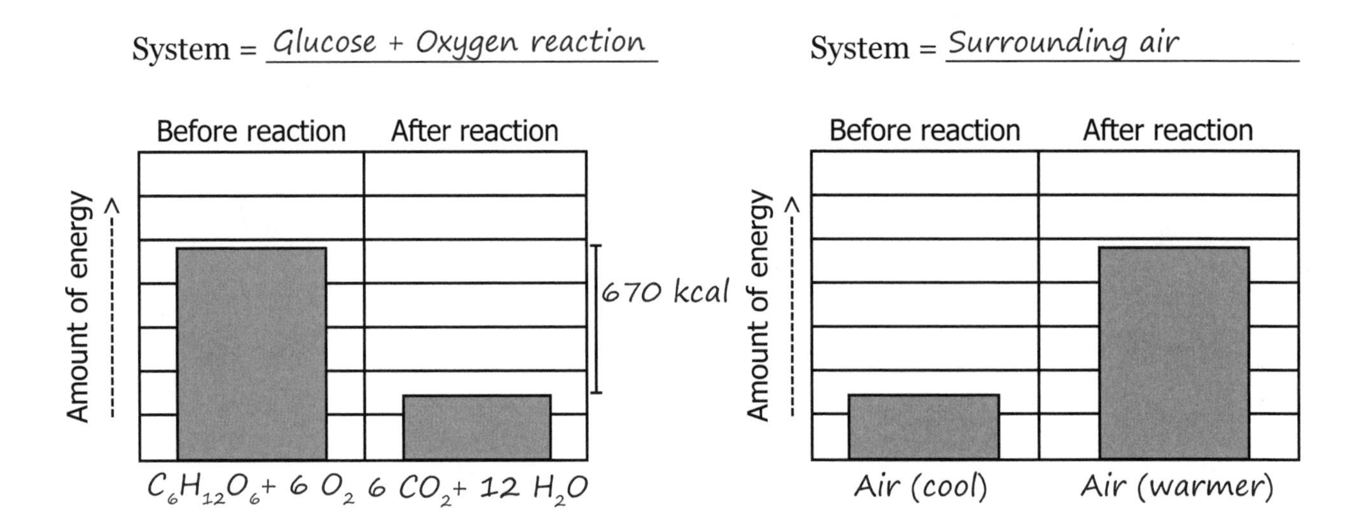

5. Complete the energy-transfer models below to represent the energy changes within each system and the direction of energy transfer between systems.

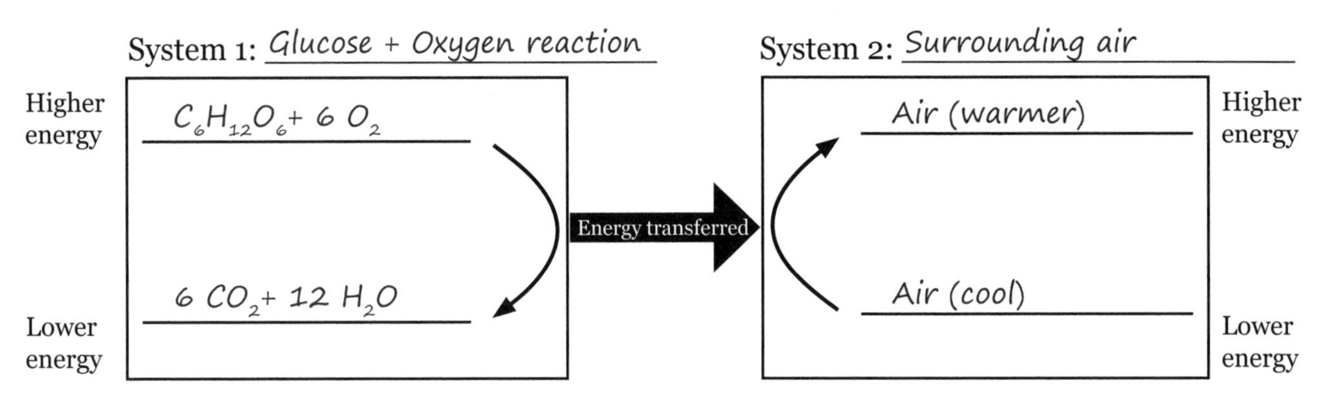

6. Now you will watch another video of the donut reacting with oxygen. In this video, the oxygen will come from liquid oxygen that is 100% oxygen. This means that there are more oxygen molecules available. The donut is the same size in this video as it was in the first video.

Teacher Talk and Actions

Show the video *Reaction Between Glucose (in Donut) and Nearly Pure Oxygen* here.

7. Record your observations below. Focus on observations that provide evidence that energy is being released by the reaction.

I saw a bright flame and heard popping sounds and the donut bounced around.

8. How do your observations from the reaction of the donut with ~100% oxygen differ from the reaction of the donut with ~20% oxygen?

The reaction with 100% oxygen was more violent than the reaction with 20% oxygen. The flame was bigger and it was louder.

9. In which demonstration was more energy transferred? Cite evidence.

More energy was transferred when the donut reacted with 100% oxygen than with air containing 20% oxygen. We know this because more light was emitted, more sound was emitted, and we observed a greater change in the motion of the donut when the reaction occurred with 100% oxygen.

10. What could explain the different observations you made about energy changes?

The main difference between the reactions is that more oxygen molecules were available to react when liquid oxygen was present. If more oxygen molecules were available, then more molecules of glucose from the donut could react.

11. When scientists reacted glucose ($C_6H_{12}O_6$) and oxygen (O_2) in the reaction chamber of a calorimeter, they determined that 670 kcal were released per mole of glucose. Add this data to the bar graphs you drew on the previous page.

Teacher Talk and Actions

The 670 kcals is the difference in the heights of the bars in the chemical reaction system bar graphs, not the actual height of either bar.

Activity 4: Energy Transfer to the Surrounding Environment

In the previous activities, you observed indicators of energy being transferred from a chemical reaction system to another system that interacts with the chemical reaction system and constructed models of the energy transfer between these two systems. Someone looking at the energy-transfer models we drew might interpret them to mean that all the energy released by the chemical reaction is transferred to the second system. Let's revisit each example and see if we can find evidence that energy is also transferred somewhere else.

Procedures and Questions

1. Watch the video of hydrogen and oxygen reacting again. In Activity 1 we focused on the energy transferred to the bottle. This time focus on observations that provide evidence that energy is also transferred somewhere else.

I can see light and hear sound. This is evidence that energy is transferred out of the bottle to the surrounding environment.

2. Think about the other two reaction systems we experienced in Activities 2 and 3. Did you make any observations that would provide evidence that energy is transferred to multiple places? If so, describe them below.

In both cases (the donut and candle reacting with oxygen), I saw light. This is evidence that energy is transferred to the surrounding environment. The light travels out of the chemical reaction system in all directions. Anything in the surrounding environment that absorbs that light increases in energy. I also felt the air around the glass beaker in addition to the glass beaker get warmer when the candle was burning. This is evidence that energy is transferred to both the beaker and the surrounding air.

3. Now go back to the energy-transfer diagrams you drew in Activities 1, 2, and 3 and revise them to better match your observations.

Students should go back and change the energy-transfer arrows to show that energy is transferred to System 2 and also to the surrounding environment.

Teacher Talk and Actions

Show the video *Reaction Between Hydrogen and Oxygen* here.

Activity 5: Comparing the Amounts of Energy Released by Different Chemical Reactions

In the previous activities, you observed indicators of energy being released from a chemical reaction system to the surrounding system and represented this energy transfer using bar graphs and energy-transfer models. In each example, your models showed that the amount of energy in the reaction system after the reaction was less than the amount of energy in the reaction system before the reaction. When you first drew bar graphs to represent the amount of energy in the reactants and the products, you had to guess what the difference in the heights of the bars should be.

When you added lines to represent the energy differences, you used the energy calculated from calorimetry measurements. However, you didn't check to see if the same number of molecules of fuel (hydrogen, candle wax, and glucose) reacted in each case. If we want to compare the amount of energy released when different fuels react with oxygen, we need to compare the amount of energy released when the **same number of molecules** of each fuel react with oxygen. Table 2.3 shows how much energy is released when 1 mole (6.02×10^{23} molecules) of various fuels react with oxygen.

Table 2.3. Energy Released During Reactions of Different Fuels With Oxygen

Chemical Reaction With Oxygen	Energy Released (kcal/mole)
Carbohydrates + $O_2 \rightarrow CO_2 + H_2O$	
Glucose ($C_6H_{12}O_6$)	670
Sucrose ($C_{12}H_{22}O_{11}$)	1,348
Fat/Fatty acid + $O_2 \rightarrow CO_2 + H_2O$	
Palmitic acid ($C_{16}H_{32}O_2$), a fatty acid	2,385
Protein/Amino acid + $O_2 \rightarrow CO_2 + H_2O$ + N-containing product	
Glycine ($C_2H_5NO_2$), an amino acid	233
Glutamine ($C_5H_{10}N_2O_3$), an amino acid	615
Hydrocarbon + $O_2 \rightarrow CO_2 + H_2O$	
Candle wax (paraffin $C_{11}H_{25}$)	1,776
Octane (C_8H_{18}), gasoline	1,317
Propane (C_3H_8)	531
Other substances + $O_2 \rightarrow CO_2 + H_2O$	
Ethanol (C_2H_6O)	311
Lactic acid ($C_3H_6O_3$)	327
Hydrogen + Oxygen ($H_2 + \frac{1}{2} O_2 \rightarrow H_2O$)	68

Teacher Talk and Actions

The values in the table are the difference in the heights of the bars not the actual height of either bar.

1. Examine the values in Table 2.3 and respond to the questions.

   a. Based on the data in the table, what substance is the best fuel? Why?

   Palmitic acid, because it releases the most energy per mole.

   b. What relationship do you notice between the number of carbon atoms making up the molecule and the amount of energy released when it reacts with oxygen?

   The more carbon atoms the molecule has, the more energy is released when the molecule reacts with oxygen.

   c. Compare the amount of energy in Table 2.3 for the reaction of hydrogen with oxygen to the differences in height you represented on your energy graphs in Activity 1 and explain the difference.

   The value in Table 2.3 lists 68 kcal/mole as the amount of energy released per mole of H_2 that reacts with oxygen, whereas the value we used in Activity 1 was 136 kcal, which is twice the amount of energy released because twice as many molecules of H_2 reacted.

2. You may be wondering how the information in Table 2.3 relates to the information about calories on your food cards. You know that foods are composed of a variety of proteins, fats, and carbohydrates and different amounts of each type of molecule. So, foods are mixtures of substances and, therefore, we need a different unit to compare the amount of energy in different foods. Food scientists chose the unit kilocalories per gram (kcal/g). The food cards you examined in Chapter 1 all provided information about the number of Calories in a serving size. A Calorie is the same as a kcal, but a serving size is much more than 1 gram. Why do you think food cards use Calories in a serving size rather than Calories in a gram?

   Reporting Calories per serving size is more useful for calculating total number of Calories consumed in a meal or in a day. Because a gram of food is much smaller than a serving size, we would need to multiply the number of kcal/g times the number of grams to figure out how many kcal we actually ate.

3. The table below shows the amount of energy released when 1 gram of three different foods each reacted with oxygen in a calorimeter.

Chemical Reaction With Oxygen	Energy Released (kcal/g)
Baked donut	3.43
Margarine	7.14
Egg white	0.52*

*Egg white is 10% protein and 90% water.

What patterns do you see across the types of foods (donut, margarine, and egg white)?

More energy is released when fats react with oxygen than when carbohydrates or proteins do.

Teacher Talk and Actions

In this context, we are referring to the Calories listed on food labels, which is actually a kilocalorie or 1,000 Calories.

Science Ideas

The activities in this lesson were intended to help you understand an important idea about energy changes during chemical reactions. Read the idea below. Look back through the lesson. In the space provided after the science idea, give evidence that supports the idea.

Science Idea #16: A chemical reaction system is defined by the atoms that make up the reacting molecules. Defining the system this way keeps the amount of matter constant so that we can focus on inputs and outputs of energy.

Evidence:

We defined the candle system to include the atoms involved in the chemical reaction—the C and H atoms making up the candle and the O atoms making up the oxygen gas. We found that the all the atoms ended up either as carbon dioxide or water molecules. When we kept all those atoms in a "closed" system, we found that mass was conserved because atoms were conserved. By not allowing any atoms/matter to enter or leave the system, we could focus on energy changes. If we hadn't keep all the atoms in the system, then we wouldn't have known whether atoms that entered or left the system "carried" some energy with them.

Science Idea #17: Some chemical reactions release energy. An increase in temperature or motion in surrounding systems or the detection of light or sound is evidence that energy was released during the chemical reaction, which means that the chemical reaction system has less energy.

Evidence:

We detected an increase in temperature of the surrounding system when copper sulfate reacted with iron. This is evidence that the chemical reaction gave off energy.

We saw light when the donut and oxygen reacted. This is evidence that the reaction gave off energy.

Tables 2.3 and 2.4 provided data about how much energy was released from various chemical reactions. The data were based on measuring how much the temperature of surrounding water increased when each substance reacted with oxygen in a calorimeter.

Science Idea #18: Energy changes within a system and transfers of energy between systems usually result in some energy being released to the surrounding environment.

Evidence:

In each example we observed, we saw evidence (e.g., the production of light and sound and an increase in temperature) that energy was transferred from the chemical reaction system to one or more other systems (e.g., the bottle system and the surrounding environment in the case of the hydrogen and oxygen reaction).

Teacher Talk and Actions

Remind students that science ideas are accepted principles that are based on a wide range of observations, not just the small number of examples they have observed.

Pulling It Together

Work on your own to answer these questions. Be prepared for a class discussion.

1. Now that you have had more experience with chemical reactions, how would you answer the Key Question: **How can we tell if chemical reactions involve changes in energy?**

We can observe evidence of energy increases in the surrounding system: The temperature increases, motion increases, or light or sound is produced. We know that increases of energy in one system must correspond to decreases in energy in another system. So, if we observe evidence of an increase in the energy of the surrounding system, then the energy must have come from another system in contact with it. In the examples we observed, a chemical reaction system was the only thing in contact with the surrounding system, so we can conclude that the chemical reaction system transferred energy to the surrounding system. Since the chemical reaction transferred energy, the chemical reaction system must have less energy after the reaction occurred and therefore the products of the reaction must have less energy than the reactants.

2. A student makes the claim that the energy released when fuel burns comes only from the fuel source. Do you agree or disagree with this claim? Write an explanation to support your claim about where the energy comes from using relevant evidence and science ideas. Then draw an energy-transfer model that is consistent with the evidence and science ideas.

Question	Does the energy released when fuel burns come only from the fuel source?
Claim	*The energy released when fuel burns comes from the chemical reaction system (which contains both fuel and oxygen molecules), not just the fuel source.*
Science Ideas	*• Science Idea #14: When the energy of one system increases, the energy of another system decreases.* *• Science Idea #15: When two systems interact and one system decreases in energy while the other system increases in energy, we say that energy was transferred from the first system to the second.* *• Science Idea #16: A chemical reaction system is defined by the atoms that make up the reacting molecules.* *• Science Idea #17: The detection of light or sound is evidence that energy was released during the chemical reaction.*

Teacher Talk and Actions

If students need help getting started, suggest that they consider how the examples of the donut burning were similar and how they were different.

Evidence	When the donut (mainly glucose) reacted with oxygen in the air (air is 20% oxygen), light was given off to the surroundings. When the scientist reacted the donut with liquid oxygen (evaporates to a gas with much more than 20% oxygen), we observed even more light and a more violent reaction (the donut jumped around in the beaker).
Models	An energy-transfer model shows that energy is transferred from the chemical reaction system to the surrounding system.

Explanation:

The energy released when fuel burns comes from the chemical reaction system, not just the fuel source. The chemical reaction system includes all the atoms involved in the chemical reaction. For example, when the glucose in the donut reacted with oxygen in the air, the chemical reaction system included the atoms from the glucose molecules and the oxygen molecules that rearrange to form carbon dioxide molecules and water molecules. During this reaction, we saw the production of light, which indicates that energy is being transferred to the surrounding system. When the scientist reacted the donut with liquid oxygen, we observed a brighter flame and a more violent reaction (the donut jumped around in the beaker). This is evidence that more energy is transferred to the surroundings during the liquid oxygen reaction than the atmospheric oxygen reaction. If the energy were just coming from the fuel (the glucose in the donut), we would have seen the same amount of energy released in each case because the size of the donut remained the same in each case. This means that the energy must have come from the combination of oxygen and glucose molecules, not from just the glucose molecules. We can represent this transfer of energy using an energy-transfer diagram like the one below. We include all the reactant and product molecules in the system box on the left to illustrate that the energy is being released from the chemical reaction system (fuel + oxygen) not from the fuel alone.

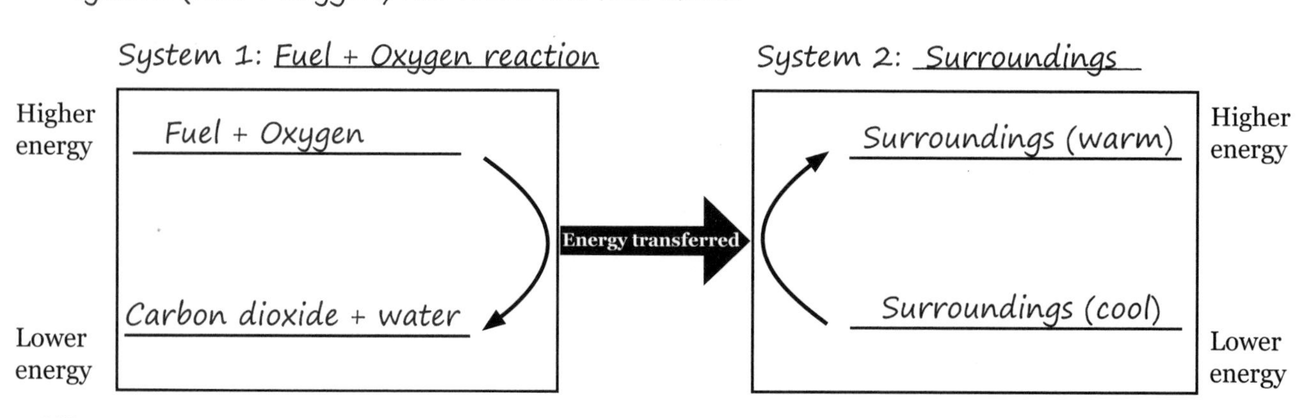

Teacher Talk and Actions

3. What indicators have you observed to suggest that chemical reactions that occur in the bodies of animals release energy? Explain/justify your answer with evidence/examples.

[NOTE: The purpose of this question is to elicit students' ideas about indicators that processes in animals release energy. Students may know that people increase in motion and temperature when they exercise and might conclude that these are both indicators that energy is being released by chemical reactions. Some students might list digestion of carbohydrate, fat, or protein polymers or cellular respiration as possible chemical reactions because they feel warm right after a meal. Other examples include fireflies and marine organisms that give off light, moving flagella and cilia of single-celled animals, sperm swimming, chromosomes moving to opposite ends of a cell during cell division, birds chirping, and dogs barking. Ideally, the class would come up with a wide range of examples that span molecular, cellular, and organism levels of biological organization. It would be great if students mention an electric eel, even though giving off electric current hasn't been listed as an indicator. In the next lesson, students will observe that energy from a battery can supply the energy needed to split water molecules into hydrogen and oxygen molecules.]

Teacher Talk and Actions

Closure and Link to subsequent lessons:

In this lesson, you observed several chemical reactions that released energy and represented the energy changes and energy transfers using bar graphs and energy-transfer models. You may be wondering whether all chemical reactions release energy. In the next lesson, we will observe several other chemical reactions to answer this question.

Lesson Guide
Chapter 2, Lesson 2.4
Energy Changes During Other Chemical Reactions

Focus: This lesson focuses on energy-requiring chemical reactions and energy transfers between those systems and their surroundings and on using energy-transfer models to represent them.

Key Question: Do all chemical reactions release energy?

Target Idea(s) Addressed

Some chemical reaction systems require an input of energy. A decrease in temperature or motion in surrounding systems, the decrease in energy of a battery connected to the system in a complete circuit, or the absorption of light or sound is evidence that energy was absorbed during the chemical reaction, which means that the chemical reaction system now has more energy. (Science Idea #19)

Energy is transferred from the sun to plants by light during the process of photosynthesis. Because an input of energy is required for photosynthesis to occur, the arrangement of atoms in product molecules (one glucose and six oxygen molecules) must have more energy than the arrangement of the same atoms in reactant molecules (six carbon dioxide and six water molecules). (Science Idea #20)

Science Practice(s) Addressed

Analyzing and interpreting data
Developing and using models
Constructing explanations
Carrying out investigations

Materials

Activity 1: *Per group of students:* new 9-volt battery; used 9-volt battery; condiment cup; 2 lead thumbtacks; salt water; *per class:* Battery testers

Activity 2: Video: *Barium Hydroxide & Ammonium Thiocyanate*

Activity 3: Video: *Aquatic Plant in Light and Dark*

Pulling It Together Q3: Video: *Nylon* (optional)

Advance Preparation

Organize materials in advance so that each set of partners or table teams has materials when class begins. Observe videos and decide how you will use them.

Activity 1: To have a more dramatic reaction, make a salt solution (1 g/liter) in advance and use that during the class. Do not tell students that there is salt in the water.

Phenomena, Data, or Models	Intended Observations	Purpose	Rationale or Notes
Activity 1 Energy-requiring chemical reaction: $2\ H_2O \rightarrow O_2 + 2\ H_2$ Bar graphs and system boxes as models of energy changes within systems and arrows in energy-transfer models to represent energy transfer between systems	• When a new battery is connected to water in a complete circuit, bubbles form, but when the battery is disconnected, bubbles stop forming. • When a used battery (representing what would happen if we left the battery connected to the water overnight) is connected to water in a complete circuit, fewer or no bubbles form. • The change from a new to used battery is represented on a bar graph with a higher bar before the reaction and a lower bar after the reaction. • The increase in energy of the reaction system is represented on a bar graph with a lower bar before the reaction and a higher bar after the reaction AND with a system box showing an upward arrow from reactants to products. • The energy transfer from the surrounding system to the energy-requiring chemical reaction system can be represented with an energy-transfer model. • Scientists have determined that this reaction requires an input of 136 kcal this quantity is the difference in the heights of the bars on the chemical reaction system bar graph. • The amount of energy required for the chemical reaction is (approximately) the same as the amount of energy released when the reverse reaction occurs.	Provide evidence that not all chemical reactions release energy. Give students a chance to apply the bar graph representation to phenomena involving energy-requiring chemical reactions. Give students a chance to apply the science idea to infer that if the surrounding system has decreased in energy then the chemical reaction system must have increased in energy. Give students a chance to apply the energy-transfer model to phenomena involving energy-requiring chemical reactions.	Students should already know that a chemical reaction occurs in each example and what the reactants and products are based on evidence obtained in Chapter 1. The change in energy of the surrounding system is represented first because this is what students have evidence for.

Phenomena, Data, or Models	Intended Observations	Purpose	Rationale or Notes
Activity 2 Energy-requiring chemical reaction: $2\ NH_4SCN + Ba(OH)_2 \rightarrow Ba(SCN) + 2\ H_2O + 2\ NH_3$ Bar graphs and system boxes as models of energy changes within systems and arrows in energy-transfer models to represent energy transfer between systems.	• When ammonium thiocyanate and barium hydroxide react with each other, water in the surrounding system freezes. • The freezing (decrease in temperature) of the surrounding system is represented on a bar graph with a higher bar before the reaction and a lower bar after the reaction. • The increase in energy of the reaction system is represented on a bar graph with a lower bar before the reaction and a higher bar after the reaction AND with a system box showing an upward arrow from reactants to products. • The energy transfer from the surrounding system to the energy-requiring chemical reaction system can be represented with an energy-transfer model. • Scientists have determined that this reaction requires an input of 24 kcal and this quantity is the difference in the heights of the bars on the chemical reaction system bar graph.	Provide evidence that not all chemical reactions release energy. Give students a chance to apply the bar graph representation to phenomena involving energy-requiring chemical reactions. Give students a chance to apply the science idea to infer that if the surrounding system has decreased in energy, then the chemical reaction system must have increased in energy. Give students a chance to apply the energy-transfer model to phenomena involving energy-requiring chemical reactions.	

Phenomena, Data, or Models	Intended Observations	Purpose	Rationale or Notes
Activity 3 Energy-requiring chemical reaction: $6\ CO_2 + 6\ H_2O \rightarrow C_6H_{12}O_6 + 6\ O_2$ Bar graphs and system boxes as models of energy changes within systems and arrows in energy-transfer models to represent energy transfer between systems.	• Aquatic plants produce more oxygen bubbles with a higher-intensity light than with a lower-intensity light. • The increase in energy of the reaction system is represented on a bar graph with a lower bar before the reaction and a higher bar after the reaction AND with a system box showing an upward arrow from reactants to products. • The energy transfer from the surrounding system to the energy-requiring chemical reaction system can be represented with an energy-transfer model. • Scientists have determined that this reaction requires an input of 670 kcal; this quantity is the difference in the heights of the bars on the chemical reaction system bar graph.	Provide evidence that not all chemical reactions release energy. Give students a chance to apply the bar graph representation to phenomena involving energy-requiring chemical reactions. Give students a chance to apply the science idea to infer that if the surrounding system has decreased in energy then the chemical reaction system must have increased in energy. Give students a chance to apply the energy-transfer model to phenomena involving energy-requiring chemical reactions. Give students a chance to use the data to revise their bar graph model of energy change during photosynthesis to include the difference in the heights of the bars on the graph.	In the energy-transfer model for photosynthesis, the energy-releasing processes in the Sun are deliberately black-boxed.
Activity 4 Data table showing the net amounts of energy required to form one mole of product	• The amount of energy required for a chemical reaction to occur depends on the molecules that react. • Forming glucose and oxygen from carbon dioxide and water (photosynthesis) requires 670 kcal/mole. • Forming various biological polymers (protein, carbohydrate, and fat) all require a net input of energy–between one and 7.3 kcal to link two monomers together).	Introduce quantitative data on the energy required to form various carbon-based substances that are required for building body structures.	

Phenomena, Data, or Models	Intended Observations	Purpose	Rationale or Notes
Pulling It Together Q3: Predict and explain a phenomenon: nylon formation	• The chemical reaction between hexamethylenediamine and adipic acid to form nylon and water molecules should require energy (because the surroundings decrease in energy).		This reaction is included in the middle school book *Toward High School Biology*.
Pulling It Together Q4: Model a phenomenon: protein synthesis	• Students brainstorm chemical reactions that occur in animals' bodies that release energy.	Gives teacher an opportunity to see what students think about energy changes and transfers that occur during protein synthesis.	By now students should be using system boxes with curved arrows to represent energy changes within a system.

Lesson 2.4—Energy Changes During Other Chemical Reactions

What do we know and what are we trying to find out?

In the previous lesson, we observed three different chemical reactions that release energy. When each reaction occurred, we observed indicators that energy was transferred from the reaction system to the surrounding system. For example, when the candle wax reacted with oxygen, the temperature of the surrounding air increased and light was produced.

The increase in temperature and appearance of light provided evidence that energy was transferred to the surrounding system. Think about other chemical reactions you have observed. Can you think of any that do not release energy?

In this lesson, you will observe some other chemical reactions and begin thinking about the Key Question. (There is no need to respond in writing now.)

Key Question: Do all chemical reactions release energy?

Teacher Talk & Actions

This is an opportunity to find out if students know about energy-requiring reactions or processes such as like cold packs. Some students may think that coldness is transferred from the cold pack to the surroundings instead of energy being transferred from the surroundings to the cold pack. If we measured the temperature of the surrounding air, we would find that the temperature of the air surrounding the cold pack decreases as the chemical reaction occurs inside the cold pack. Similarly, as the chemical reaction in the cold pack occurs, the cold pack feels cold to us because our hands transfer energy to the chemical reaction in the cold pack.

Activity 1: Chemical Reaction Between Water Molecules

Materials

For each team of students
New 9-volt battery
Used 9-volt battery
Condiment cup
2 lead thumbtacks
Water

For each class
1–2 battery testers

In Lesson 2.1, you observed that a toy car moved faster when connected in a circuit to a "new" battery but the toy car moved slower or did not move at all when connected to a "used" battery. In Lesson 1.4, you observed that bubbles of hydrogen and oxygen gas formed when water was connected in a circuit to a 9-volt battery, but you focused only on changes in matter. In this activity, you will examine the chemical reaction again, but now you will focus on changes in energy. Recall that when an electric current is passed through water, a chemical reaction occurs that forms oxygen gas and hydrogen gas. The chemical equation for the reaction is as follows:

$$2\,H_2O \rightarrow O_2 + 2\,H_2$$

Procedures and Questions

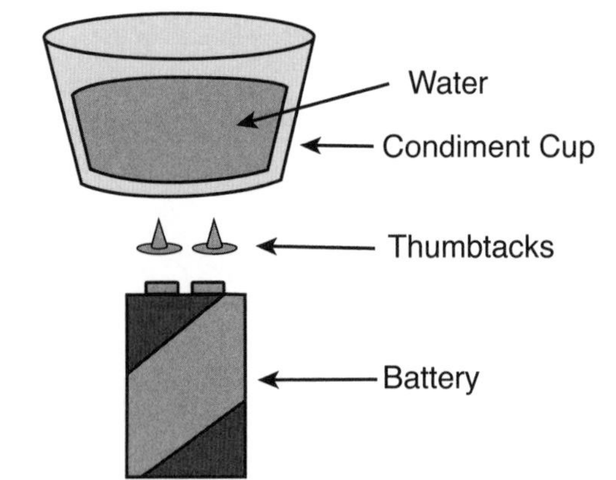

1. Place the cup upside down on the table and punch the thumbtacks through the bottom of the cup, making sure that they do not touch each other but that they are close enough so that they will line up with the leads of the battery.

2. Put some water in the cup.

3. Carefully place the cup with the thumb tacks on a new 9-volt battery so that the thumbtacks line up with the leads of the battery.

4. Record your observations about this chemical reaction below. Focus on observations that provide evidence that energy changes are occurring during the chemical reaction.

Bubbles form only when the thumbtacks touch the leads of the battery.

Teacher Talk & Actions

Safety Notes

1. Wear safety goggles, nonlatex aprons, and thermal gloves during the setup, hands-on, and take down segments of this activity.

2. Use caution when using sharp tools and materials.

3. Immediately wipe up any spilled water on the floor to avoid a slip-and-fall hazard.

4. Properly clean up and dispose of waste materials.

5. Wash your hands with soap and water immediately after completing this activity.

5. Now place the cup of water with the thumbtacks on a used 9-volt battery.

6. Record your observations below. Focus on observations that provide evidence that energy changes are occurring during the chemical reaction.

Bubbles don't form as rapidly (or at all), even when the thumbtacks touch the leads of the battery.

7. During this reaction, does anything appear to end up with either more energy or less energy than it had at the beginning? Cite evidence to support your answer.

Perhaps the battery has less energy. We observed that little or no gas was produced when we placed the cup on the "used" battery. I have observed with flashlights and battery-operated toys that after a battery has been used for a while, it is no longer able to power an electrical device.

Representing energy changes and transfers

8. If you were to leave the cup of water on the new battery for a long time and then test the battery with a battery tester, you would see that the needle would no longer be in the "good" region. Sketch a bar graph to represent the energy change in the battery. Draw a taller bar when the battery system has more energy and a shorter bar when the battery system has less energy. In the parentheses, describe how the system is different before and after the reaction.

System = Battery

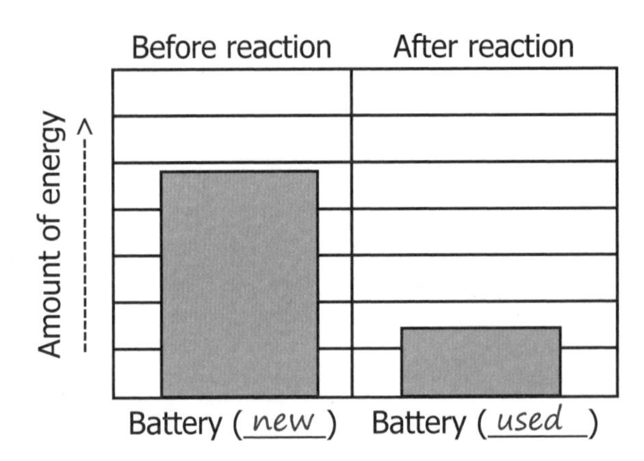

9. Based on the bar graph you drew of the battery system, what do you think happened to the amount of energy in the chemical reaction system? (Assume that the battery system and the chemical reaction systems are the only two systems interacting in the case.)

Because the battery system decreased in energy, the chemical reaction system must have increased in energy.

Teacher Talk & Actions

10. Draw an energy-transfer model to represent the energy changes and transfers you observed. Use the blank graphs above each box to draw bars representing the amount of energy in each system before and after the reaction to help you complete the energy-transfer model.

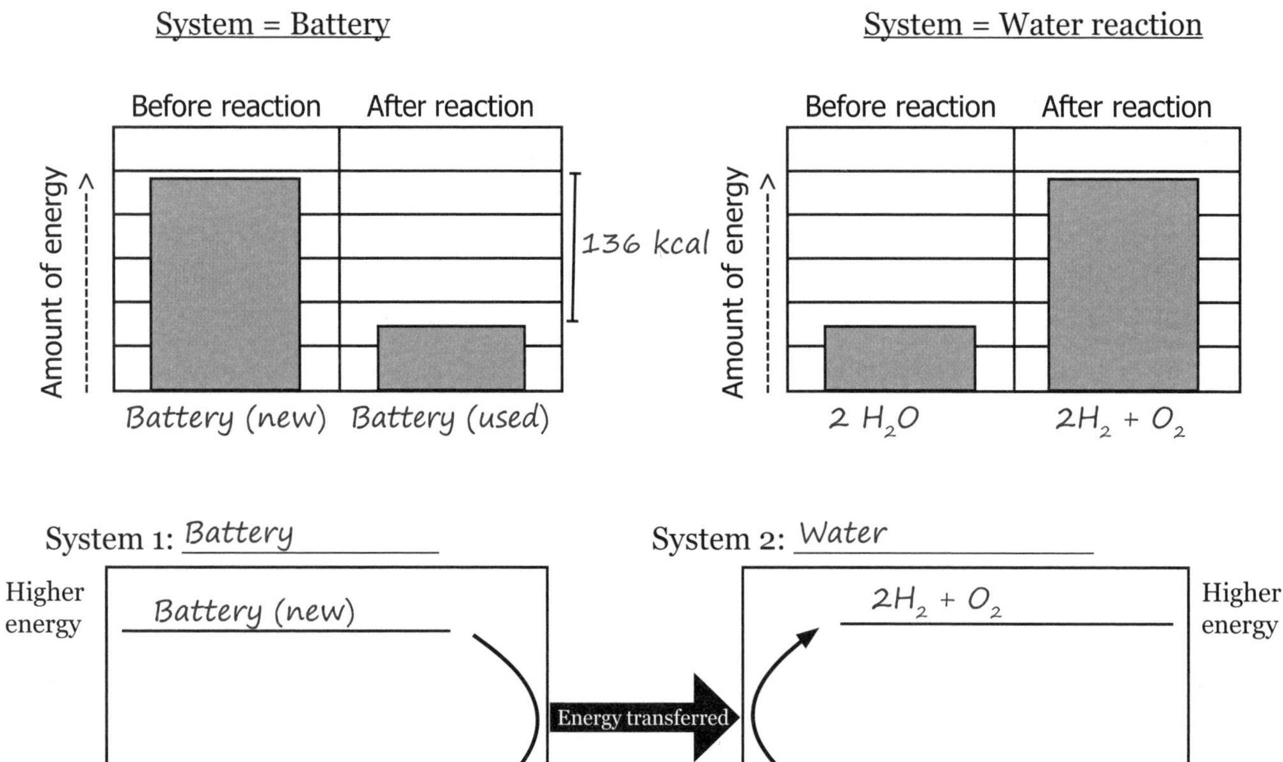

System 1: *Battery*

System 2: *Water*

Higher energy

Battery (new)

Battery (used)

Lower energy

Energy transferred

$2H_2 + O_2$

$2\ H_2O$

Higher energy

Lower energy

11. When scientists carried out this water reaction inside the reaction chamber of a calorimeter, they detected a decrease in the temperature of the surrounding water instead of an increase. They used the temperature change to calculate the energy change caused by the reaction. They determined that the reaction $2\ H_2O \rightarrow 2\ H_2 + O_2$ requires an input of 136 kcal. That means that for every 2 moles of H_2O that react to form 2 moles of H_2 and 1 mole of O_2, 136 kcal must be added to the system. Add this data to the bar graphs above.

12. Look back at Activity 1 in Lesson 2.3. How does the amount of energy released when 2 moles of H_2 and 1 mole of O_2 react to form 2 moles of H_2O compare to the amount of energy required to form 2 moles of H_2 and 1 mole of O_2 from 2 moles of H_2O?

The amount of energy released by the $2\ H_2 + O_2 \rightarrow 2\ H_2O$ reaction is equal to the amount of energy required by the $2\ H_2O \rightarrow 2\ H_2 + O_2$ reaction.

Teacher Talk & Actions

The value here is the difference in the heights of the bars in the chemical reaction system bar graphs, not the actual height of either bar.

Activity 2: Chemical Reaction Between Ammonium Thiocyanate and Barium Hydroxide

In this activity, we will look at the chemical reaction between ammonium thiocyanate and barium hydroxide and look for indicators of energy changes. Recall from Lesson 1.3 that ammonium thiocyanate and barium hydroxide react to form barium thiocyanate, water, and ammonia. The chemical equation for the overall reaction is:

$$Ba(OH)_2 + 2\ NH_4SCN \rightarrow Ba(SCN)_2 + 2\ H_2O + 2\ NH_3$$

Procedures and Questions

Observing the reaction

1. Watch the video of the ammonium thiocyanate and barium hydroxide reaction.

2. Record your observations about this chemical reaction below. Focus on observations that provide evidence that energy changes are occurring during the chemical reaction.

The water under the beaker freezes.

3. During this reaction, does anything appear to end up with either more energy or less energy than it had at the beginning? Cite evidence to support your answer.

The surroundings (water on the plank of wood) have less energy because the temperature of the water decreased. My evidence is that the water froze.

Representing energy changes and transfers

4. Sketch bar graphs to represent the energy changes you observed during the chemical reaction for the surrounding system (water under the beaker) and the chemical reaction system. Draw a taller bar when the system has more energy and a shorter bar when the system has less energy. Below each bar, describe how the system is different before and after the reaction.

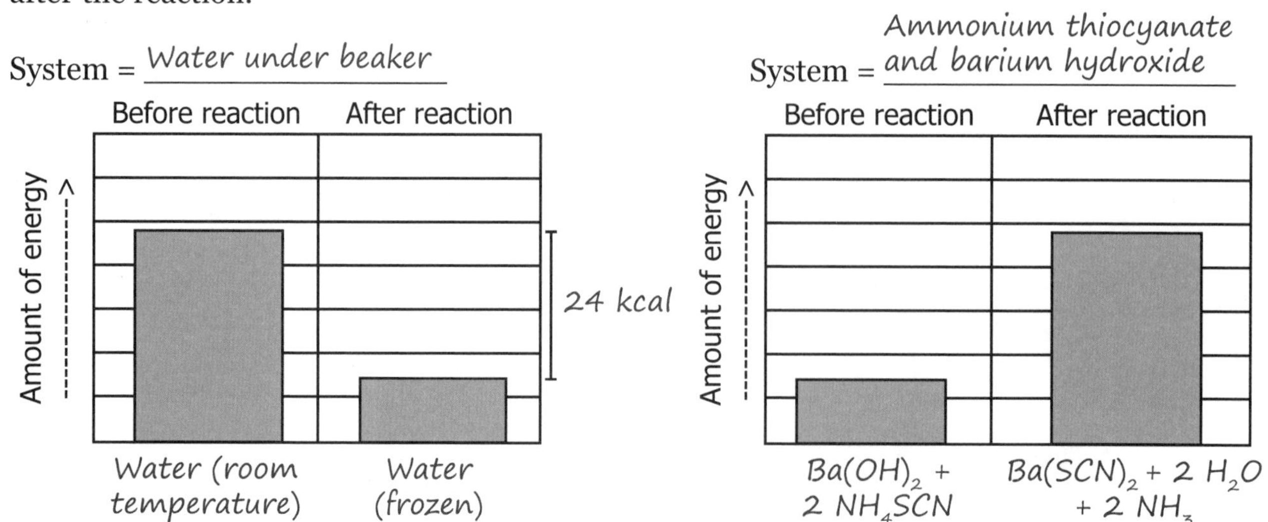

Teacher Talk & Actions

Show the video *Barium Hydroxide & Ammonium Thiocyanate* here.

5. Construct an energy-transfer model that describes the energy transfer between the chemical reaction system and the water system.

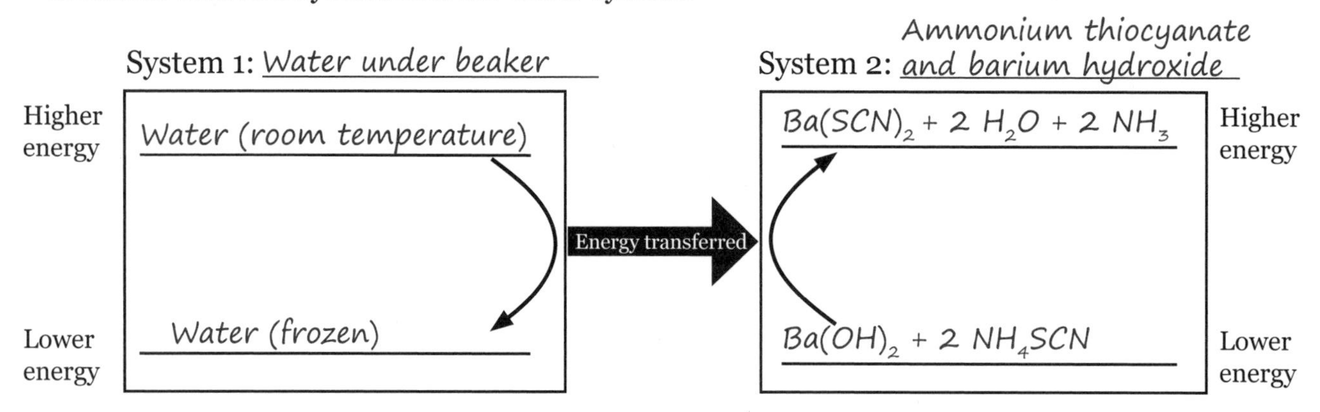

6. When scientists reacted ammonium thiocyanate and barium hydroxide in the reaction chamber of a calorimeter, they determined that an input of 24 kcal was required per mole of barium thiocyanate formed. Add this data to the bar graphs above.

Teacher Talk & Actions

The value here is the difference in the heights of the bars in the chemical reaction system bar graphs, not the actual height of either bar.

Activity 3: Chemical Reaction Between Carbon Dioxide and Water

In this activity, we will look at the chemical reaction between carbon dioxide and water that forms glucose and oxygen and look for indicators of energy changes. Recall that plants produce glucose and oxygen from carbon dioxide and water during a process called photosynthesis. The chemical equation for the overall process is as follows:

$$6\,CO_2 + 6\,H_2O \rightarrow C_6H_{12}O_6 + 6\,O_2$$

Procedures and Questions

Observing the reaction

1. Watch the video *Aquatic Plant in Light and Dark.*

2. Record your observations below. Focus on observations that provide evidence that energy changes are occurring during the chemical reaction.

When light was shone on the plant, bubbles formed. When the light was turned off, fewer bubbles formed.

3. Scientists wanted to study the relationship between the intensity of light and the amount of oxygen produced during photosynthesis. Each minute, they measured the amount of oxygen produced in a very small unit of volume called microliters (µl). Their findings are summarized in Table 2.4 and the graph on the next page:

Table 2.4. The Amount of Oxygen Produced Under Different Intensity Light

Time (min.)	Oxygen Produced Under Low-Intensity Light (µl)	Oxygen Produced Under Medium-Intensity Light (µl)	Oxygen Produced Under High-Intensity Light (µl)
1	0.05	0.29	0.49
2	0.23	1.12	2.06
3	0.51	2.41	4.36
4	0.85	3.99	7.34
5	1.15	5.86	10.79
6	1.46	7.72	14.30
8	1.99	11.52	21.74
10	2.60	15.33	28.94
12	3.16	19.09	36.28
15	4.09	24.74	47.56

Teacher Talk & Actions

Show the video *Aquatic Plant in Light and Dark* here.

The absorption of light is an important observation here.

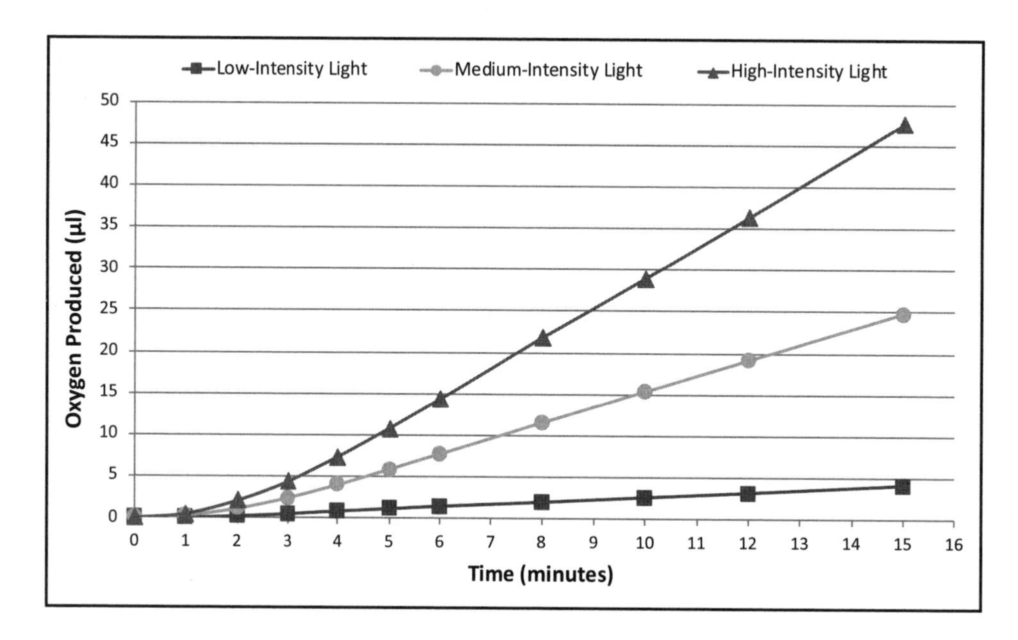

4. Under what light intensity was the most oxygen produced?

High-intensity light

5. What conclusions about the amount of oxygen produced under the different intensities of light can you make from this data?

The higher the light intensity, the more oxygen was produced by the plant.

The longer the plant was exposed to light of each intensity, the more oxygen was produced.

6. How are the data about the relationship between light intensity and oxygen production by plants that are shown in the graph similar to the data you collected in Lesson 2.1, p. 88, about the effect of light intensity on a solar-powered toy car? Your answer should compare energy changes and transfers in the two different phenomena.

In both cases, increasing the amount of energy transferred from the light source to the object caused a greater change in the energy of the object. For the solar car, the greater amount of energy transferred from a light source caused a greater change in the motion of the car. For the aquatic plant, the greater amount of energy transferred from a light source caused more oxygen to be produced from a chemical reaction.

Teacher Talk & Actions

Representing energy changes and transfers

7. Sketch a bar graph to represent the energy changes that occur during the chemical reaction for the photosynthesis reaction system. Draw a taller bar when the system has more energy and a shorter bar when the system has less energy.

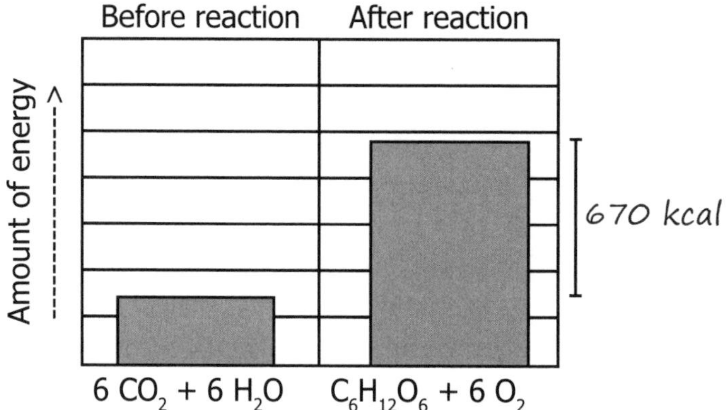

8. Scientists have determined that an input of 670 kcal is required per mole of glucose formed. Add this data to the bar graph above.

9. Construct an energy-transfer model that describes the energy transfer between the Sun system and the chemical reaction system. (*Note:* The energy-releasing processes that occur in the Sun are complex and won't be discussed in this unit, so you don't have to complete a box to account for what is happening in the Sun.)

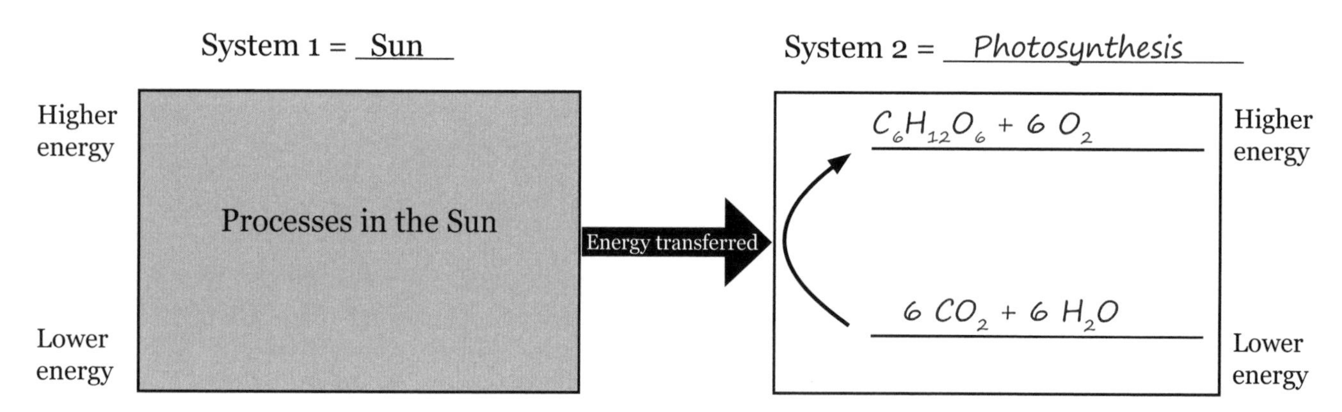

Teacher Talk & Actions

Activity 4: Comparing the Amounts of Energy Required by Chemical Reactions

In the previous lesson, you looked at tables that showed how much energy is released when certain substances react with oxygen (Table 2.3). Below is a similar table but this time it shows how much energy is required by certain chemical reactions per mole of product formed (indicated by an asterisk).

Table 2.5. Energy-Requiring Chemical Reactions

Chemical Reaction	Energy Required (kcal/mole)
Photosynthesis: $6\ CO_2 + 6\ H_2O \rightarrow C_6H_{12}O_6 + 6\ O_2$	670
Breaking water: $H_2O \rightarrow H_2 + ½\ O_2$	68
Protein synthesis: 2 Amino acids → Dipeptide + H_2O	2–4
Carbohydrate synthesis:	
2 Glucose → Maltose + H_2O	1.0
Glucose + fructose → Sucrose + H_2O	3.6
Fat synthesis: 3 Fatty acid + glycerol → Triglyceride +3 H_2O	7.3

Note: Protein and carbohydrate synthesis involve linking many amino acids or carbohydrate monomers together. The value in the table gives the energy required for each link.

Procedures and Questions

1. Examine the values in Table 2.5. Which reaction requires the most energy per mole? Which reaction requires the least?

The amount of energy required to synthesize one mole of glucose is much more than the amount of energy required to synthesize one mole of a dipeptide (plus a mole of water molecules) from amino acids and to synthesize a mole of maltose and water molecules from glucose monomers.

2. Based on data in Table 2.5, estimate how much energy would be required to synthesize the enzyme amylase, a protein made up of about 100 amino acids.

Table 2.5 shows that it takes 2-4 kcal/mole to link two amino acids together. Linking about 100 amino acids together would take between 200-400 kcal/mole.

Teacher Talk & Actions

This is the minimum amount of energy required when any two amino acids are linked together in the laboratory. Much more energy is required to ensure that the amino acids are accurately sequenced to form a protein. As students will see in Lesson 2.8, building proteins in cells requires a lot of energy!

Science Ideas

The activities in this lesson were intended to help you understand important ideas about energy changes during chemical reactions. Read the ideas below. Look back through the lesson. In the space provided after the science ideas, give evidence that supports each one.

Science Idea #19: Some chemical reaction systems require an input of energy. A decrease in temperature or motion in surrounding systems, a decrease in the energy of a battery connected to the system in a complete circuit, or the absorption of light or sound is evidence that energy was absorbed during the chemical reaction, which means that the chemical reaction system now has more energy.

Evidence:

The chemical reaction that produced hydrogen and oxygen from water required energy because (a) the reaction only occurred when a new battery was connected to the reaction, and (b) the battery had less energy after the reaction occurred. Therefore, the battery must have transferred energy to the chemical reaction system.

The production of glucose and oxygen from water and carbon dioxide required an input of energy because more oxygen was produced with higher-intensity light than with lower-intensity light. (Note: Because the solar car moved more with higher-intensity light, we have evidence the car absorbed energy from the bulb. Because the aquatic plant produced more oxygen with higher-intensity light, we have evidence that the plant absorbed energy from the bulb.)

Science Idea #20: Energy is transferred from the Sun to plants by light during the process of photosynthesis. Because an input of energy is required for photosynthesis to occur, the arrangement of atoms in product molecules (1 glucose and 6 oxygen molecules) must have more energy than the arrangement of the same atoms in reactant molecules (6 carbon dioxide and 6 water molecules).

Evidence:

The increased motion of the solar-powered car provided evidence that energy can be transferred by light. The correlation between increased light intensity and increased oxygen production provided evidence that energy is required for photosynthesis to occur and that energy from the Sun was absorbed by the water plant. If light was absorbed, energy must have been transferred to the plant, so the products of the photosynthesis reaction system must have more energy than the reactants.

Teacher Talk & Actions

Remind students that science ideas are accepted principles that are based on a wide range of observations, not just on the small number of examples they have observed. Because scientists have observed photosynthesis in all plants they have studied, the science idea generalizes to all plants.

Pulling It Together

Work on your own to answer these questions. Be prepared for a class discussion.

1. Now that you have had more experiences with chemical reactions, answer the Key Question: **Do all chemical reactions release energy?**

Not all chemical reactions release energy. In this lesson, we saw three reactions that required an input of energy. In the case of forming hydrogen and oxygen from water, a battery transferred the required energy. In the case of the formation of glucose and oxygen from carbon dioxide and water, light from the Sun transferred the required energy. In the case of the reaction between barium hydroxide and ammonium thiocyanate, the surrounding system transferred energy to the reaction system as indicated by the decrease in temperature of the surrounding system.

2. Below is a bar graph that represents the amount of energy in a chemical reaction system before and after a particular chemical reaction occurs.

System = Chemical reaction

Before reaction
After reaction
Amount of energy ------->
Reactants
Products

Based on the bar graph, does the reaction release energy or require an input of of energy? Explain.

This reaction takes in energy. The bar graph shows that the reaction system has more energy after the reaction than before the reaction. This means that energy must have been transferred to the reaction system from another system.

Teacher Talk & Actions

3. Hexamethylenediamine and adipic acid are two colorless liquids that react with each other to form solid nylon.

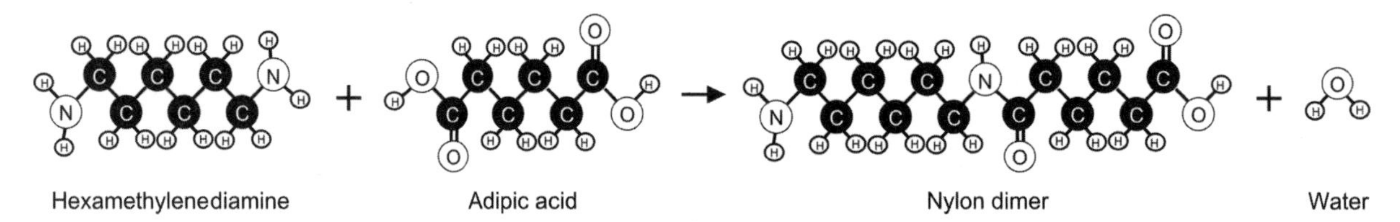

Nylon is a polymer that is used to make fabric and other kinds of materials. During nylon formation, the temperature of the surrounding air decreases. Do the products of this reaction have more, less, or the same amount of energy as the reactants? Write an explanation using models, data, science ideas, and systems thinking.

Question	Do the products of this reaction have more, less, or the same amount of energy as the reactants?
Claim	*The products of the reaction have more energy than the reactants.*
Science Ideas	*• When the energy of one system decreases, the energy of another system increases. (from Science Idea #14)* *• When two systems interact and one system decreases in energy while the other system increases in energy, energy was transferred from the energy-releasing system to the energy-requiring system (from Science Idea #15)* *• A decrease in temperature in surrounding systems is evidence that energy was absorbed during the chemical reaction and indicates that the chemical reaction system now has more energy. (from Science Idea #19)*
Evidence	*The temperature of the surroundings decreases during the reaction, which is evidence that energy is transferred from the surroundings to the chemical reaction system.*
Models	*An energy-transfer model showing that energy is transferred to the chemical reaction system from the surrounding system is consistent with the science ideas and evidence listed above.*

Teacher Talk & Actions

Show the *Nylon* video if your students are unfamiliar with this reaction. (Safety note: The professor shown in the video should have been wearing goggles, and the reaction should have been carried out in a hood.)

Explanation:

The products of the hexamethylenediamine/adipic acid reaction have more energy than the reactants because energy was transferred from the surroundings to the chemical reaction system. The decrease in temperature in the surrounding system is evidence that energy in the surrounding system decreased. We know from science ideas that (a) when the energy of one system decreases the energy of another system increases and that (b) if the energy of one system decreases while the energy of another system increases, then energy was transferred from the energy-releasing system (in this case the surrounding system) to the energy-requiring system (in this case the chemical reaction system). The energy-transfer model below is consistent with the science ideas and evidence listed.

4. We know from Table 2.5 that protein synthesis requires an input of energy. Using what you learned in Lesson 1.4, draw a model that illustrates the energy changes and transfers that occur during protein synthesis. Explain your model using science ideas.

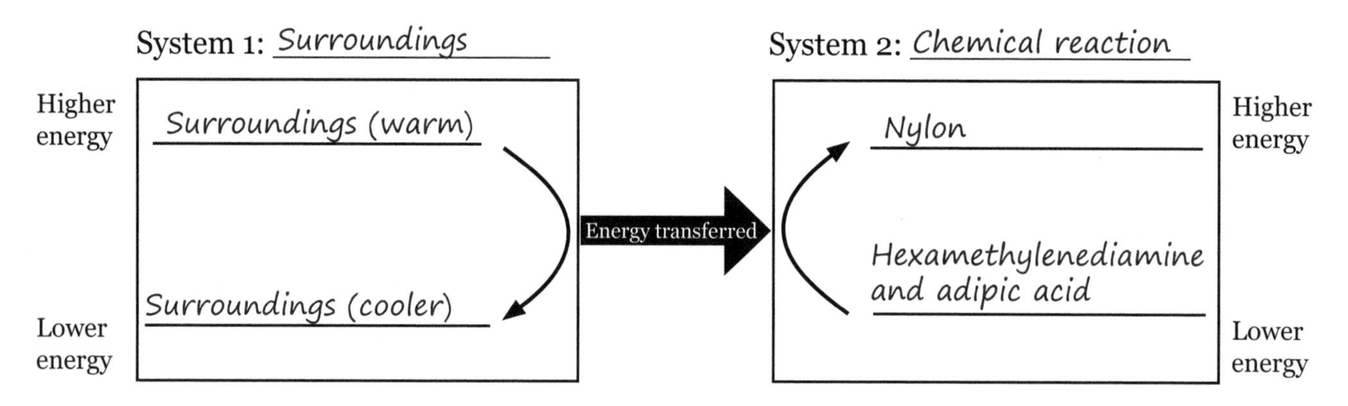

Models should include 2 systems, one that gives off energy and one that requires energy, and an energy transfer arrow that connects them. The protein-synthesizing system is on the right and should show amino acids on the lower-energy line and proteins + water molecules on the higher-energy line.

Identifying the system on the left is a challenging task but worth the effort because it connects what students have been doing in physical systems to living systems. Students don't know what chemical reaction system transfers the energy. However, based on their general sense that "food provides energy," they might suggest that the reaction of glucose, fatty acids, or even amino acids with oxygen could provide the energy. Students could cite the idea that polymers from food are broken down to monomers during chemical reactions with water molecules (Science Idea #10) and the circulatory system carries the monomers to body cells (Science Idea #11). Students could cite evidence from Table 2.3 that the reaction of glucose, fatty acids, and amino acids with oxygen all release energy. Lesson 2.7 will provide evidence they can use to answer the question with more confidence.

Teacher Talk & Actions

Closure and Link to subsequent lessons:

In this lesson, you observed several chemical reactions that required an input of energy and represented the energy changes and energy transfers using bar graphs and energy-transfer models. Using these representations, along with the idea that when the energy of one system increases, the energy of another system decreases (Science Idea #14), helped us to make sense of several phenomena. Now you know that not all chemical reactions release energy; some chemical reactions require an input of energy. Did you notice that in Lesson 2.3 we had to input a little bit of energy to get the reactions to occur? For example, the oxygen and hydrogen needed to be sparked. How can a chemical reaction that releases energy still require an input of energy? In the next lesson, we try to figure out why that input of energy was needed.

Lesson Guide
Chapter 2, Lesson 2.5
Making Sense of Energy Changes During Chemical Reactions

Focus: In this lesson, students examine models of reactants and products of chemical reactions to determine what bonds of reactant molecules must break and what bonds of product molecules must form during two different chemical reactions. Then they use data about bond energies to calculate (a) the total amount of energy required to break the bonds of reactants, (b) the total amount of energy released when the bonds of products form, and (c) the net energy change of the chemical reactions.

Key Question: Why do energy-releasing chemical reactions require an initial input of energy?

Target Idea(s) Addressed

Different arrangements of atoms are associated with different amounts of energy. Changing the arrangement of atoms requires breaking some bonds and forming other bonds. (Science Idea #21)

Energy is required to break bonds between atoms and energy is released when bonds form between atoms. Differences in the total amount of energy required to break bonds of reactant molecules and the total amount of energy released when bonds form between product molecules account for the net change in energy during chemical reactions.

- When a chemical reaction releases energy, it is because the amount of energy required to break bonds of reactant molecules is less than the amount of energy released when bonds of product molecules form.

- When a chemical reaction requires energy, it is because the amount of energy required to break bonds of reactant molecules is greater than the amount of energy released when bonds of product molecules form. (Science Idea #22)

Science Practice(s) Addressed

Developing and using models

Materials

Activity 1: Video: *Making Water (Animation), Part 2;* poster of Science Idea #21; strong toy magnets (2 per small group)

Advance Preparation

Download video and test it in advance.

Phenomena, Data, or Models	Intended Observations	Purpose	Rationale or Notes
Activity 1 Observations of models of reactants and products of the chemical reaction $2\ H_2 + O_2 \rightarrow 2\ H_2O$	• The number of hydrogen atoms is the same before and after the chemical reaction. • The number of oxygen atoms is the same before and after the chemical reaction. • The arrangement of H and O atoms is different in the reactants and products: In the reactants, each H atom is bonded to another H atom and each O atom is bonded to another O atom, but in the products one O atom is bonded to 2 H atoms. • Breaking bonds requires an input of energy (because a molecule has less energy than the same set of atoms separated). • Forming bonds gives off energy (because a molecule has less energy than the same set of atoms separated).	Give students an opportunity to use their observations as evidence for the conclusion/claim that differences in energy result from differences in the arrangement of the same atoms. This will prepare them to think about which bonds between atoms are broken and which new bonds form during the chemical reaction.	The chemical reaction $2\ H_2 + O_2 \rightarrow 2\ H_2O$ involves breaking only four bonds and forming only four bonds. This is a simple example to start with.
Activity 2 Data on average bond energy of O=O, H-H, and O-H bonds are used to calculate the total amount of energy required to break bonds of reactant molecules (2 moles H_2 + 1 mole O_2) and the total amount of energy released when bonds of 2 moles of H_2O molecules form. Data are used to calculate the total amount of energy required to break bonds of 2 moles of H_2O molecules and released when 2 moles of H_2 + 1 mole of O_2 form.	• Since the calculated 440 kcal released when 2 moles of H_2O molecules form is 118 kcal more than the calculated 322 kcal required to break the bonds of 2 moles of H_2 + 1 mole of O_2 molecules, there should be a net release of 118 kcal during the chemical reaction. • The calculation is consistent with students' observation that energy was released (as light and sound and motion) during the chemical reaction. • Similar calculations for the reverse reaction provides the basis for the prediction that energy is required to break 2 moles of H_2O molecules into 2 moles of H_2 molecules and 1 mole of O_2, which is consistent with students' observations that an input of energy (from a battery) was required.	Use the calculations to reinforce the idea that bond breaking requires energy and bond forming releases energy. See that predictions based on calculations are consistent with qualitative observations; this is an important aspect of the nature of science.	Having students carry out the calculations in a simpler system where only a few bonds are broken and formed is intended to scaffold a similar calculation for a more complex system where a greater number and variety of bonds are broken and formed.

Phenomena, Data, or Models	Intended Observations	Purpose	Rationale or Notes
Activity 3 Data on average bond energy of C-C, C-O, C-H, O-H, and O=O bonds are used to calculate the total amount of energy required to break bonds of reactant molecules (1 mole $C_6H_{12}O_6$ + 6 moles O_2) to form 6 moles CO_2 + 6 moles H_2O.	• Because the calculated 3,564 kcal released when 6 moles of CO_2 molecules + 6 moles of H_2O molecules are formed is more than the calculated 2,927 kcal required to break the bonds of 1 mole of $C_6H_{12}O_6$+ 6 moles of O_2 molecules, there should be a net release of 637 kcal during the chemical reaction. • The calculation is consistent with students' observation that energy was released (as light and sound and motion) during the chemical reaction. • Students calculate that reacting 10 grams of glucose with oxygen releases 35 kcal and reacting 100 grams of glucose with oxygen releases 350 kcal. • This calculation is consistent with students' observation that more energy was released when more molecules react.	Provide an opportunity for students to learn ideas about and calculations of energy changes to a chemical reaction in living organisms.	
Activity 4 Data on average bond energy of C-C, C-O, C-H, O-H, and O=O bonds are used to calculate the total amount of energy required to break bonds of reactant molecules (1 mole dimer + 1 mole of H_2O) to form one mole of one amino acid from the dimer and one mole of the other amino acid.	• Because the calculated 1,952 kcal released when 1 mole of alanine and 1 mole of glycine are formed is more than the calculated 1,949 kcal required to break the bonds of 1 mole of protein dimer + 1 mole of H_2O molecules, there should be a net release of 3 kcal during the chemical reaction. • When the calculation is simplified to considering only the bonds that have to break and the bonds that have to form, we calculate that 178 kcal are released when bonds of the products form and 175 kcal are required to break bonds of the reactants. This results in the same net value of 3 kcal released during the chemical reaction.	Provide an opportunity for students to see that it is not necessary to include all the bonds in the calculation of energy changes.	

Phenomena, Data, or Models	Intended Observations	Purpose	Rationale or Notes
Pulling It Together Q2: Butyric acid in butter reacts with oxygen	• Students predict that energy is released during the reaction between butyric acid and butter because they calculate that 2,004 kcal are required to break the bonds of the reactant molecules and 2,376 kcal are released when the bonds of the product molecules form.	Provide an opportunity for students to apply ideas about and calculations of energy changes to a chemical reaction in living organisms.	
Pulling It Together Q3: Protein synthesis reaction	• Students predict that energy is absorbed during protein synthesis because they calculate that 534 kcal are required to break the necessary bonds of the reactant molecules and 525 kcal are released when the bonds of the product molecules form.		

Lesson 2.5—Making Sense of Energy Changes During Chemical Reactions

What do we know and what are we trying to find out?

In the previous lessons, you saw examples of chemical reactions that release energy and examples of other chemical reactions that require an input energy. Yet even the energy-releasing chemical reactions you observed still required an input of a little bit of energy to start the reaction. For example, your teacher used a lighter to start the candle wax and oxygen reaction, and the scientist in the video used a lighter to start the reaction between hydrogen gas and oxygen gas. And the reaction between the donut and oxygen reaction required a flame to start it. Why was an input of energy needed even though these chemical reactions all release energy?

In this lesson, you will learn why chemical reactions, even energy-releasing reactions, need an initial input of energy and begin thinking about the Key Question. (There is no need to respond in writing now.)

Key Question: Why do energy-releasing chemical reactions require an initial input of energy?

Teacher Talk and Actions

Activity 1: Energy and the Breaking and Forming of Bonds

In this activity, we will look more closely at how atoms of reactants and products are arranged to see if the arrangement of atoms can give us ideas about why energy-releasing chemical reactions nonetheless require an input of energy to occur.

Procedures and Questions

Atom rearrangement

1. Watch the video about the oxygen and hydrogen reaction we observed in Lesson 2.3 and answer the following questions.

a. Compare the models of the molecules before and after the reaction.

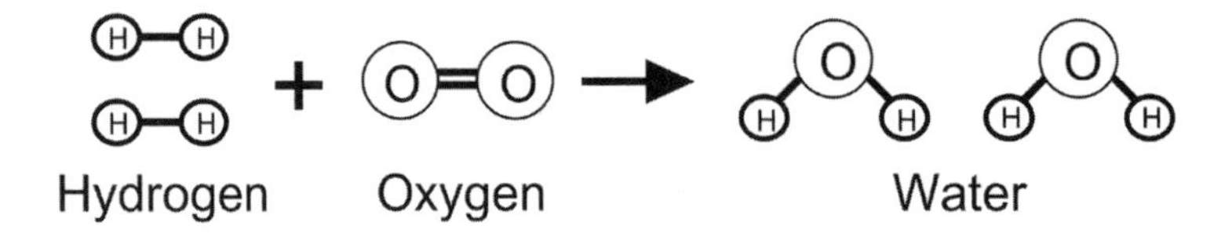

b. What do you notice about the type of atoms that make up these molecules?

There are the same types of atoms before and after the reaction.

c. What do you notice about the number of each type of atom before and after the reaction?

There are the same number of each type of atom before and after the reaction.

d. What do you notice about the way in which the atoms are connected to one another?

The way in which the atoms are connected is different before and after.

2. We learned that this chemical reaction gives off energy to the surroundings, and we drew bar graphs, like the ones below, that showed that the system had more energy before the reaction than after the reaction.

System = Chemical Reaction

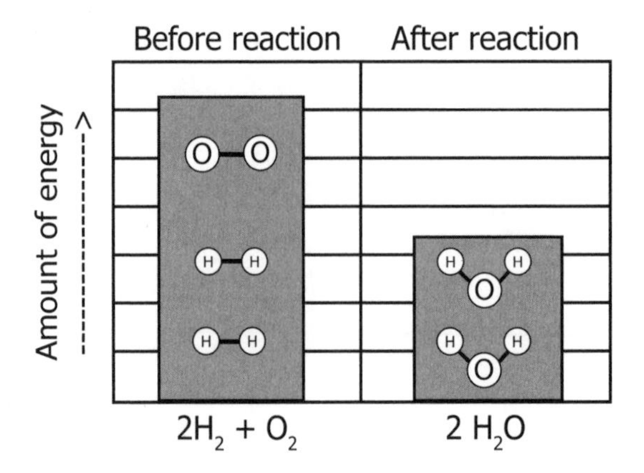

Teacher Talk and Actions

Show the video *Making Water (Animation), Part 2* here.

3. Based on the similarities and differences in the models, what do you think accounts for the difference in the amounts of energy associated with the reactants and products?

If the only thing that is different about the molecules of reactants and products is how the atoms are arranged (connected, bonded), the energy differences may be due to differences in the arrangement of atoms.

Bonds and energy

You just observed that the number and type of bonds between atoms of the reactants is different from the number and type of bonds between atoms of the products. All chemical reactions involve breaking bonds between atoms and forming new bonds. For product molecules to form, bonds between atoms of reactant molecules must be broken. **Breaking bonds between any two atoms always requires an input of energy.** Separate atoms have more energy (are less stable) than those same atoms bonded together.

4. The bar graph below represents the energy associated with hydrogen molecules and oxygen molecules and the energy associated with the same set of hydrogen and oxygen atoms separated. Draw a line on the graph to represent the amount of energy required to break the bonds of 2 H_2 + O_2.

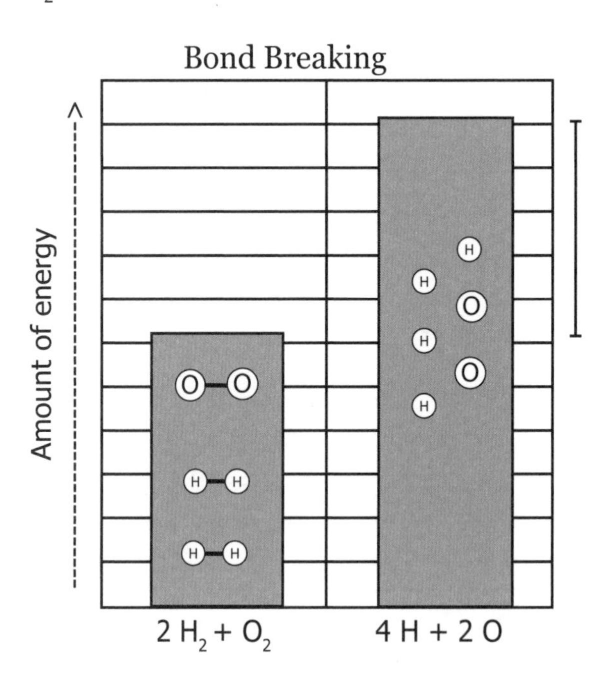

5. Once separated, the H and O atoms bond to form H_2O molecules. **Forming bonds between any two atoms always releases energy.** As with bond breaking, the energy difference is because a molecule made up of atoms bonded together has less energy (is more stable) than the same set of atoms separated. The bar graph below represents the energy of the separate H and O atoms (left bar) and the energy of the atoms bonded as water molecules (right bar). Draw a line on the graph to represent the amount of energy released when 4 H atoms + 2 O atoms.

Teacher Talk and Actions

A correct understanding of changes in bond energy is necessary to explain phenomena about energy changes during chemical reactions at the atomic/molecular scale (i.e., whether a reaction releases or requires a net input of energy results from differences between total energy required to break bonds between atoms of reactants and the total energy released when bonds form between atoms of products). Students can then reconcile their molecular scale explanations with explanations at the macroscopic scale (i.e., whether or not a reaction releases or requires a net input of energy depends on changes in motion and temperature and the emission or absorption of light or sound).

Most high school and university students have considerable difficulty appreciating that energy is released when bonds form. Many students think, incorrectly, that energy is stored in bonds and is released when those bonds break. Often this is the result of students' experiences with physical models such as ball-and-stick models or LEGOs, in which forming bonds does appear to require energy. The physical models are useful for representing atom rearrangement during chemical reactions, but they misrepresent the energy changes when bonds form.

We strongly urge you to let students play with pairs of magnets so that they can feel the magnets requiring energy to pull them apart but releasing energy (which pulls students' hands together) as the magnets get close to each other. Students can also observe the sound that the magnets make as they come together, another indicator of a release of energy.

4H + 2 O are "hypothetical" intermediates—the actual mechanism of the reaction is likely more complicated. However, the hypothetical intermediates are fine for the purposes of determining the net energy change in the reaction.

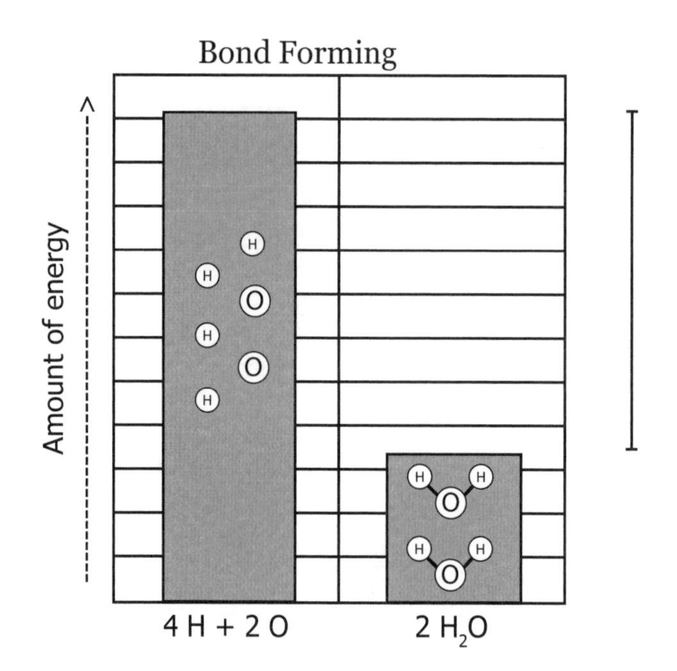

6. Let's combine the two graphs to represent the two steps—breaking bonds between atoms of reactants and forming bonds between atoms to make products. Draw a bar graph for the oxygen and hydrogen chemical reaction system that includes bars for the reactants ($2\ H_2 + O_2$), the separated atoms ($4\ H + 2\ O$), and the products ($2\ H_2O$).

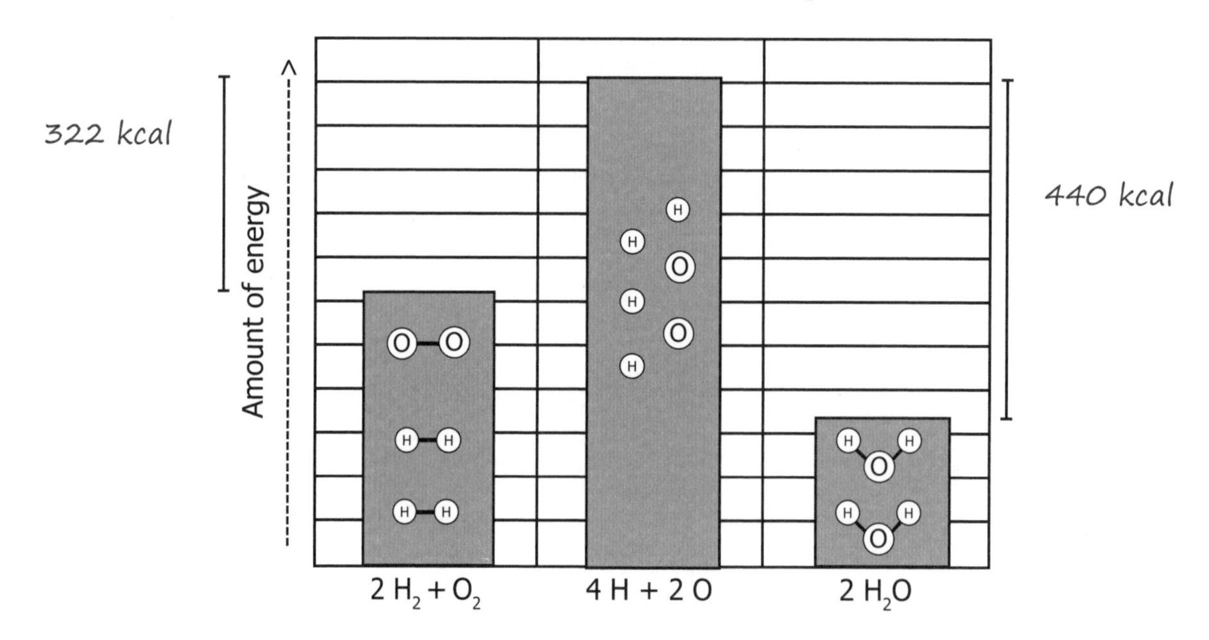

Teacher Talk and Actions

Activity 2: Bond Energies

Scientists have determined the average amount of energy that is required to break particular bonds between atoms and the average amount of energy that is released when particular bonds form. We call these values **bond energies**. We can use bond energies to calculate the amount of energy required to break the bonds of the reactant molecules and the amount of energy released when the bonds of the product molecules form. We can then compare these two values to predict if, overall, the reaction would be expected to release energy or require an input of energy.

In this activity, we will use bond energies of H-H, O=O, and O-H bonds to estimate the net energy change when hydrogen and oxygen molecules react to form water molecules and compare our estimation to experimental measurements of what actually happens. We will then repeat the process to estimate the net energy change for the reverse reaction and compare our estimation to experimental data.

Procedures and Questions

1. Table 2.6 shows bond energies for bonds between pairs of C, O, H, and N atoms. Notice that bonds between different pairs of atoms are associated with different bond energies. What do you notice about the relationship between the amount of energy required to break a bond and the amount of energy released when that bond forms?

Table 2.6. Average Bond Energies

Type of Bond	Average Amount of Energy Required to Break Bond (kcal/mole)	Average Amount of Energy Released When Bond Forms (kcal/mole)
C-H	98	98
O-H	110	110
H-H	103	103
C-C	80	80
C-O	85	85
C-N	65	65
O=O	116	116
C=O	187	187
C=C	145	145

Note: Remember that a *mole* is standard number of molecules (6.02×10^{23}).

Oxygen and hydrogen reaction

2. Let's first predict the net energy change for the reaction $O_2 + 2\ H_2 \rightarrow 2\ H_2O$. Use the data from Table 2.6 and the models below to respond to the questions.

H–H + O=O → H–O–H H–O–H

H–H

Hydrogen Oxygen Water

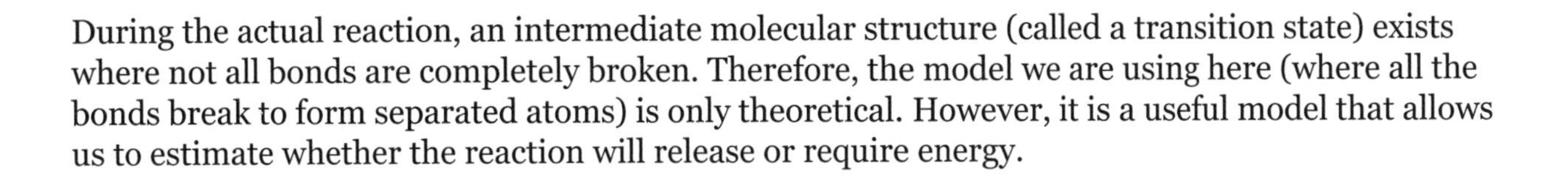

Teacher Talk and Actions

During the actual reaction, an intermediate molecular structure (called a transition state) exists where not all bonds are completely broken. Therefore, the model we are using here (where all the bonds break to form separated atoms) is only theoretical. However, it is a useful model that allows us to estimate whether the reaction will release or require energy.

Students could highlight bonds broken with one color and new bonds formed with another.

a. What types of bonds connect atoms of the reactant molecules (O_2 and 2 H_2)?

Oxygen atoms of oxygen molecules are connected by O=O bonds, and hydrogen atoms of hydrogen molecules are connected by H-H bonds.

b. What types of bonds connect atoms of the product molecules (2 H_2O)?

Oxygen and hydrogen atoms of water molecules are connected by O-H bonds.

c. For each type of bond, fill in the tables below.

Reactants:

Type of bond	Number of bonds broken	Average bond energies (kcal/mole)	Total amount of energy (kcal)
O=O	1	116	116
H-H	2	103	206
Total amount of energy required to break all bonds of the reactant molecules (kcal)			322

Products:

Type of bond	Number of bonds formed	Average bond energies (kcal/mole)	Total amount of energy (kcal)
O-H	4	110	440
Total amount of energy released when all bonds of product molecules form (kcal)			440

3. Go back to the bar graph of the chemical reaction that you drew in Activity 1 (Step 6) and write the energy required to break the bonds of the reactants and the energy released when the bonds of products form on the graph.

4. Will the chemical reaction 2 H_2 + O_2 → 2 H_2O release energy or require a net input of energy? How much energy? Justify your answers.

The chemical reaction will release 118 kcal for every 2 moles of H_2O that form. 322 kcal must be added to break bonds of 2 moles H_2 and 1 mole O_2 and 440 kcal is released when bonds form between 2 moles of H_2O. The difference is 118 kcal and it will be released because 2 moles of H_2O has less energy than 2 moles H_2 + 1 mole O_2.

5. Is your answer consistent with your observations of the oxygen + hydrogen reaction?

Yes, in the previous activities we saw indicators of energy being released (transferred to the surrounding systems) as light, sound, and motion when hydrogen and oxygen reacted to form water. The bond energy calculation also showed that the chemical reaction releases energy.

Teacher Talk and Actions

Be sure to emphasize that the 322 kcal represents the total amount of energy **required** to **break** all bonds of the reactant molecules (kcal). Therefore, it represents the difference in the heights of bars not the actual height of the reactant bar.

Be sure to emphasize that the 440 kcal represents the total amount of energy **released** when all bonds of product molecules **form.** Therefore, it represents the difference in the bars not the actual height of the bars.

The graph should show that the difference in the heights of bars 1 and 2 is 322 kcal, and the difference in the heights of bars 2 and 3 is 440 kcal.

If students ask why their calculated value of 118 kcal is higher than the measured value of 68 kcal/mole in Table 2.3, ask them to compare the number of moles of H_2 that react in the two cases. They should notice that their calculation is based on twice as many moles of reactants and products, which should release twice as much energy. So their calculated value would be 59 kcal/mole H_2. The difference between 59 and 68 is because bond energy for O-H bonds is an average value. Breaking/forming the first O-H bond in water involves a different amount of energy than breaking/forming the second O-H. So, the calculated values won't be exact.

If students are not familiar with the idea of *net,* pause here to review.

By convention, the net energy change is calculated by subtracting the energy of reactants from the energy of the products: Energy (products) – Energy (reactants) = net energy change. If the Energy (products) is greater than the Energy (reactants), the net energy change will be positive, meaning that the reaction required a net input of energy. If the Energy (products) is less than Energy (reactants), then the net energy change will be negative, meaning that the reaction released energy. We have found that most students have an easier time visualizing what is going on graphically than mathematically.

Water reaction

6. Now let's try calculating the change in bond energy for the reverse reaction, which we showed requires an input of energy:

$$2\ H_2O \rightarrow 2\ H_2 + O_2$$

Using the molecular models below and the bond energies in the table below, calculate the net amount of energy either released or absorbed by this reaction system.

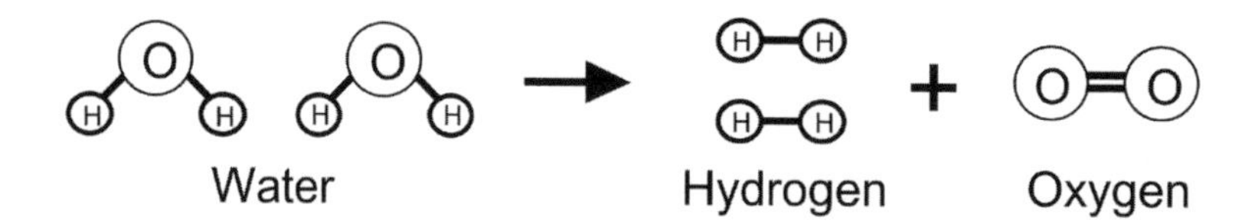

Reactants:

Type of bond	Number of bonds broken	Average bond energies (kcal/mole)	Total amount of energy (kcal)
O-H	4	110	440
Total amount of energy required to break all bonds of reactant molecules (kcal)			440

Products:

Type of bond	Number of bonds formed	Average bond energies, kcal/mole	Total amount of energy (kcal)
O=O	1	−116	116
H-H	2	−103	206
Total amount of energy released when all bonds of product molecules form (kcal)			322

7. Draw a bar graph to represent the energy changes in this reaction and write the energy required to break the bonds of the reactants and the energy released when the bonds of products form on the graph.

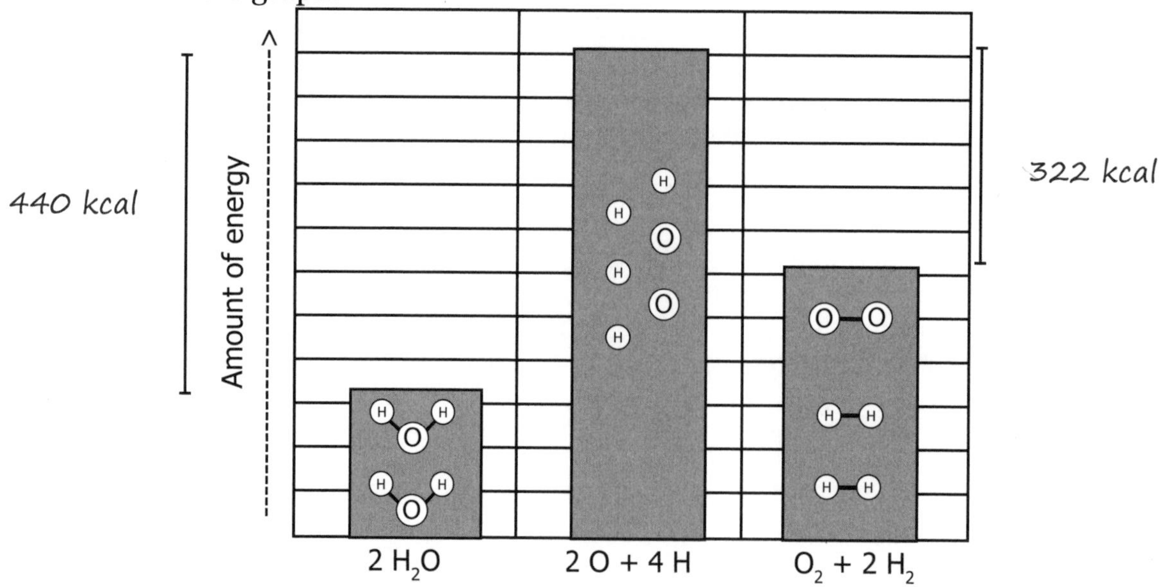

Teacher Talk and Actions

8. Will the chemical reaction 2 $H_2O \rightarrow 2 H_2 + O_2$ release energy or require a net input of energy? How much energy? Justify your answers.

The chemical reaction will require a net input of 118 kcal for every 2 moles of H_2O that react. 440 kcal must be added to break bonds of 2 moles of H_2O and 322 kcal is released when bonds form between 2 moles H_2 and 1 mole O_2. The difference is 118 kcal and it will be a net input because 2 moles of H_2O has less energy than 2 moles H_2 + 1 mole O_2.

9. Is the answer you calculated consistent with your observations of the water reaction?

Yes, both our calculation and our conclusions from the water decomposition experiment indicated that the reaction required energy. We observed that (a) water didn't decompose to form hydrogen and oxygen unless the reaction system was connected to a new battery, and (b) the battery had less energy after it had been used for a while. From these observations. we concluded that the battery transferred energy to the reaction system, which meant that the energy of the products (hydrogen and oxygen molecules) was greater than the energy of the reactants (water molecules). Our calculation showed the same thing.

Teacher Talk and Actions

Activity 3: Energy and the Breaking and Forming of Bonds During a More Complicated Reaction

The example reactions we examined during the previous activity involved breaking and forming only a few bonds. In this activity, we will try using the same approach to chemical reactions that involve breaking and forming many bonds. The reaction we will look at is the reaction between glucose and oxygen. We discussed this reaction in Lesson 2.3 when we investigated the burning of a donut. During this reaction, one molecule of glucose reacts with six molecules of oxygen to form six molecules of carbon dioxide and six molecules of water. The chemical reaction shown below uses molecular models to represent the reactants and products.

Procedures and Questions

1. Study the models below to determine which bonds of the reactant molecules will be broken. For each row in the table below, record the total number of the type of bond that will be broken and calculate the total amount of energy that will be required to break them.

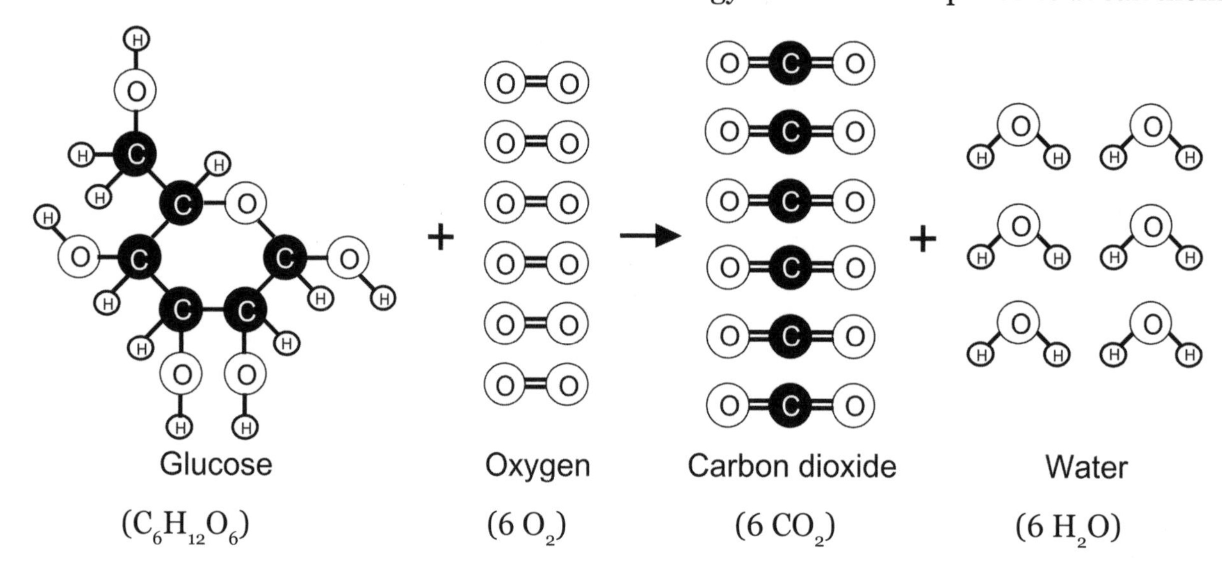

Reactants:

Type of bond	Number of bonds broken	Average bond energies (kcal/mole)	Total amount of energy (kcal)
C-C	5	80	400
C-O	7	85	595
C-H	7	98	686
O-H	5	110	550
O=O	6	116	696
Total amount of energy required to break all bonds of reactant molecules (kcal)			2,927

Teacher Talk and Actions

In Lessons 2.3 and 2.5, cellular respiration is referred to as a chemical reaction, which could lead students to conclude that all the bond breaking and bond forming occurs all at once. In Lesson 2.7, students will be told that cellular respiration is a carefully controlled multistep process. The unit does not present the steps involved in cellular respiration.

It is important to note that the reaction between glucose and oxygen is an ongoing process in which bond breaking and bond forming does not occur all at once, but instead involves many steps. In combustion, the process is not orderly, but in cellular respiration, the process is carefully controlled and occurs in discrete steps. In both cases, we can summarize the process with the overall chemical reaction.

2. Use the models on the previous page to determine which bonds will form to make the product molecules. For each row in the table below, record the total number of the type of bond that will be formed and calculate the total amount of energy that will be released when they form.

Products:

Type of bond	Number of bonds formed	Bond energy (kcal/mole)	Total amount of energy (kcal)
C=O	12	187	2,244
O-H	12	110	1,320
Total amount of energy released when all bonds of product molecules form (kcal)			3,564

3. Draw a bar graph to represent the energy changes in this reaction. Label each bar on the graph with the types of atoms and/or molecules it represents. Draw lines on the graph to indicate the amount of energy required to break the bonds of reactants and the amount of energy released when the bonds of products form.

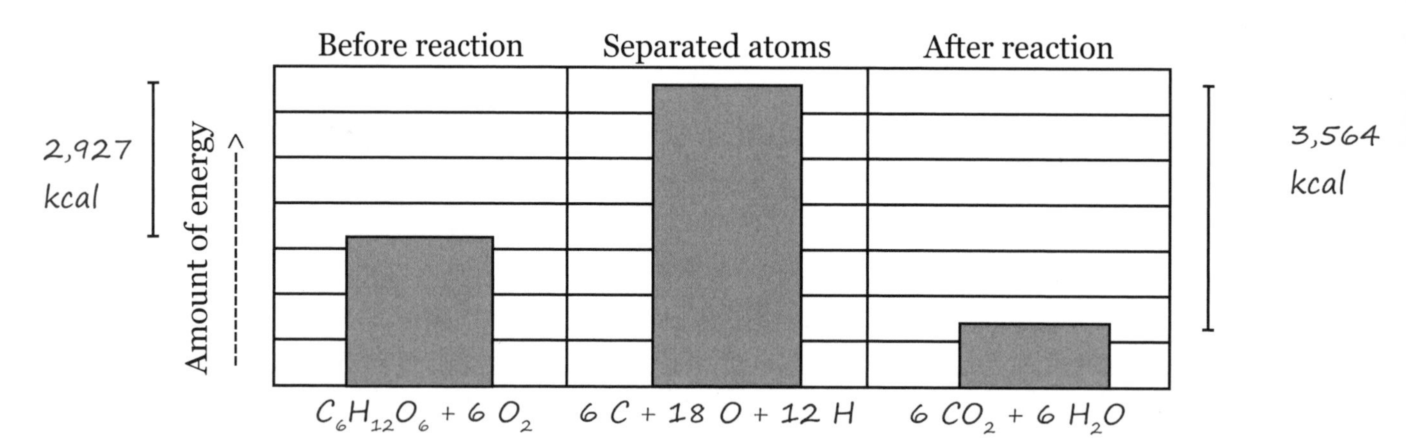

4. Will the chemical reaction $C_6H_{12}O_6 + 6\ O_2 \rightarrow 6\ CO_2 + 6\ H_2O$ release energy or require a net input of energy? How much energy? Justify your answers.

There is a net release of energy because more energy is released when new bonds form than was required to break bonds. 2,927 kcals are required to break the bonds of 1 mole of glucose and 6 moles of oxygen and 3,564 kcals are released when bonds of 6 moles of carbon dioxide molecules and 6 moles of water molecules are formed. The difference is 637 kcal, which means that the reaction will release energy because 1 mole of glucose and 6 moles of oxygen have more energy than 6 moles of carbon dioxide and 6 moles of water.

5. Is your calculation consistent with your observations of the reaction between the donut and oxygen?

Yes, both our calculation and our conclusions from the donut/oxygen reaction video indicated that the reaction released energy. We observed light and sound during the reaction which is evidence that energy was transferred from the reaction system to the surroundings. This means that the energy of the products (carbon dioxide and water molecules) was less than the energy of the reactants (glucose and oxygen molecules). Our calculation showed the same thing.

Teacher Talk and Actions

6. Based on the calculation for the glucose + oxygen reaction, what would you predict that the calculation for photosynthesis would show and why?

I would predict that the calculation for photosynthesis would show that the reaction requires an input of 637 kcals because the photosynthesis reaction is the reverse of the glucose + oxygen reaction.

7. Look back at the graph you drew in Question 3 for the reaction between glucose and oxygen ($C_6H_{12}O_6 + 6\ O_2 \rightarrow 6\ CO_2 + 6\ H_2O$). Complete the bar graph below to represent the energy changes during photosynthesis ($6\ CO_2 + 6\ H_2O \rightarrow C_6H_{12}O_6 + 6\ O_2$). Label each bar on the graph with the types of atoms or molecules it represents. Draw lines on the graph to show the amount of energy required to break the bonds of the reactants and the amount of energy released when the bonds of products form.

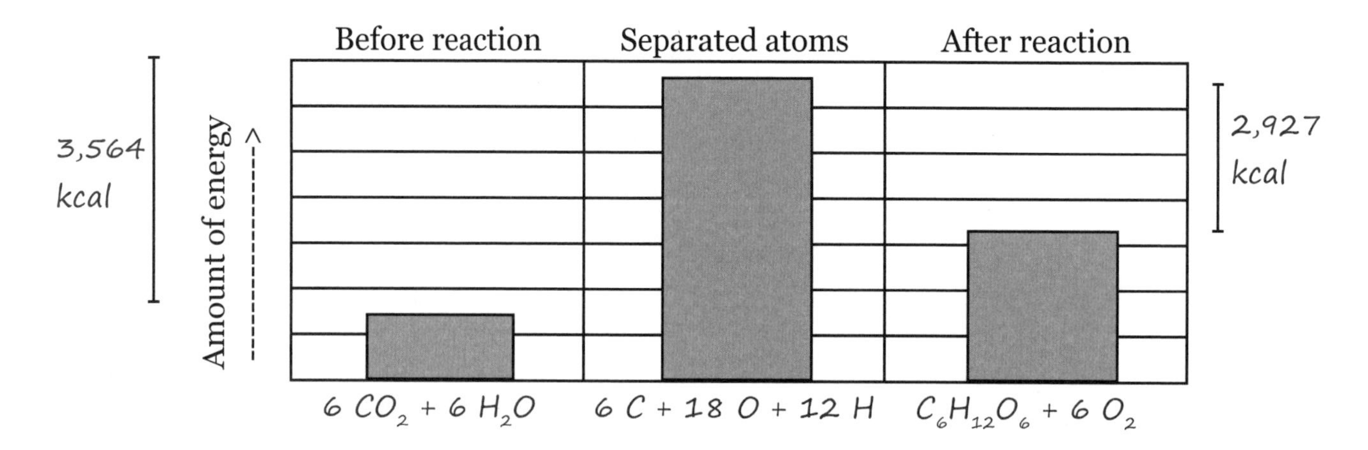

8. Your answer to Question 6 indicates how much energy would be required to produce 1 mole of glucose and 6 moles of oxygen from 6 moles of carbon dioxide and 6 moles of water. How much energy do you predict would be required to produce twice as much glucose and oxygen from twice as much carbon dioxide and water? Justify your prediction based on observations you made in earlier lessons.

I think twice as much energy (2 x 637 = 1,274 kcal) would be required to produce 2 moles of glucose and 12 moles of oxygen. My prediction is consistent with our observation in Lesson 2.4 that the aquatic plant produced more oxygen with a higher-intensity light than with a lower-intensity light.

Teacher Talk and Actions

Activity 4: Simplifying Our Energy Calculations

In the previous example, we imagined a hypothetical intermediate consisting of separate atoms and then included the breaking of each bond between atoms of the reactant molecules and the making of each bond between atoms of the product molecules in our calculations. In Lesson 1.5, you observed that only a few bonds break and form when food proteins react with water to make amino acids monomers and only a few bonds break and form when amino acids react to form new protein polymers plus water molecules. In this activity, we'll investigate whether we can simplify our calculations to include only the bonds that have to break and the bonds that have to form.

Procedures and Questions

Breaking and making all the bonds

1. First let's do the calculation by breaking and making all the bonds in the molecules. Study the models below to determine which bonds of the reactant molecules will be broken. Record the total number of each bond to be broken in the table below. Then calculate the total amount of energy for each row.

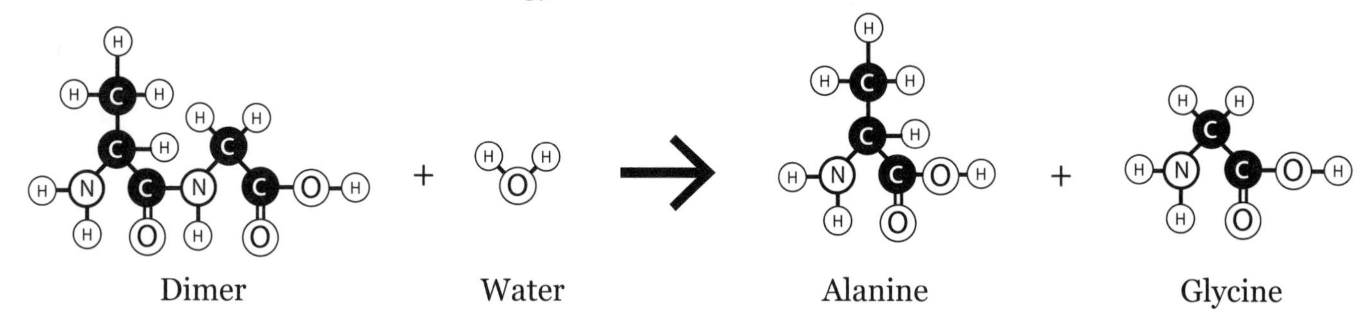

Reactants:

Type of bond	Number of bonds broken	Average bond energies (kcal/mole)	Total amount of energy (kcal)
C-C	3	80	240
C-O	1	85	85
C-H	6	98	588
O-H	3	110	330
N-H	3	93	279
C-N	3	65	195
C=O	2	116	232
Total amount of energy required to break all bonds of reactant molecules (kcal)			1,949

Teacher Talk and Actions

2. Use the models to determine which bonds will form to make the product molecules. Record the total number of each type of bond formed in the table below. Then calculate the total amount of energy for each row.

Products:

Type of bond	Number of bonds formed	Bond energy (kcal/mole)	Total amount of energy (kcal)
C-C	3	80	240
C-O	2	85	170
C-H	6	98	588
O-H	2	110	220
N-H	4	93	372
C-N	2	65	130
C=O	2	116	232
Total amount of energy released when all bonds of product molecules form (kcal)			1,952

3. Will this chemical reaction release energy or require a net input of energy? How much energy? Justify your answers.

The chemical reaction will release 3 kcal for every mole of protein dimer that reacts. 1,949 kcal must be added to break bonds of mole of dimer and mole of H_2O, and 1,952 kcal are released when bonds form between the product molecules. The difference is 3 kcal and it will be a net input because 1 mole of protein dimer + 1 mole of H_2O has less energy than 1 mole of alanine + 1 mole of glycine.

4. Is your answer to Question 3 consistent with data in Table 2.5 in Lesson 2.4, Activity 4, page 132?

Table 2.5 indicates that 2-4 kcal/mole are required to produce a dimer from two amino acids. That means that about 2-4 kcal/mole will be released when a dimer is digested to two amino acids. All of the information on our energy tables provides evidence that if a chemical reaction requires a certain amount of energy then the reverse of that reaction should release that same amount of energy.

Simplifying the bond energy calculations

5. Compare the models of the reactant and product molecules. Notice that they are very similar and only a few atoms rearrange during the reaction. If we did the calculation using only the unique bonds that break and form instead of using all the bonds, would we get the same answer as in Step 3? Explain.

I think we will get the same answer because in the first calculation we broke the same kinds of bonds we made. This is redundant because it means we are counting the energy needed to make and break the same bonds. The simplification will focus on just the bonds that are different between the reactant and product molecules.

Teacher Talk and Actions

6. In the tables below, indicate the minimum number of each bond that must be broken and the minimum number of each bond that must be formed and calculate the total amount of energy in each row.

Reactants:

Type of bond	Number of bonds broken	Average bond energies (kcal/mole)	Total amount of energy (kcal)
C-C	0	80	0
C-O	0	85	0
C-H	0	98	0
O-H	1	110	110
N-H	0	93	0
C-N	1	65	65
C=O	0	116	0
Total amount of energy required to break all bonds of reactant molecules (kcal)			175

Products:

Type of bond	Number of bonds formed	Bond energy (kcal/mole)	Total amount of energy (kcal)
C-C	0	80	0
C-O	1	85	85
C-H	0	98	0
O-H	0	110	0
N-H	1	93	93
C-N	0	65	0
C=O	0	116	0
Total amount of energy released when all bonds of product molecules form (kcal)			178

7. According to the simplified calculation, what is the amount of energy released or absorbed during this reaction?

The reaction will release 3 kcals per mole of protein dimer.

8. How does this compare to the answer you got for Step 3?

This is the same value we obtained from the full calculation in Step 3.

Teacher Talk and Actions

Science Ideas

The activities in this lesson were intended to help you understand important ideas about changes in bond energy during chemical reactions. Read each science idea and list evidence from the chemical reactions you observed in this lesson that supports the idea.

Science Idea #21: Different arrangements of atoms are associated with different amounts of energy. Changing the arrangement of atoms requires breaking some bonds and forming other bonds.

Evidence:

We observed that the reaction between hydrogen and oxygen gave off light and increased the motion of the bottle, providing evidence that the reaction gave off energy. Since energy was transferred from the chemical reaction system to the plastic bottle, the products (water molecules) must have less energy than the reactants (hydrogen + oxygen). Models (based on data) showed us that both reactants (O_2 + 2 H_2) and products (2 H_2O) were made up of 4 H atoms and 2 O atoms. Since the only thing different was how the O and H atoms were arranged/linked together, the different arrangements must be the reason that the reactants (O_2 + 2 H_2) and products (2 H_2O) have different amounts of energy.

Science Idea #22: Energy is required to break bonds between atoms and energy is released when bonds form between atoms. Differences in the total amount of energy required to break bonds of reactant molecules and the total amount of energy released when bonds form between product molecules account for the net change in energy during chemical reactions.

- When a chemical reaction releases energy, it is because the amount of energy required to break bonds of reactant molecules is less than the amount of energy released when bonds of product molecules form.
- When a chemical reaction requires energy, it is because the amount of energy required to break bonds of reactant molecules is greater than the amount of energy released when bonds of product molecules form.

Evidence:

From our calculations of the bond energies of 2 H_2O and of 2 H_2 + O_2, we saw that it took more energy to break the bonds of 2 H_2O than was released when 2 H_2 + O_2 formed, which is why the reaction required a net input of energy.

From our calculations of the bond energies of $C_6H_{12}O_6$ + 6 O_2 and of 6 CO_2 + 6 H_2O, we saw that it took less energy to break the bonds of $C_6H_{12}O_6$ + 6 O_2 than was released when 6 CO_2 + 6 H_2O formed, which is why the reaction had a net release of energy.

Teacher Talk and Actions

Any of the energy-releasing or energy-requiring reactions can be used as an example here.

Pulling It Together

Work on your own to answer these questions. Be prepared for a class discussion.

1. Now that you have learned about bond energies, answer the Key Question: **Why do energy-releasing chemical reactions require an initial input of energy?**

Energy-releasing chemical reactions require an initial input of energy because breaking the bonds between the atoms of the reactant molecules always requires energy. For the reaction to occur, some bonds of the reactant molecules must first be broken. Only then can they bond to other atoms, which releases energy. So even if the energy released is more than the initial input of energy, the initial input of energy is still required.

2. Butyric acid is a fatty acid that makes up a triglyceride that makes up ~4% of butter. Butyric acid can react with oxygen to form carbon dioxide and water as shown in the models below. Use the models and bond energies tables to calculate the net change in bond energy and determine whether energy is released or absorbed during the reaction.

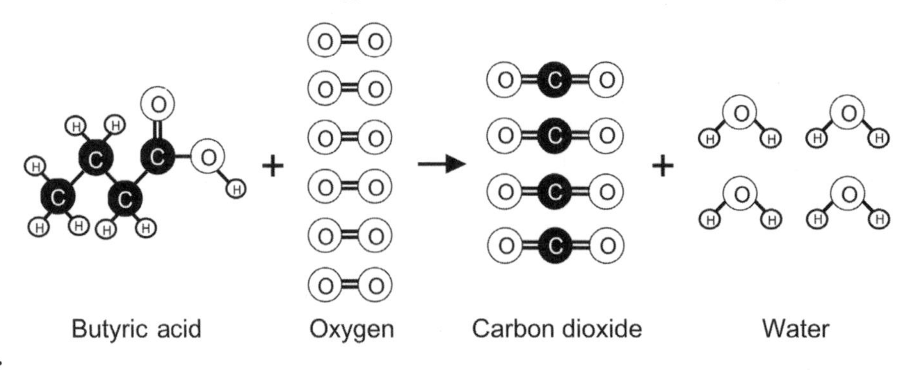

Reactants:

Type of bond	Number of bonds broken	Average bond energies (kcal/mole)	Total amount of energy (kcal)
C-C	3	80	240
C-O	1	85	85
C-H	7	98	686
O-H	1	110	110
C=O	1	187	187
O=O	6	116	696
Total amount of energy required to break all bonds of reactant molecules (kcal)			2,004

Products:

Type of bond	Number of bonds formed	Average bond energies (kcal/mole)	Total amount of energy (kcal)
C=O	8	187	1,496
O-H	8	110	880
Total amount of energy released when all bonds of product molecules form (kcal)			2,376

Teacher Talk and Actions

When butter goes rancid, butyric acid is liberated from the triglyceride by hydrolysis and causes an unpleasant smell.

Draw a bar graph to represent the energy changes in this reaction.

System = Butyric Acid and Oxygen

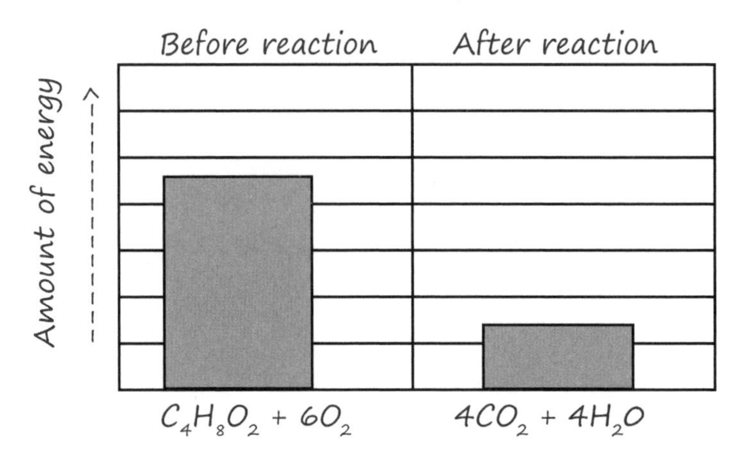

Is energy released or absorbed by the chemical reaction system? Why?

Energy is released because the amount of energy released when the bonds of the product molecules form is greater than the amount of energy required to break the bonds of the reactant molecules.

3. In Chapter 1, you learned that humans build the proteins that make up their body structures from amino acids from food. In the example below, four amino acids react to form a small piece of a protein polymer and three water molecules. Use the models and bond energies tables to calculate the net change in bond energy and determine whether energy is released or absorbed during the reaction. First, put a line through the minimum number of bonds that need to be broken in the reactants. Then circle the new bonds that are formed in the products.

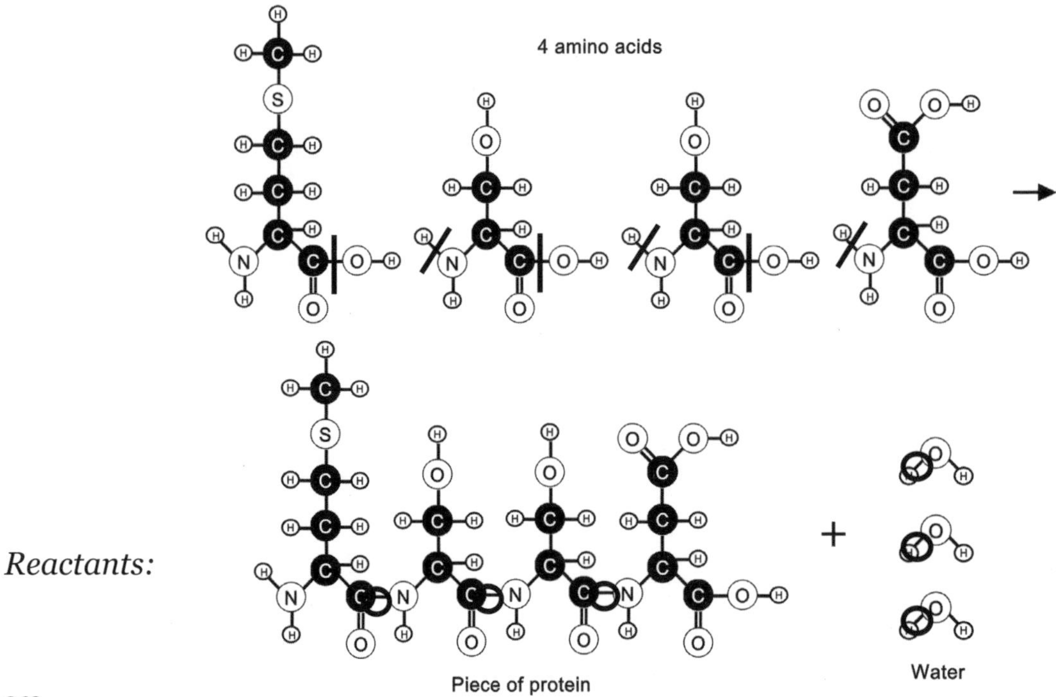

Teacher Talk and Actions

It's fine if students include a middle bar in the graph to represent the energy associated with the separated atoms. If they do, the bar should be higher than the bars for both the reactants and products and should be labeled underneath with "4 C + 8 H +12 O."

Type of bond	Number of bonds broken	Average bond energies (kcal/mole)	Total amount of energy required (kcal)
C-O	3	85	255
N-H	3	93	279
Total amount of energy required to break all bonds of reactant molecules (kcal)			534

Products:

Type of bond	Number of bonds formed	Average bond energies (kcal/mole)	Total amount of energy released (kcal)
C-N	3	65	195
O-H	3	110	330
Total amount of energy released when all bonds of product molecules form (kcal)			525

Draw a bar graph to represent the energy changes in this reaction.

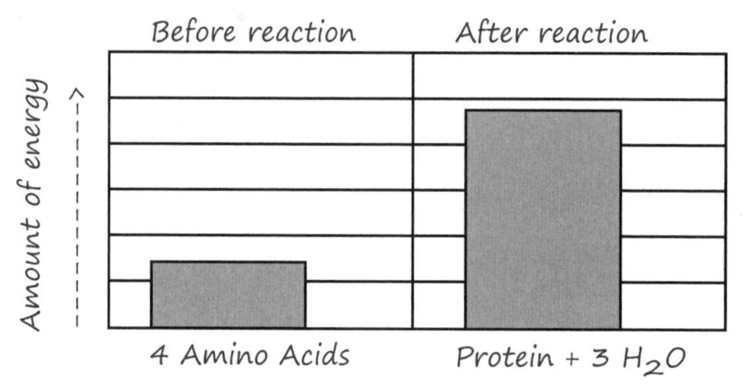

Is energy released or absorbed by the chemical reaction system? Why?

Energy is absorbed because the amount of energy released when the bonds of the product molecules form is less than the amount of energy required to break the bonds of the reactant molecules.

4. During a class on energy changes and chemical reactions, the teacher asks students if they think that all chemical reactions require an input of energy. One student raises her hand and says, "Yes, all chemical reactions require an input of energy." Another student raises his hand and says, "No, not all chemical reactions require an input of energy because we observed several chemical reactions that gave off energy, like burning glucose." The teacher says that both students are correct. Why would the teacher say this?

The first student is correct if she is thinking about the initial step in a chemical reaction. All chemical reactions involve breaking bonds between atoms, and this requires an input of energy.

The second student is correct if he is thinking about the net energy change during a chemical reaction. Some reactions release energy, even though some energy must be added at the beginning to break the bonds between atoms of reactant molecules. We determined that the amount of energy required to break bonds between atoms of glucose and oxygen is less than the amount of energy released when the bonds of the products (carbon dioxide and water) form.

Teacher Talk and Actions

To link their calculations back to an important phenomenon, ask students:

When do you think our bodies use this chemical reaction?

When building proteins

What does this tell you about whether protein-building releases or requires energy?

Building proteins requires energy. In the reaction shown, the protein built was just 4 amino acids, whereas the proteins used to build body structures are made up of hundreds of amino acids. So, building body proteins probably requires a lot of energy.

Note that the calculated amounts of energy grossly underestimate the actual energy needed for protein synthesis because they do not reflect the energy that cells must expend to get the amino acid sequence right. Doing so requires that cells maintain a protein-synthesizing system (the ribosomes) that ensures that each amino acid added to the protein is the one coded for by the RNA that was transcribed from the DNA.

Closure and Link to subsequent lessons:

In this lesson, you used models and the idea that an input of energy is always required to break bonds between atoms and energy is always released when bonds form between atoms (Science Idea #22) to explain why we need to ignite a candle or a donut and why we need to spark a mixture of hydrogen and oxygen gas, even though all three reactions release a lot of energy. Because all chemical reactions involve both breaking and making bonds, all chemical reactions require some energy (to break bonds) and all chemical reactions release energy (as bonds form). When we refer to a reaction as being energy-releasing or energy-requiring we actually mean that there is a net amount of energy released or a net amount of energy required. Given this, we could modify Science Ideas #17 and #19 to be clearer that we are referring to a net change in energy. So, Science Idea #17 would say "Some chemical reaction systems release a net amount of energy." And Science Idea #19 would say "Some chemical reaction systems require a net input of energy." In the next lesson, we will consider how a reaction that releases a net amount of energy can provide the energy for a reaction or process that requires a net input of energy.

Lesson Guide
Chapter 2, Lesson 2.6
Coupling Energy-Releasing to Energy-Requiring Chemical Reactions

Focus: Students consider how an energy-releasing system can provide the energy for an energy-requiring system and use an energy-transfer model to predict the relationship between systems in three examples that increase in complexity and level of abstraction: energy transfer by a seesaw from one circus performer to another, energy transfer by battery leads and wires from the chemical reaction inside the battery to the energy-requiring water splitting reaction, and energy transfer from the energy-releasing combustion of gasoline to increase a car's motion.

Key Question: Can the energy released by an energy-releasing chemical reaction provide the energy for an energy-requiring chemical reaction? If so, how?

Target Idea(s) Addressed

Energy released during a chemical reaction/process can be used to "drive" other reactions/processes that require a net input of energy if

a. the amount of energy released by the first reaction is more than the amount of energy required by the second reaction/process, and

b. a mechanism, or "coupling device/machine," is available to transfer the required amount of energy from the energy-releasing to the energy-requiring process. (Science Idea #23)

Science Practice(s) Addressed

Developing and using models

Materials

Activity 1: *Per class:* Video: *Circus Performers*

Advance Preparation

Download video and test it in advance.

Phenomena, Data, or Models	Intended Observations	Purpose	Rationale or Notes
Activity 1 Circus performers use a seesaw to propel each other into the air.	• When a performer lands on the raised side of the seesaw, the person standing of the other side is propelled up into the air.	Give the students an introduction to a coupling mechanism. Give the students a chance to discuss the necessity of a coupling mechanism in order to transfer energy. Give the students a chance to discuss how much energy the first system needs to transfer in order for the process in the second system to occur.	The seesaw is a simple system where the transfer of energy is easily observed.
Activity 2 Chemical reaction occurs inside of the battery.	• Because energy is released by the battery when it was used to make the car move, the reactants of the reaction must have more energy than the products of the reaction. • An energy-transfer model can be drawn to represent the energy-releasing chemical reaction on the left side transferring energy to the energy-requiring chemical reaction on the right side. • The thumbtacks are necessary to transfer the energy from the battery system to the water-reaction system. • Because batteries tend to get warmer when they transfer energy to other systems, there is some energy that is also transferred to the surrounding air system.	Give students an opportunity to investigate the coupling of two reactions in a familiar phenomenon.	
Activity 3 The reaction of octane and oxygen is coupled to the motion of a car.	• Students develop an energy-transfer model that couples the energy-releasing reaction between octane and oxygen to the car's wheels.	Give the students the opportunity to apply the energy-transfer model to phenomena involving an energy-releasing chemical reaction and a car.	

Phenomena, Data, or Models	Intended Observations	Purpose	Rationale or Notes
Pulling It Together Q3: Explain a phenomenon in which oxygen is produced in a simple physical system ($2\ H_2O \rightarrow 2\ H_2 + O_2$)	More oxygen gas forms when water is connected in a complete circuit to a fresh battery than to a used battery.	Give students an opportunity to apply ideas about energy changes within systems and energy transfer between systems in a simple physical system.	
Pulling It Together Q4: Explain a phenomenon in which oxygen is produced in a complex biological system (Photosynthesis in pond weed)	Pond weed produces more oxygen gas when grown under a 100 w light bulb than under a 40 w light bulb.	Give students an opportunity to apply ideas about energy changes within systems and energy transfer between systems in a complex biological system.	

Lesson 2.6—Coupling Energy-Releasing to Energy-Requiring Chemical Reactions

What do we know and what are we trying to find out?

In previous lessons, we saw that some chemical reactions have a net release of energy and other chemical reactions require a net input of energy. We also saw that a battery or sunlight can transfer energy to energy-requiring reactions/processes. For example, we saw that batteries transferred energy to the toy car and to the water reaction system.

In this lesson, we will revisit the battery and water activity to see if we can learn more about transferring energy between systems and begin thinking about the Key Question. (There is no need to respond in writing now.)

> **Key Question: Can the energy released by an energy-releasing chemical reaction provide the energy for an energy-requiring chemical reaction? If so, how?**

Teacher Talk and Actions

Activity 1: Coupling

How can we transfer energy from one system to another? **Coupling** is the word we use to describe the transfer of energy released by one system to another system. For example, think of circus performers on a seesaw, as shown in the diagram below. The process of moving one performer (Person 1) from the ground up into the air requires an input of energy because that person goes from not moving to moving. The process of landing gives off energy because the person goes from moving to not moving. When Person 1 lands, the seesaw "couples" the energy-releasing system (Person 1) to the energy-requiring system (Person 2). When Person 1 lands on the raised side of the seesaw, energy is transferred by the movement of the seesaw to Person 2. This results in Person 2 being propelled into the air without having to jump.

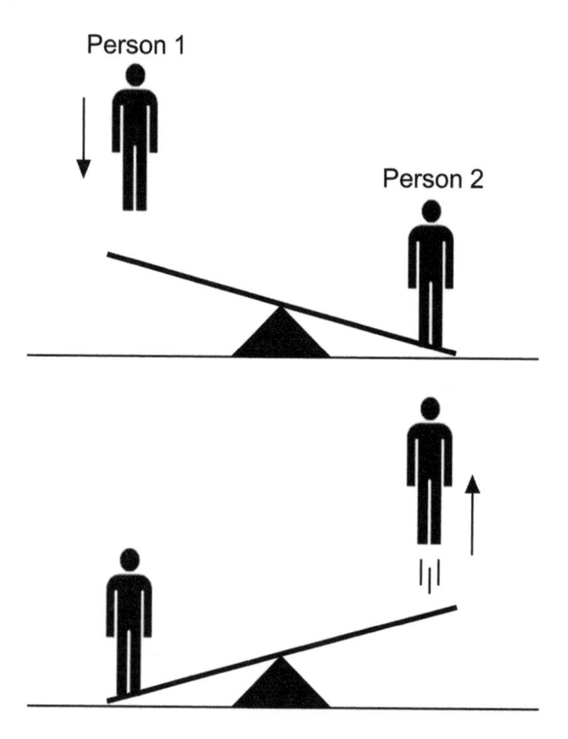

In this case, the seesaw is a **coupling device.** A coupling device is an object or system that enables the transfer of energy between two systems. We can represent the transfer of energy using the following model.

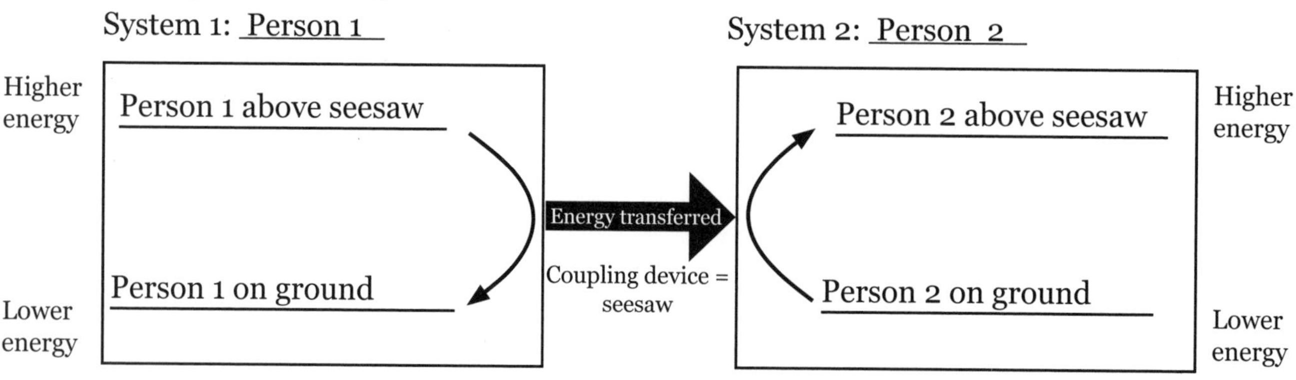

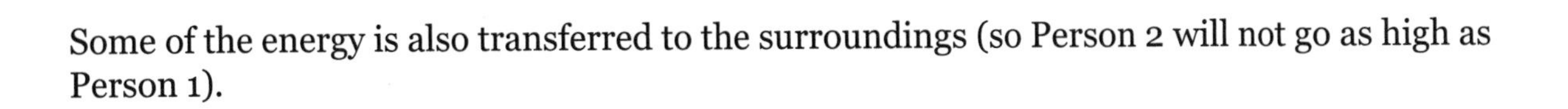

Teacher Talk and Actions

Some of the energy is also transferred to the surroundings (so Person 2 will not go as high as Person 1).

Draw students' attention to the addition of the coupling device in this model.

Procedures and Questions

1. Watch the video of the circus performers and record your observations below.

When one or two circus performers landed on the raised side of the seesaw, the person standing on the other side was shot up into the air.

2. What do you think would happen if we removed the seesaw as illustrated below?

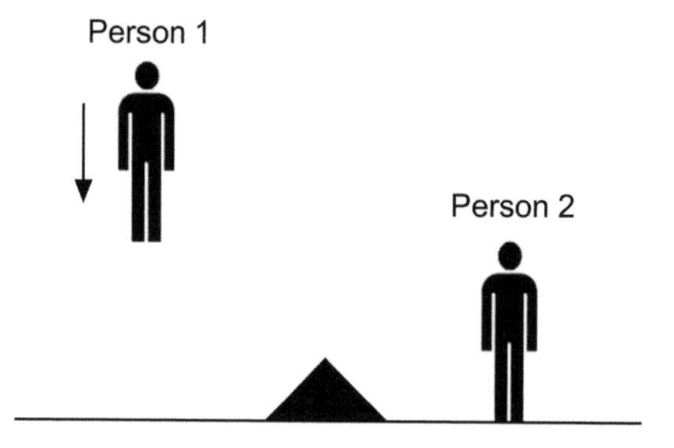

Person 1 would land on the ground but energy would not be transferred to Person 2. All the energy would be transferred to the ground. Person 2 would not move, because there is no device that can transfer the energy from Person 1 to Person 2.

3. What can you conclude about the necessity of a coupling device?

A coupling device is necessary in order to transfer the energy. If there is not a coupling device, the two systems do not interact and therefore cannot transfer energy to each other.

4. You may have noticed that when Person 2 is propelled into the air, he or she does not reach the height that Person 1 started out at. Why do you think this is?

Perhaps not all of the energy from Person 1 is transferred to Person 2. When energy is transferred from one system to another, typically there is some energy that gets transferred to the surrounding system. (Science Idea #18)

Teacher Talk and Actions

Show the video *Circus Performers* here.

If the video is played several times, students can make many observations about the relationship between the mass of the performers on the left and the mass and height of the performers on the right. For example, it takes two performers in System 1 to lift a heavier person to the same height or a lighter person higher. Keeping track of the relative masses and relative heights can provide evidence that energy released by System 1 must be sufficient to provide the energy needed by System 2.

5. Now consider the case where Person 1 was a child who had less mass than Person 2 as illustrated below. What do you think would happen in this case?

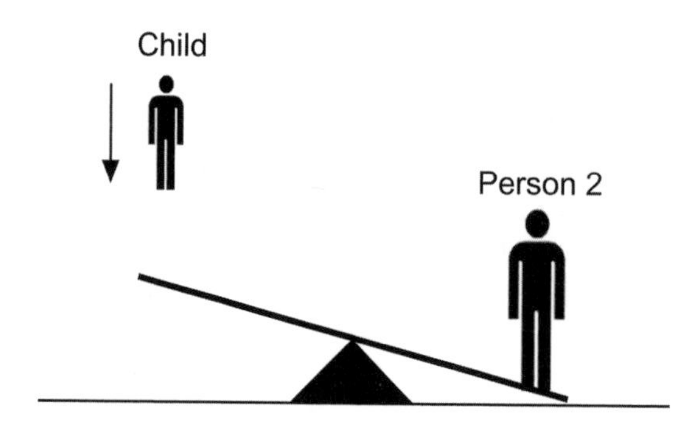

Person 2 will likely not go anywhere because the child has less mass than the person. The amount of energy a person has when moving depends on the mass of the person. So, the child would not have as much energy as Person 1 did when landing on the raised side.

6. What can you conclude about the amount of energy that must be released by System 1 compared to the amount of energy required by System 2 in order for the process in System 2 to occur?

The amount of energy released by System 1 cannot be less than the amount of energy required by System 2. If the amount of energy released by System 1 is less than the amount of energy required by System 2, the process in System 2 will not be able to occur.

7. In Step 4, we talked about the fact that some of the energy from Person 1 is transferred to the surroundings. What does that tell you about the amount of energy released by System 1 compared to the amount of energy required by System 2? Explain.

Because some energy is transferred to the surroundings, not all of the energy released by System 1 is available for use by System 2; therefore, the amount of energy released by System 1 has to be greater than the amount of energy required by System 2.

Teacher Talk and Actions

Activity 2: Coupling Chemical Reactions

In the water-breaking reaction, the 9-volt battery transfers energy to the chemical reaction system. Let's take a look at what is happening inside the battery. Inside the battery, there is actually a chemical reaction occurring. In some batteries, the reaction is between zinc and manganese dioxide and the reaction equation is as follows:

$$Zn + 2\ MnO_2 \rightarrow ZnO + Mn_2O_3$$

Procedures and Questions

1. Based on what you have learned so far, do you think that this reaction has a net input or output of energy?

Because the battery provides the energy required by the $2\ H_2O \rightarrow 2\ H_2 + O_2$ reaction, the $Zn + 2\ MnO_2$ reaction must have a net output of energy. Also, we saw that the battery transferred energy to the toy car and made it move.

2. Draw bar graphs to represent the amount of energy in the battery before and after the reaction.

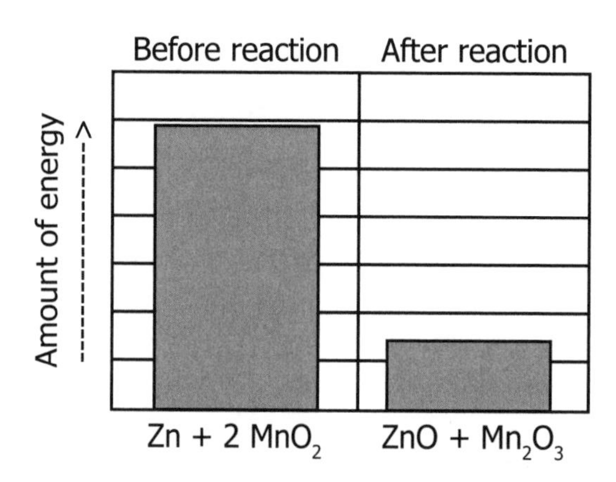

3. Do you think the products (ZnO and Mn_2O_3) of this reaction have *more, less,* or *the same* amount of energy as the reactants (Zn and MnO_2)? Explain why you think so.

Less. Since energy is released during the reaction, the amount of energy associated with the products (ZnO and Mn_2O_3) must be less than the amount of energy associated with the reactants (Zn and MnO_2).

4. Discuss with your group how you think the energy from the Zn and MnO_2 reaction could be transferred to the water. Write your thoughts below:

Energy is transferred when an electric current is passed through the water. That only happens when there is a complete circuit through which current can flow. When the leads on the battery touch the thumbtacks (which are in contact with the water), the circuit is complete. (The bubbles stopped forming when the leads were disconnected from the thumbtacks.) The leads and thumbtacks act as the "coupling mechanism" that transfers the energy from the battery to the water.

Teacher Talk and Actions

If students are having trouble accepting that this is a chemical reaction (atoms are rearranged), point out that there are no bonds between Zn and O in the reactants but there are in the products.

5. Complete the energy-transfer model below to represent how the battery system is coupled to the water system. The energy-releasing process should be in the box on the left and the energy-requiring process should be in the box on the right.

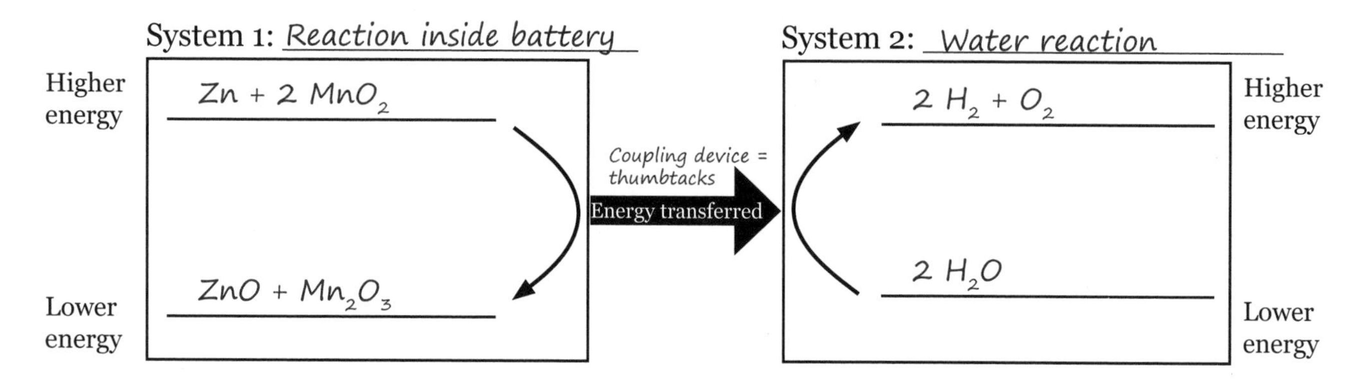

6. What do you think would happen if you didn't put thumbtacks through the bottom of the cup of water? How do you think that would affect the energy transfer?

I don't think the water-breaking reaction would occur without the thumbtacks. There would be no energy transfer because there would be no way to get the energy from the battery to the water. (The circuit would not be complete.)

7. Have you noticed that a battery tends to get warm when it is used to power a device? What does this mean about the energy being transferred from the battery? Is all of it being transferred to the water-reaction system or is some of it being transferred elsewhere?

Not all of the energy from the battery is being transferred to the water-reaction system. The increase in temperature of the battery is evidence that some energy is being transferred to the battery casing and the surrounding air.

8. Revise the energy-transfer model in Step 5 to reflect your answer in Step 7.

Students could revise their diagram in a variety of ways. They could put in a third box for the surrounding environment or just split the arrow and label it "energy transferred to environment." See facing page for one example.

Teacher Talk and Actions

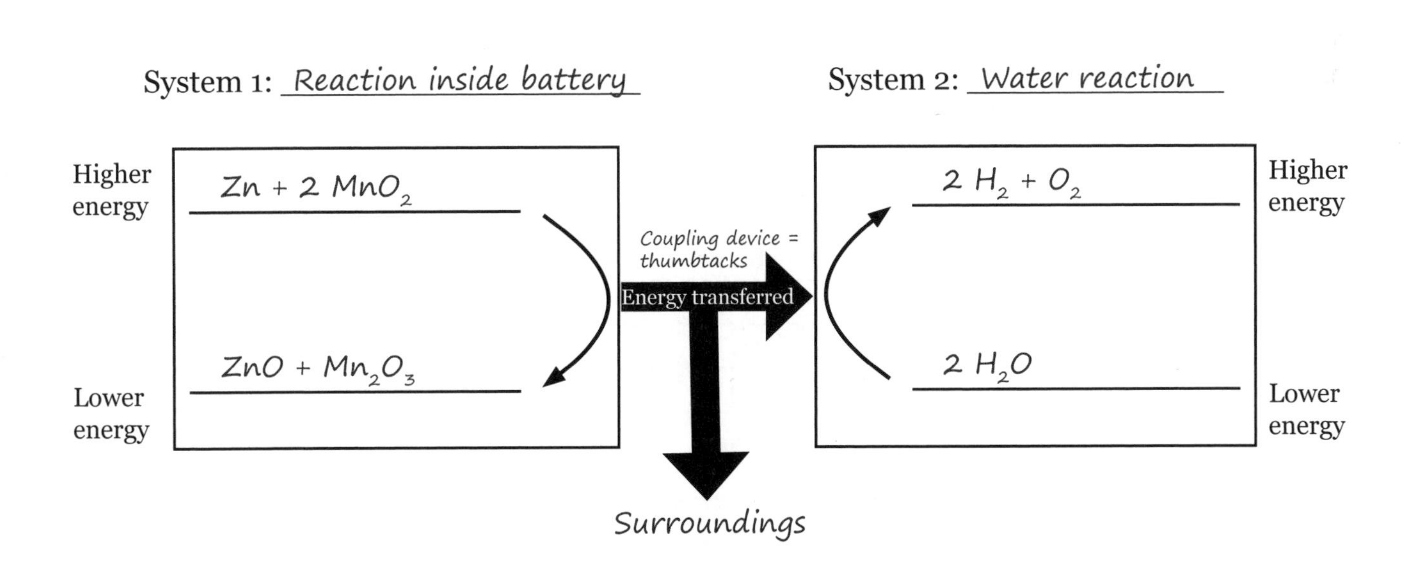

Activity 3: Coupling Chemical Reactions to Motion

We'll now examine how the energy released during a chemical reaction is coupled to the motion of a car. In a car, gasoline reacts with oxygen to form carbon dioxide and water. The chemical equation for the reaction between gasoline (which is primarily made up of octane, C_8H_{18}) and oxygen is as follows:

$$2\ C_8H_{18} + 25\ O_2 \rightarrow 16\ CO_2 + 18\ H_2O$$

This reaction releases approximately 2,633.4 kcals of energy for every two moles of octane reacted.

Procedures and Questions

1. Do you think the product molecules (16 CO_2 and 18 H_2O) of this reaction have *more, less,* or *the same* amount of energy as the reactant molecules (2 C_8H_{18} and 25 O_2)? Explain why you think so.

Because energy is released during the reaction, the amount of energy associated with the products (16 CO_2 and 18 H_2O) is less than the amount of energy associated with the reactants (2 C_8H_{18} and 25 O_2).

2. Draw bar graphs to represent the amount of energy in the system before the reaction and the amount of energy in the system after the reaction.

System = Gasoline + Oxygen reaction

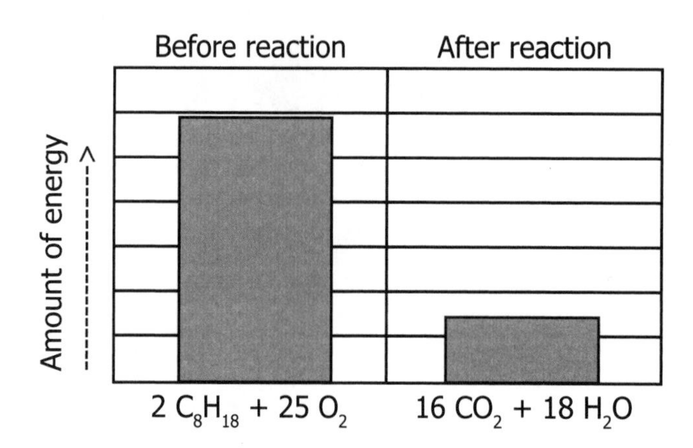

Teacher Talk and Actions

3. Complete the energy-transfer model below to represent how the reaction between octane and oxygen is coupled to increase a car's speed. The energy-releasing process should be in the box on the left and the energy-requiring process should be in the box on the right. Question 4 will help you think about the coupling device.

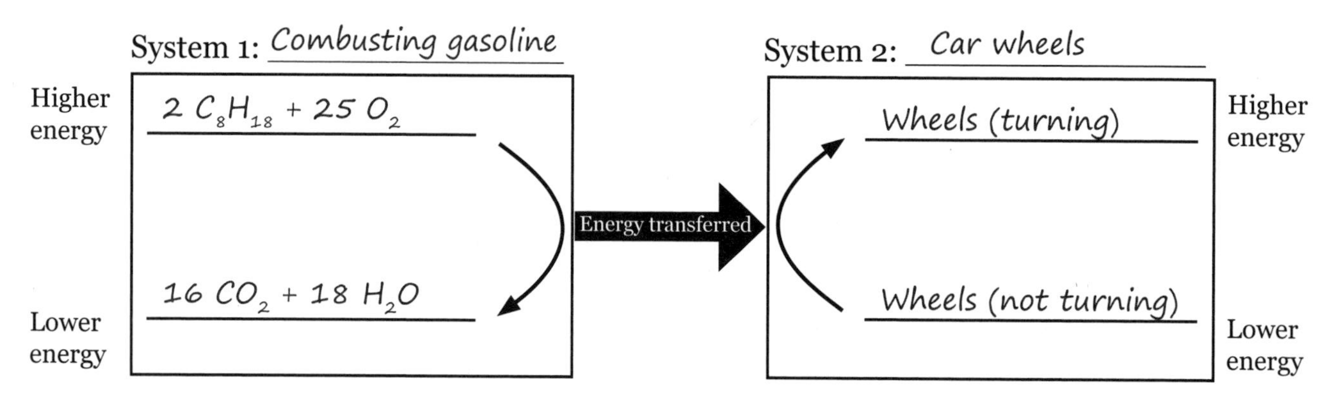

4. In this example, we are coupling the octane and oxygen reaction to the motion of an object. The ability to transfer energy from a chemical reaction system to make a car move was a major breakthrough of the Industrial Revolution.

Discuss with your group what the coupling mechanism between System 1 and System 2 could be. (Hint: What part of the car couples the reaction between gasoline and oxygen to the movement of the car?)

The coupling device is the car's engine. (Some students may know that the heat from the combustion of gasoline expands a gas that moves a piston, which then turns the drive shaft connecting the wheels.)

Teacher Talk and Actions

After students have discussed how energy from the combustion of gasoline could be used to increase a car's speed, you can show them an example of a combustion engine. (Several animated examples can be found at *www.animatedengines.com.*)

Science Ideas

The activities in this lesson were intended to help you understand an important idea about coupling chemical reactions. Read the idea below. Look back through the lesson. In the space provided after the science idea, give evidence that supports the idea.

Science Idea #23: Energy released during a chemical reaction/process can be used to "drive" other reactions/processes that require a net input of energy if

a. the amount of energy released by the first reaction is more than the amount of energy required by the second reaction/process, and

b. a mechanism, or "coupling device/machine," is available to transfer the required amount of energy from the energy-releasing to the energy-requiring process.

Evidence:

A chemical reaction inside the battery released energy that was transferred to the water by the thumbtacks.

A chemical reaction inside the car's engine released energy that was transferred to the wheels by the motor.

The person who landed on the raised side of the seesaw needed to have enough mass so that enough energy could be transferred to raise the person standing on the other side of the seesaw.

Teacher Talk and Actions

Pulling It Together

Work on your own to answer these questions. Be prepared for a class discussion.

1. Now that you have learned about coupling, answer the Key Question: **Can the energy released by an energy-releasing chemical reaction provide the energy for an energy-requiring chemical reaction? If so, how?**

Yes, the energy released by an energy-releasing reaction can provide the energy for an energy-requiring process. In order for this to happen there needs to be something to couple the two reactions/processes together. Just like the seesaw was needed to transfer the energy from the person falling to the person rising, the battery system could transfer energy to the water system if the two systems were connected in a complete circuit using the thumbtacks. A car engine is also a coupling device that transfers energy from the combustion of gasoline to the motion of the car wheels.

2. Why is a coupling mechanism needed to transfer energy from one chemical reaction to another?

Somehow the two reactions/processes need to "touch" each other so the energy released by one reaction/process can be transferred to the energy-requiring reaction/process.

- *Both people "touched" the seesaw, so the seesaw could transfer the energy from the person falling to the other person.*
- *The leads and thumbtacks connected in a complete circuit allows the reaction inside the battery and the reaction that splits water into hydrogen and oxygen to "touch."*
- *The car engine "touches" the combustion of gasoline and the shaft that turns the car wheels.*

When the coupling mechanism touches both systems, it can transfer the energy from one system to the other. If there is no coupling mechanism, the energy from the energy-releasing reaction will be transferred to the surrounding air (so the air will warm up).

3. Write a complete explanation for why more oxygen gas is formed when water is connected in a complete circuit to an unused battery versus a used battery.

More oxygen gas forms when water is connected in a complete circuit to an unused battery than to a used battery because (a) energy is required to produce oxygen gas (and hydrogen gas) from water (System 2), (b) the energy comes from the battery (or from an energy-releasing chemical reaction in the battery) (System 1), and (c) an unused battery can transfer more energy than a used battery can transfer. [A more sophisticated response would add the following mechanism for why: As the battery transfers energy to the water reaction system, reactants in the battery (e.g., $Zn + 2\ MnO_2$) are converted to products (e.g., $ZnO + Mn_2O_3$). When enough of the reactants have been converted to products, the chemical reaction system in the battery can no longer transfer energy to the water reaction system, and the production of oxygen gas in System 2 slows down or stops.]

Teacher Talk and Actions

Some students may know that the piston is the part of the engine that transfers force from the expanding gas to the crankshaft.

A complete explanation should make all three points (a, b, and c) included in the ideal student answer. Encourage students to use energy transfer models to help them think through their explanations. Students constructed an energy transfer model for the water reaction in Activity 2 of this lesson. When discussing student responses, ask students to list evidence for each point (a through c).

• For a, students can cite evidence from Table 2.5 in Lesson 2.4, Activity 4 (SE p. 132) that focuses on the "energy required" column.

• For b, students can cite evidence that oxygen gas is only produced when the system is connected to a battery system.

• For c, students can cite their observations of the battery-powered toy car (i.e., car moves more with fresh versus used battery) as evidence that the fresh battery can transfer more energy than the used battery.

If students need additional support, consider posting a chart with one column listing a, b, and c as claims and another column listing evidence for each claim.

4. Write a complete explanation for why an aquatic plant produces more oxygen gas when grown under a 100 w light bulb than under a 40 w light bulb.

An aquatic plant produces more oxygen gas when grown under a 100 w light bulb than under a 40 w light bulb, because (a) energy is required for an aquatic plant to produce oxygen gas (and glucose) from carbon dioxide and water (System 2), (b) the energy comes from the light bulb (System 1), and (c) the 100 w light bulb can transfer more energy to the aquatic plant system than the 40 w light bulb can. [A mechanistic explanation goes beyond NGSS expectations.]

5. Propose an energy-transfer diagram for the motion of an athlete. Start by constructing the right side (the system receiving the energy). For the box on the left (the system providing the energy), think about what processes could possibly provide the energy. Then explain why you selected that process.

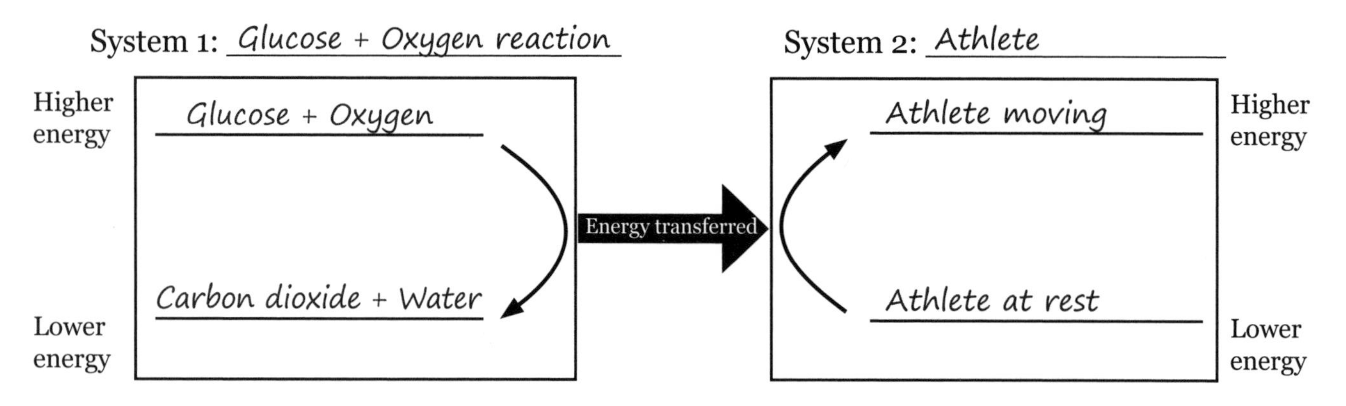

Teacher Talk and Actions

Students constructed an energy transfer model for photosynthesis in Lesson 2.4, Activity 3. When discussing student responses, ask students to list evidence for each point (a through c).

- For a, students can cite evidence from Table 2.5 in Lesson 2.4, Activity 4 (SE p. 132) that focuses on the "energy required" column.

- For b, students can cite evidence that an aquatic plant only produces oxygen gas in the light, not in the absence of light.

- For c, students can cite the experimental data in Lesson 2.4, Activity 3 (SE p. 130) that shows oxygen production increases with light intensity.

If students need additional support, consider posting a chart with one column listing a, b, and c as claims and another column listing evidence for each claim.

Question 5 is designed to get students to start thinking about what they will be learning in Lesson 2.7. Students may have trouble completing the System 1 box. Encourage them to use Table 2.5 to get ideas for energy-releasing reactions and rule out ones that would not be feasible given that the reaction must take place inside the body.

Students may recognize that the athlete gets hot when moving. These students could incorporate this observation into their model.

Closure and Link to subsequent lessons:

In this lesson, we learned that energy released during a chemical reaction in a system (System 1) can be used to power a chemical reaction in another system that requires a net input of energy (System 2) as long as (a) System 1 releases more energy than System 2 needs and (b) a coupling mechanism like the seesaw exists that transfers energy from System 1 to System 2. Why must System 1 always release more energy than system 2 needs? System 1 must always release more energy because some of the energy will always be transferred as heat to its surroundings. (For example, the energy that warms the battery is not being transferred to the water system.) In the next lesson, we will apply what we have learned to help us figure out how an athlete obtains energy to move and grow.

Lesson Guide
Chapter 2, Lesson 2.7
Coupling Chemical Reactions to Motion and Growth in Living Organisms

Focus: Students apply their understanding of coupled systems and evidence from experiments on the effects of exercise on carbon dioxide production, oxygen consumption, glucose consumption, and body temperature to develop an energy-transfer model for muscle contraction. Using isolated muscle-fiber proteins, students observe that glucose + oxygen do not cause fibers to shorten but an ATP solution does and use their observations plus additional data to revise their energy-transfer models. The final activity provides data from experiments on the effect of exercise on muscle growth that students use to construct an energy-transfer model linking cellular respiration to muscle growth.

Key Question: What chemical reactions in our bodies can provide the energy needed for body motion and growth? How are motion and growth coupled to this energy-releasing process?

Target Idea(s) Addressed

Cellular respiration is a chemical process during which bonds are broken between the atoms that make up glucose and oxygen molecules and bonds are formed between the atoms that make up carbon dioxide and water molecules, resulting in a net release of energy. (Science Idea #24)

Multicellular organisms need energy to move and grow. At the macroscopic level, muscle contraction moves the bones that the muscles are attached to. Cellular respiration provides energy for muscle contraction (motion) and for building muscles during training (growth). (Science Idea #25)

Multicellular organisms need a way to obtain oxygen for cellular respiration and eliminate carbon dioxide produced. The respiratory system, including the lungs, takes in air rich in oxygen that passes into blood vessels and gives off air rich in carbon dioxide that has come from blood vessels. (Science Idea #26)

Some of the energy given off during cellular respiration is used to produce ATP and water from ADP and inorganic phosphate. Energy released as heat is used to maintain body temperature. (Science Idea #27)

The reaction between ATP and water to form ADP and inorganic phosphate provides the energy input for essential energy-requiring processes in living systems, such as muscle contraction and the synthesis of polymers for growth and repair. (Science Idea #28)

Science Practice(s) Addressed

Analyzing and interpreting data
Developing and using models
Constructing explanations
Obtaining, evaluating, and communicating information

Materials

Activity 1: *Per group of students:* Limewater test solution; 2 50 ml beakers; 2 straws

Activity 2: Video: *Muscle Contraction*

Pulling It Together: Videos: *Hamster* and *Maize Plants Growing*

Advance Preparation

Organize materials in advance so that each team has materials when class begins. Observe videos to see how each part is integrated with student text.

Phenomena, Data, or Models	Intended Observations	Purpose	Rationale or Notes
Activity 1			
Limewater test before and after exercise	• Limewater turns cloudy faster after 5 minutes of vigorous exercise than after 5 minutes at rest.	Provide evidence for the reactants (glucose + oxygen) and one of the products (carbon dioxide) of cellular respiration.	
Experimental data showing effect of exercise on oxygen uptake of thigh muscle	• The amount of oxygen taken up increases with exercise time, with the greatest rate of increase being from 10–20 seconds.	Give students opportunities to analyze experimental data and use the data to draft a model.	
Experimental data showing the effect of mild, moderate, and heavy intensity exercise on glucose uptake by thigh muscle over time	• Glucose uptake increases over time for all three conditions. More glucose is taken up during moderate-intensity than low-intensity exercise and more glucose is taken up during high-intensity exercise than during moderate-intensity exercise.		
Experimental data showing the effect of exercise on thigh muscle temperature	• Thigh-muscle temperature increases with exercise time. Temperature increases at a constant rate over 180 seconds (three minutes), whereas oxygen uptake doesn't increase at a constant rate over the same timeframe.		

Phenomena, Data, or Models	Intended Observations	Purpose	Rationale or Notes
Activity 2 Video of muscles contracting and moving bones and of muscle fibers shortening	• The tennis player's arms move because contraction of muscles moves the bones they are attached to.	Give students an opportunity to interpret experimental data and use the data to revise an energy-transfer model for muscle contraction.	
Video of sliding filaments	• Muscles contract as filaments making up muscle fibers slide past one another.	Give students an opportunity to use mathematics to revise their energy-transfer model for muscle contraction.	
Effect of glucose + oxygen on isolated muscle fibers	• Isolated muscle fibers (containing actin and myosin protein filaments) do not shorten when glucose + oxygen is added.		
Effect of ATP + H_2O on isolated muscle fibers	• However, isolated muscle fibers do shorten when a solution of ATP is added.		

Phenomena, Data, or Models	Intended Observations	Purpose	Rationale or Notes
Activity 3 Experimental data on the amount of protein synthesized per hour in biceps muscles of experimental arm (that engaged in resistance training) and control arm (that did not engage in resistance training)	• More protein is synthesized in the arm that engaged in resistance training than in the control arm.	Give students an opportunity to learn the practice of developing and revising a model in a complex biological system (the human body). Give students opportunities to analyze experimental data, draw conclusions, and use conclusions to revise a model. Give students an opportunity to use an evidence-based model to construct an explanation for how proteins in food become muscle proteins.	
Pulling It Together Q2: Brown fat cells help infants to stay warm	Brown fat cells carry out cellular respiration but don't transfer the energy released to the ATP/ADP system.	Give students an opportunity to apply ideas about energy transfer to the environment as heat to understand how fat cells work.	

Phenomena, Data, or Models	Intended Observations	Purpose	Rationale or Notes
Pulling It Together Q3: Weight loss	Weight loss (decrease in measured mass) occurs when more carbon atoms leave the system (as CO_2) than enter the system (as carbon-containing polymers from food).	Give students an opportunity to apply ideas about atom rearrangement and conservation to weight loss (i.e., a decrease in measured mass of a living organism).	Students' ability to reason about the reverse of weight gain, perhaps assisted by drawing models of before and after, is an indication of their understanding of why changes in measured mass don't violate the law of conservation of atoms.
Pulling It Together Q4: Hamster on a wheel	Models are useful for making sense of energy changes and transfers when a hamster runs on a wheel.	Give students an opportunity to apply ideas about and models of energy changes within systems and energy transfer between systems to a complex phenomenon they observed in Lesson 1.2.	Students don't yet know why the spinning wheel makes the light bulb glow, which they will learn if they take physics. Interested students can look up "electromagnetic induction" on the internet.

Phenomena, Data, or Models	Intended Observations	Purpose	Rationale or Notes
Pulling It Together Q5: Plant growth	Models are useful for making sense of energy changes and transfers when a plant grows.	Give students an opportunity to apply ideas about and models of energy changes within systems and energy transfer between systems to plant growth.	Challenges the misconception that plants obtain energy for growth *directly* from the Sun.
Pulling It Together Q6: Dangerous diet pills	Models are useful for making sense of energy changes and transfers when a substance in a (now banned) diet pill prevents energy from being transferred from cellular respiration to the ATP/ADP system.	Give students an opportunity to apply ideas about and models of energy changes within systems and energy transfer between systems to plant growth.	

Lesson 2.7—Coupling Chemical Reactions to Motion and Growth in Living Organisms

What do we know and what are we trying to find out?

In the previous lesson, you saw that the energy released by one system can be used to make something happen in another system if two conditions are met: (1) the amount of energy released from the first system must be sufficient to supply the energy needs of the second system, and (2) a coupling device must connect the two systems so that the energy can be transferred from one system to the other. Both conditions must be met. When you observed the video of the circus performers, you decided that if the seesaw was removed then the energy from the falling circus performers couldn't be used to move another performer.

You also observed that the energy released by a chemical reaction inside a new battery could transfer energy to interacting water molecules to form hydrogen and oxygen molecules if the battery was connected in a complete circuit to the water system. In this example, the chemical reaction in the battery released sufficient energy and was coupled to the water-breaking reaction using thumbtacks to complete the circuit. You also decided that if the performer that lands on the raised end of the seesaw had less mass than the performer on the other end, the performer on the other end would not be propelled into the air.

We know that exercise must require energy, because exercise involves an increase in body motion. When people run, their arms and legs move because their muscles are contracting. We also know that the chemical reactions between amino acid monomers to form protein polymers (and water molecules) require energy. That means that building muscle requires energy, because building muscle involves making new muscle proteins. Athletes in training need a source of energy for moving their bodies during their athletic event and for building their muscles before the event. In this lesson, we'll focus first on an athlete's motion and then on making muscle.

Answer the Key Questions to the best of your knowledge. Be prepared to share your ideas with the class.

Key Questions: What chemical reactions in our bodies can provide the energy needed for body motion and growth? How are motion and growth coupled to this energy-releasing process?

Teacher Talk and Actions

Activity 1: What Chemical Reaction Provides Energy for Body Motion During Exercise?

Materials
For each team of students
Limewater test solution (turns cloudy in presence of CO_2)
Two 50 ml beakers
Two straws

We know that as we exercise our muscles move, and movement requires energy. But where does that energy come from? We can start to think about this question by creating a draft energy-transfer model.

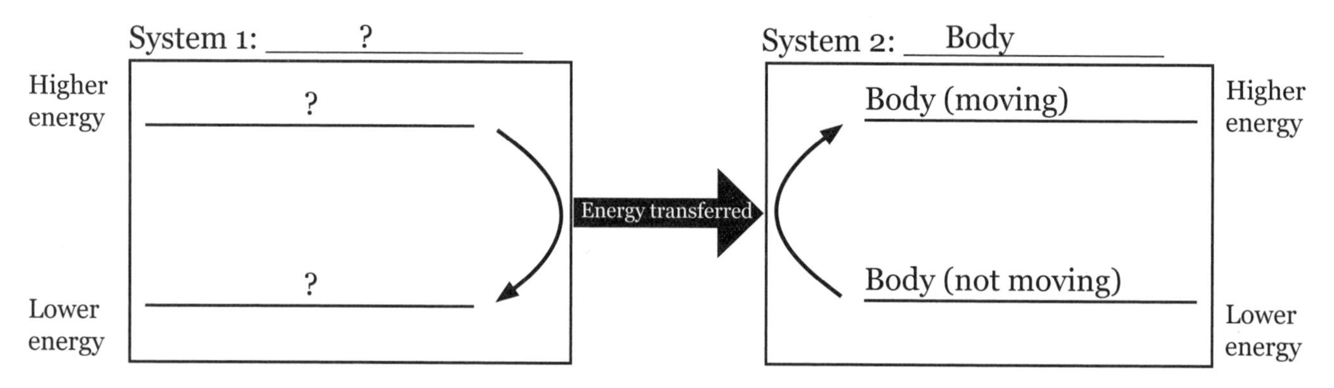

We have put question marks in System 1 of our energy-transfer model because we don't know where the energy for body movement is coming from. Our goal is to complete the model.

Procedures and Questions

As a class, discuss the following questions:

1. What are some chemical reaction(s) that occur in the body that might provide the energy needed for exercise?

   Table 2.3, p. 117, includes several energy-releasing chemical reactions, e.g., carbohydrates + $O_2 \rightarrow CO_2 + H_2O$, fat/fatty acids + $O_2 \rightarrow CO_2 + H_2O$. Since these reactions release energy, the reactants have a higher amount of energy than the products.

2. Represent one possible reaction in the System 1 box of the energy-transfer model.

   The downward arrow indicates that energy is released as reactants form products, so the reactants (e.g., glucose + oxygen) should be listed on the top line of the System 1 box and the products ($CO_2 + H_2O$) should be listed on the bottom line.

Teacher Talk and Actions

Students should look at Table 2.3 (SE p. 117) for ideas.

In this activity, we'll collect data and examine data from four investigations to help us figure out what chemical reaction could be providing the energy necessary for body motion during exercise and then represent this chemical reaction on our energy-transfer model.

Investigating the effect of exercise on carbon dioxide production

Several of the chemical reactions proposed produce carbon dioxide, so it makes sense to see whether the amount of carbon dioxide produced increases during exercise. In this investigation, you will compare the amount of carbon dioxide produced after resting to the amount of carbon dioxide produced after exercising.

3. Fill two 50 ml beakers with 25 ml of limewater solution. Label one "after resting" and the other "after exercise." After sitting at rest for 5 minutes, blow through a straw into the 50 ml "after resting" beaker with limewater and measure the number of seconds it takes until the limewater turns cloudy. Record the time in the table below.

4. Exercise for 5 minutes (e.g., jumping jacks, skipping rope, running) and then blow through a straw into the "after exercise" limewater solution, measure the number of seconds it takes until the limewater turns cloudy, and record the time in the table below.

Experimental Condition	Seconds It Takes Limewater to Turn Cloudy
After 5 min. at rest	31–41
After 5 min. of exercise	19–29

5. Compare the results of the limewater test for both conditions.

Limewater turned cloudy faster after exercise than after being at rest. That means more carbon dioxide was produced after exercising than after resting.

6. What evidence does this provide about the chemical reaction(s) that provide energy for exercise?

Carbon dioxide is a product of the chemical reaction.

Teacher Talk and Actions

Investigating the effect of exercise on oxygen consumption

Several of the chemical reactions listed in Table 2.3 involve a fuel reacting with oxygen, so it makes sense to see whether the amount of oxygen consumed increases during exercise. Let's look at data from an experiment that investigated the effect of exercise on oxygen consumption.

7. Read about the experiment that scientists conducted, examine the data they collected, and answer the questions that follow with your group.

Experiment 1: Effects of Exercise on Oxygen Uptake

Scientists recruited five healthy men who ranged in age from 22 to 25 years. The men were asked to perform a single-legged knee extension exercise for 180 seconds. Scientists sampled blood every 10 seconds from a vessel carrying blood to the muscle and from another vessel carrying blood away from the muscle and measured the amount of oxygen contained in each sample. The amount of oxygen consumed was calculated by subtracting the amount of oxygen in the blood leaving the muscle from the amount of oxygen in the blood entering the muscle. They repeated the study three times and averaged the measurements for each time point.

Figure 1 shows the average rate of oxygen consumption by the quadriceps (thigh) muscle before and during exercise.

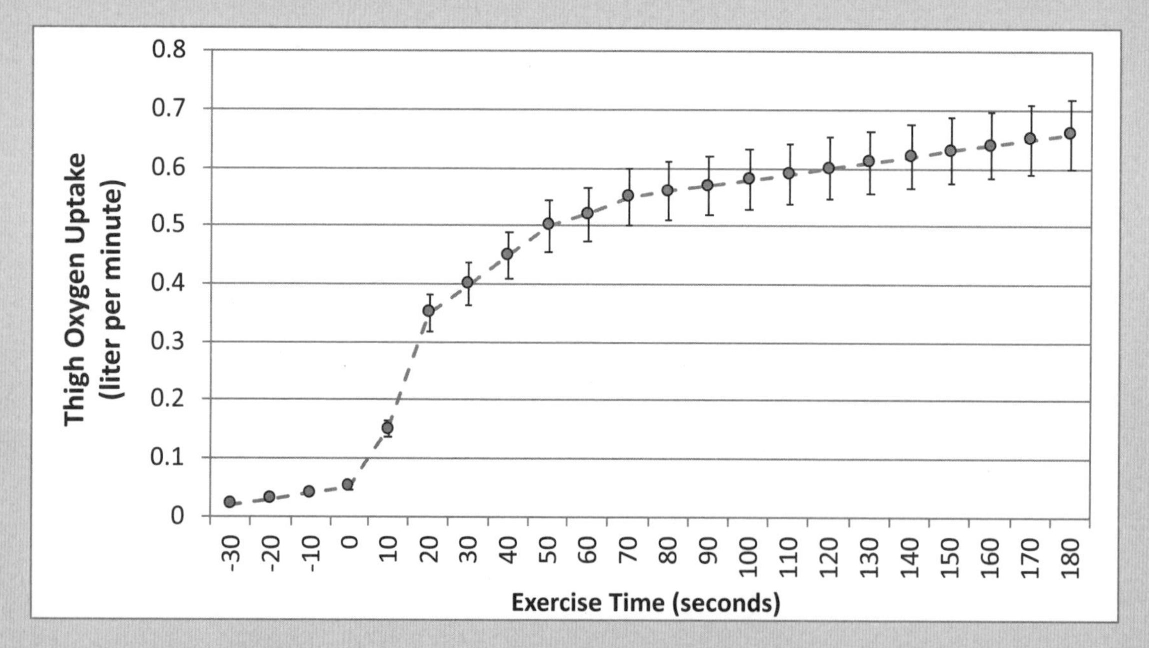

Figure 1. Oxygen consumption by the quadriceps muscle before and during knee-extensions

Source: Krustrup, P., J. González-Alonso, B. Quistorff, and J. Bangsbo. 2001. Muscle heat production and anaerobic energy turnover during repeated intense dynamic exercise in humans. *Journal of Physiology* 536 (3): 947–956.

Teacher Talk and Actions

8. Citing specific data from the graph, describe what happens to oxygen uptake of the muscle during exercise.

The amount of oxygen uptake increases from 0-180 seconds after starting to exercise. It increases the most between 10 and 20 seconds. (During the first 10 seconds, the amount of oxygen used increases from 0 to .15 liters/minute (an increase of .15 liters). Between 10 and 20 seconds, the amount of oxygen used increases from .15 to .35 liters/minute (an increase of .20 liters). Between 20 and 30 seconds it increases from .35 to .40 liters/minute (an increase of .05 liters).

9. Do the data provide evidence to support the chemical reaction you proposed in Question 2? Explain.

The data show that more oxygen was used during exercise than at rest, which provides evidence that oxygen is a reactant of the chemical reaction that provides energy for exercise.

Investigating the effect of exercise on glucose uptake

To find out whether glucose was the other reactant of the chemical reaction that provides energy for exercise, scientists measured the amount of glucose taken up from blood by muscles during exercise.

10. Read about the experiment that scientists conducted, examine the data they collected, and answer the questions that follow with your group.

Experiment 2: Glucose Uptake by Leg Muscles

Scientists measured the amount of glucose taken up by leg muscles every 10 minutes during mild, moderate, and heavy intensity exercise.

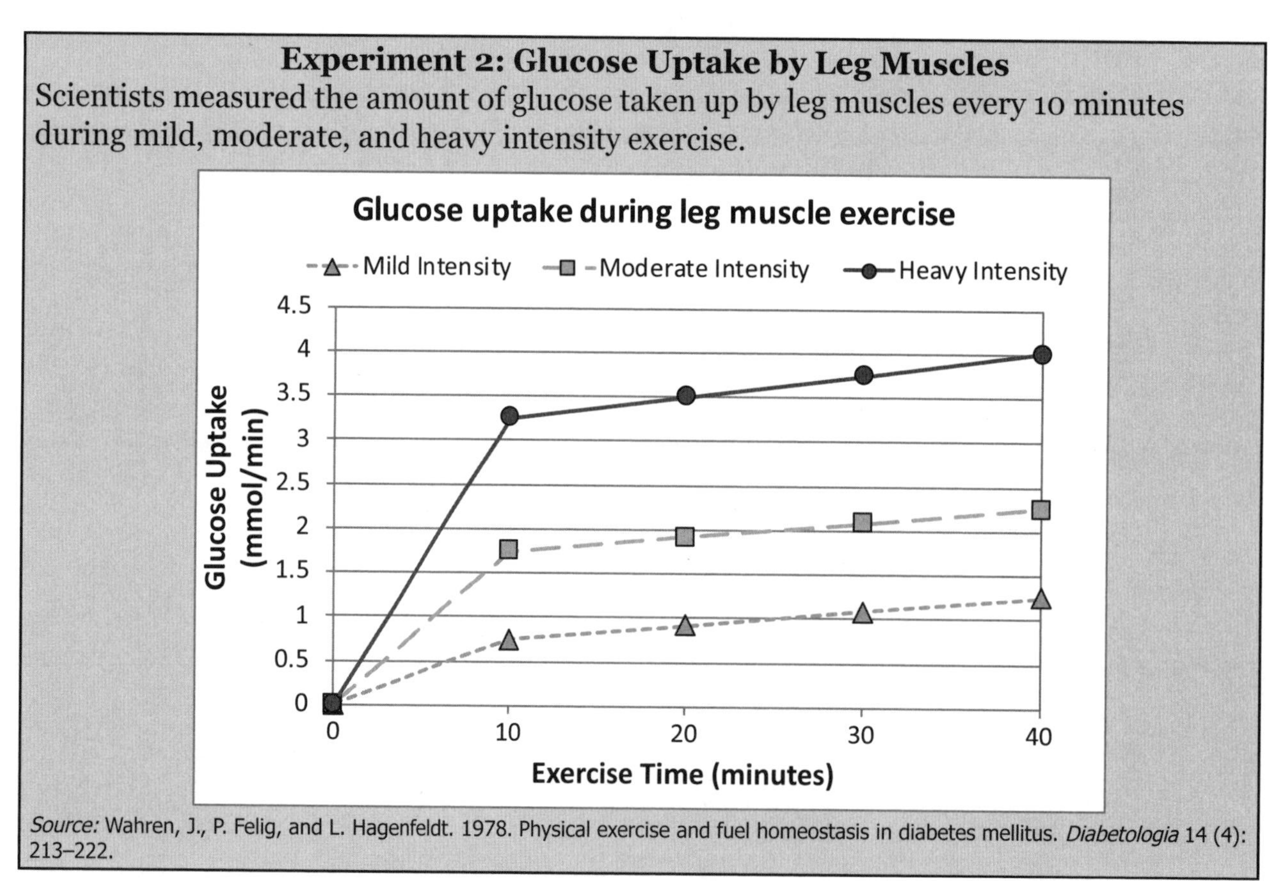

Source: Wahren, J., P. Felig, and L. Hagenfeldt. 1978. Physical exercise and fuel homeostasis in diabetes mellitus. *Diabetologia* 14 (4): 213–222.

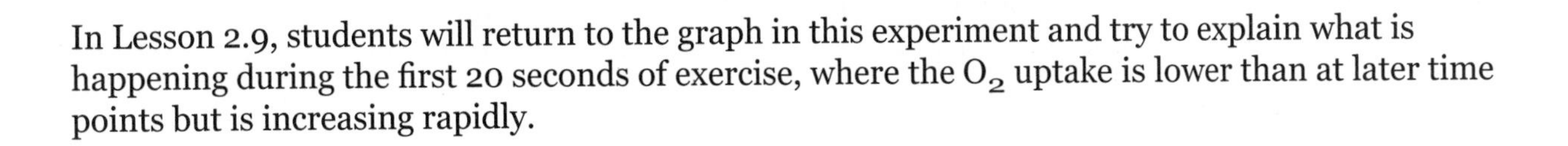

Teacher Talk and Actions

In Lesson 2.9, students will return to the graph in this experiment and try to explain what is happening during the first 20 seconds of exercise, where the O_2 uptake is lower than at later time points but is increasing rapidly.

If students say it increases, ask "When does it increase the most?"

The bars represent the range of the average measurements.

11. Using the graph, describe what happens to glucose uptake during exercise.

The amount of glucose taken up increases over time during mild, moderate, and heavy intensity exercise. At each time point (10 min., 20 min., etc.) more glucose is taken up during moderate versus light-intensity exercise and even more glucose is taken up during heavy-intensity exercise.

12. Do the data provide evidence that glucose is a reactant of the chemical reaction involved during exercise? Explain.

Yes. The data show that the muscle takes up more glucose with exercise time and intensity, suggesting glucose is used during exercise and, therefore, is a reactant of the chemical reaction involved during exercise.

13. Based on evidence from the investigations and what you know about chemical reactions, you can now propose a chemical reaction that is involved in exercise and justify your proposal.

a. Start by writing the names of reactants and products you think you know, citing evidence.

Reactants: glucose and oxygen (uptake of both increased with increasing exercise)

Product: carbon dioxide (production increased with increasing exercise)

b. Write the chemical formulas for each substance and start to write a chemical equation for the reaction.

Glucose ($C_6H_{12}O_6$) + oxygen (O_2)$\rightarrow$ carbon dioxide (CO_2)

c. Using models and your knowledge of matter conservation, predict what other product might form and justify your prediction.

The formulas show reactant molecules (glucose) contain H atoms but not the product carbon dioxide. So, there must be a product that contains H atoms. The glucose + oxygen reaction in Table 2.3, Lesson 2.3, shows that water (H_2O) is typically produced in reactions between carbon-containing molecules and oxygen.

d. Write an equation for the chemical reaction you think is involved in exercise.

$C_6H_{12}O_6 + 6\ O_2 \rightarrow 6\ CO_2 + 6\ H_2O$

Investigating the effect of exercise on body temperature

Data in Table 2.3 show that when one mole of glucose ($C_6H_{12}O_6$) reacts with O_2 in a calorimeter to produce CO_2 and H_2O, 670 kcal of energy is released. If we have correctly identified the reactants and products of the chemical reaction that occurs in the body, then energy should be released as the chemical reaction occurs. Is energy released when the chemical reaction occurs in the body? If so, then we would have more evidence that the chemical reaction that occurs in the body during exercise has the same reactants and

Teacher Talk and Actions

Many students have difficulty understanding the concept of *rate,* because it is a ratio. It may be helpful to unpack the question into smaller questions (e.g., What happens to the rate of glucose uptake during 40 minutes of mild-intensity exercise? During 40 minutes of high-intensity exercise?) Compare the amounts of glucose used by the leg muscle after 10 minutes of mild-, moderate-, and high-intensity exercise.

At this point, students have evidence from Table 2.3 that the reaction releases energy in a calorimeter, but they don't have evidence that it releases energy in the human body or that it is called cellular respiration.

Once students have written the chemical formulas for reactants glucose and oxygen and the product carbon dioxide, ask "What atoms are present in reactants? In products?" Students should reason from their knowledge of atom conservation that there must be a product with H atoms. Then ask, "What substances do you know that have H atoms?" You can also remind students that they might get ideas from the table in Lesson 2.3 to get ideas. Students should notice that all the reactions in Table 2.3 produce H_2O as a product.

Remind students that a mole is a large number of molecules.

products as the chemical reaction that was carried out in a calorimeter. One indicator that energy is given off is an increase in temperature. In this investigation, you will examine data on the effect of exercise on muscle temperature.

14. Read about the experiment that scientists conducted, examine the data they collected, and answer the questions that follow with your group.

Experiment 3: Effects of Exercise on Muscle Temperature

In 2001, Peter Krustrup and his colleagues recruited five healthy men ranging in age from 22 to 25 years and obtained informed consent that satisfied the requirements of the university's human subjects approval board. Volunteers performed a single-legged knee-extension exercise while scientists recorded internal temperature of the thigh (quadriceps) muscle before and during exercise. Scientists repeated the study three times, averaged the data, and plotted the average and range in Figure 1.

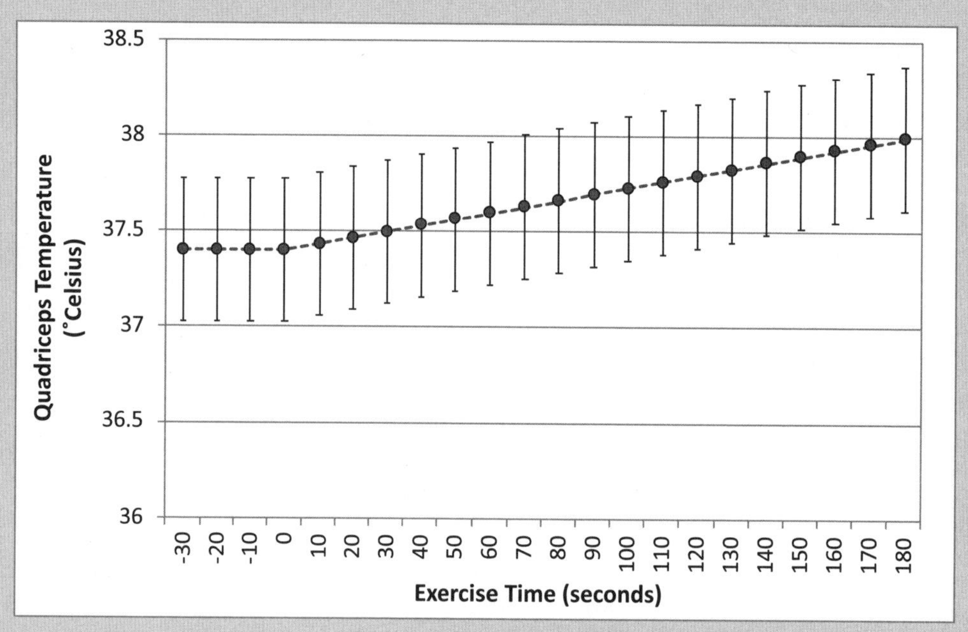

Figure 1. Temperature of the muscle before and during knee-extensions

Source: Krustrup, P., J. González-Alonso, B. Quistorff, and J. Bangsbo. 2001. Muscle heat production and anaerobic energy turnover during repeated intense dynamic exercise in humans. *Journal of Physiology* 536 (3): 947–956.

15. Using the chemical formulas for the substances shown in Table 2.3, write an equation for the chemical reaction that occurs in the body during exercise that shows that atoms are conserved.

$C_6H_{12}O_6 + 6\ O_2 \rightarrow 6\ CO_2 + 6\ H_2O$

16. Where in the body do you think this chemical reaction is occurring? Explain.

Muscle cells need the energy released by this reaction to exercise/contract, so the reaction probably occurs in all skeletal muscle cells (and probably heart muscle cells as well).

Teacher Talk and Actions

Point out that the bars on the graph represent the range of the 15 measurements (5 men × 3 times) with the midpoint (red) representing the average.

17. Living cells stay alive by using energy-releasing chemical reactions, such as the reaction between glucose and oxygen, to "drive" a variety of energy-requiring chemical reactions/processes. They do so by carrying out the glucose and oxygen reaction in a series of carefully regulated steps. The process is called **cellular respiration.** Both the word and the symbolic equation for this chemical reaction are shown below:

Cellular respiration

Glucose + oxygen → carbon dioxide + water

$$C_6H_{12}O_6 + 6\ O_2 \rightarrow 6\ CO_2 + 6\ H_2O$$

We are now ready to complete the energy-transfer model we started with at the beginning of the activity.

Based on what you have learned in this activity, complete the diagram below.

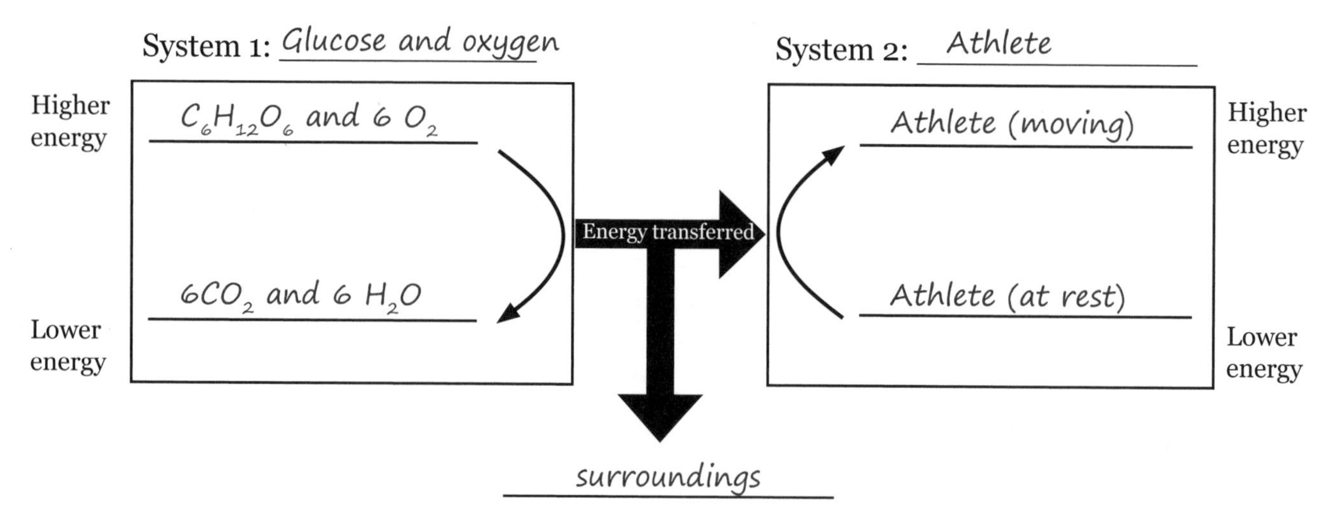

18. The energy-transfer model shows that an athlete can use energy from cellular respiration to move her body. Do you think all the energy from the glucose + oxygen reaction is used for motion? Justify your claim.

Claim: The athlete's body can't use all the energy released from the glucose + oxygen reaction for motion.

Evidence: Data from Experiment 3 (SE p. 172) showed that the leg temperature increased with exercise.

Reasoning: The data provided evidence that some of the energy from the reaction between glucose and oxygen was transferred to the surrounding muscle. Because energy is not created or destroyed the energy that increases muscle temperature isn't available for motion.

19. Compare cellular respiration to the burning donut reaction you observed in Lesson 2.3, Activity 3, page 113.

In both examples, glucose and oxygen react to form carbon dioxide and water and release energy. Both examples transfer energy to the surroundings as heat. The burning donut also transfers energy to the surroundings as light, whereas cellular respiration does not.

Teacher Talk and Actions

To avoid having students think that cellular respiration occurs only in muscle cells, ask the following questions: "Do you think any other body cells require energy? Can you think of any other cellular processes that involve motion? What about during cell division? (Chromosomes move to opposite ends of cells.) In addition to smooth muscle, what moves food down the digestive system (cilia on cells lining the gut)? Can you think of any other cells that probably have cilia (cells lining the respiratory track—that's what moves mucous up, so you can cough it out)?"

Although *Next Generation Science Standards* (*NGSS*) expect students to know by the end of grade 12 that "energy cannot be created or destroyed," *Matter and Energy in Growth and Activity* (*MEGA*) does not list it as a science idea because providing students with evidence for it is both challenging and time-consuming. Instead, *MEGA* gets at (a) energy conservation in Lesson 2.1 through Science Idea #15 (when two systems interact, and one system decreases in energy while the other system increases in energy, we say that energy was transferred from the first system to the second) and (b) dissipation in Lesson 2.3 through Science Idea #18 (energy changes within a system and energy transfers between systems usually result in some energy being released to the environment). If students are struggling to identify science ideas to justify their claim, refer to Science Idea #23 and ask, "Why does part (a) say *more*"?

If students continue to struggle, refer them to Science Idea #18 and ask, "How can this science idea help us understand why Science Idea #23 says *more*"?

For students who don't appreciate the subtle distinction between these science ideas and the *NGSS* statement, it's OK to have them use the *NGSS* statement in their explanations or justifications.

Students should add the arrow to their model showing energy transfer to the surroundings.

Activity 2: How Is the Energy of Cellular Respiration Coupled to Muscle Contraction?

We now know that cellular respiration, the chemical reaction between glucose and oxygen, provides the energy for motion. How does the body couple cellular respiration to motion? Scientists worked for nearly 100 years in laboratories throughout the world to try to figure this out. They started by making observations at the organ level of the body-system hierarchy (muscles), then zoomed in to make observations at the cell level (muscle cells or fibers), and finally carried out experiments at the molecular level (muscle proteins). Their initial observations are shown in the muscle contraction video. Let's look closely at the video to see if it will give us ideas about how cellular respiration is coupled to muscle contraction.

Procedures and Questions

1. To help you visualize how body movement is caused by muscle contraction, watch the first part of the muscle contraction video (Part I). Describe what you see happening at the organ level that enables the tennis player to hit the ball.

Muscles, such as the biceps muscle in the arm, are connected by ropelike tendons to bones. The shortening of the biceps muscle pulls on the tendons, which moves the bones.

2. Revise System 2 of the energy-transfer model you drew for Step 17 in Activity 1 to reflect what you observed at the organ level.

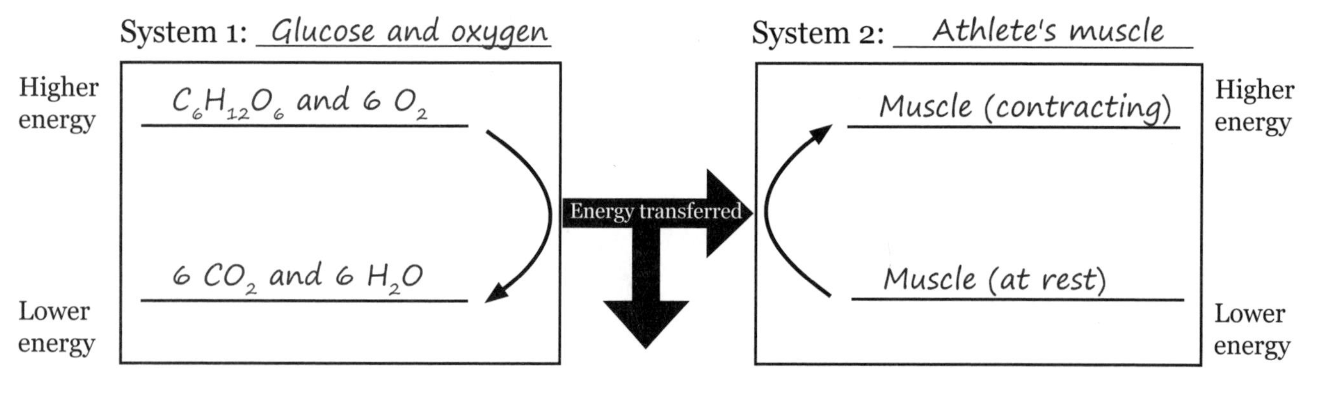

3. Part II of the video shows the muscle fibers that make up muscles and the filaments that make up the fibers. Watch the second part of the video and describe what you see happening at the fiber/cell level that enables the muscle to contract.

Muscle fibers/cells are made up of many overlapping filaments. The sliding of muscle filaments past one another causes the muscle to contract.

Teacher Talk and Actions

Show the video *Muscle Contraction, Part I,* "Muscle Contraction Causes Body Motion," pausing to let students record their observations.

Show the video *Muscle Contraction, Part II,* "Sliding Filaments Cause Muscle Contraction," pausing to let students record their observations.

4. Scientists discovered that it is possible to separate the muscle-fiber proteins actin and myosin from all other substances present in live muscle cells without destroying the structure and orientation of the proteins. This enabled scientists to carry out a variety of experiments on muscle proteins to try to understand what happens at the molecular level when muscle fibers contract. They could add different substances to the protein system and observe whether any of the substances caused the fibers to shorten. Based on what you know so far, what substances would you try first? Explain.

I would try adding glucose + oxygen, because the experiments in Activity 1 showed that glucose and oxygen are the reactants of an important energy-releasing reaction in the body.

5. Modify System 2 in your energy-transfer model to represent what you think could be happening at the molecular level.

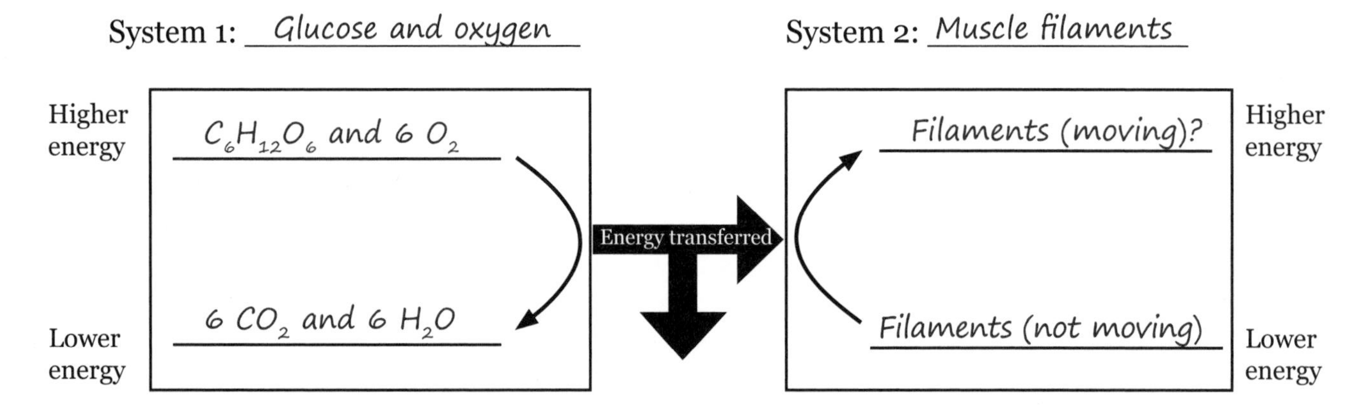

6. Scientists tried dripping a solution of glucose on the isolated muscle proteins to see if the fibers would shorten. Since oxygen was present in the air, both reactants of cellular respiration would be available to the muscle-fiber proteins. Watch the video (Part III) and describe what happens. Modify your model to represent your observations.

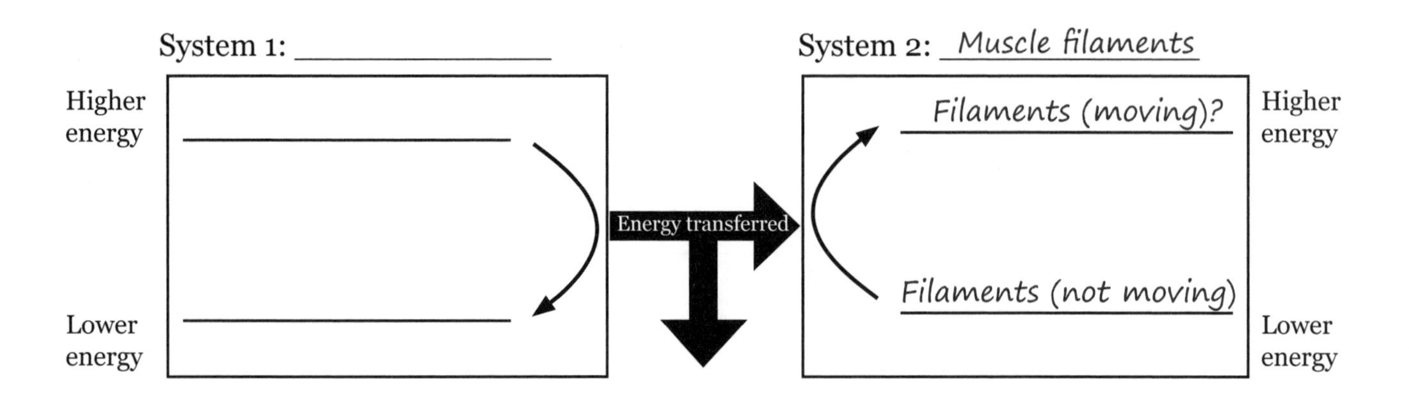

Teacher Talk and Actions

Show the video *Muscle Contraction, Part III,* “Dripping Glucose on Muscle Fibers,” pausing to let students record their observations.

Based on the video, students can erase glucose + oxygen in the System 1 box because this chemical reaction doesn’t appear to transfer energy directly to the muscle protein system. They can leave the energy-transfer arrow from System 1 to System 2 because something must be transferring energy to make the muscle proteins slide/pull. If they leave the energy-transfer arrow to the surroundings, ask them why they think so. They should cite Science Idea #18 as the reason that energy is always transferred to the surroundings.

7. The scientists' observation that the muscle filaments didn't move/slide in the presence of glucose and oxygen provided evidence that the glucose + oxygen reaction does not transfer energy directly to the muscle filament system. Perhaps the glucose + oxygen system transferred its energy to another system that then transferred energy to the muscle filament system. Scientists reasoned that if that other system was somehow lost or destroyed when the muscle filaments were isolated, then maybe adding back the missing system would cause the muscle filaments to contract just like whole muscle. To try to find out what that missing system might be, scientists tried adding back different substances found in muscle. One of the substances they tried was adenosine triphosphate (ATP), the molecule you examined in Lesson 1.3. Scientists had made two important observations about ATP that made it a good molecule to try:

- The reaction ATP + H_2O → ADP + P_i releases about 10 kcal/mol. That meant that the reaction could provide the energy required for actin and myosin to move past one another.

- ATP was found in freshly prepared resting rabbit muscle. When the muscle was electrically stimulated to contract continuously for 10 minutes, the concentration of ATP decreased, and the concentration of ADP increased. However, if the muscle rested for an hour, the concentration of ATP returned to normal.

The video (Part IV) shows what the scientists observed. Watch the video and describe what happens when ATP is added to the muscle-fiber protein system.

When ATP was added, the muscle fibers shortened.

8. Complete the energy-transfer model to reflect your observations.

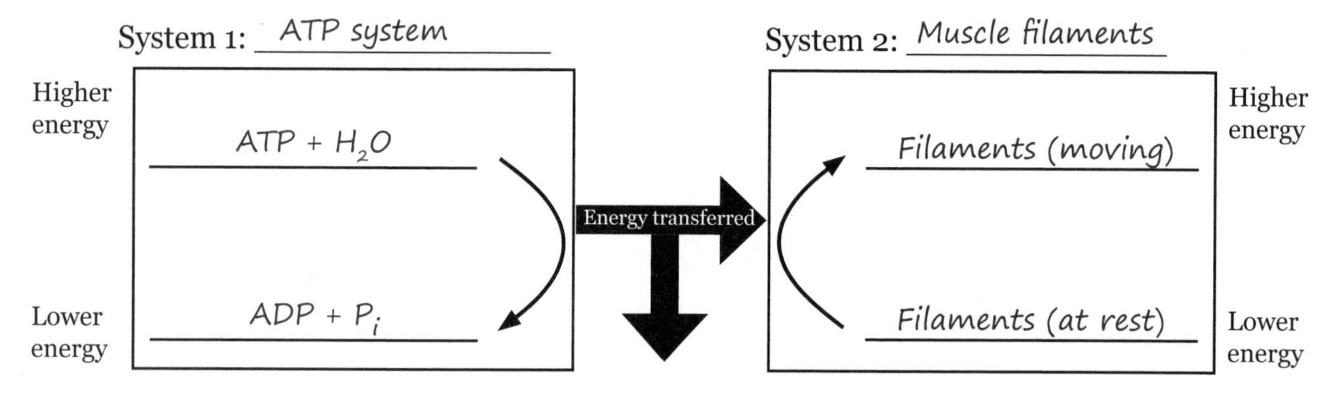

9. Scientists' observation that the concentration of ATP in live muscle returned to normal after an hour of rest suggested that muscle cells must somehow be able to produce ATP + H_2O from ADP + P_i. Since the reaction ATP + H_2O → ADP + P_i releases energy, the production of ATP + H_2O from ADP + P_i must require energy. Where could the energy come from? Scientists in Denmark, Germany, Japan, the former Soviet Union, and the United States carried out numerous experiments from 1930–1950 that provided the answer:

- The complete breakdown of a molecule of glucose in muscle and liver cells requires oxygen and correlates with the production of 32 molecules of ATP + H_2O from ADP + P_i.

Teacher Talk and Actions

Show the video *Muscle Contraction, Part IV,* “Dripping ATP on Muscle Fibers,” pausing to let students record their observations.

Teachers should guide students to (a) label System 1 ATP system, (b) put ATP + H_2O on top line and ADP + P_i on bottom line, (c) put a curved downward arrow from ATP + H_2O to ADP + P_i, and (d) add an arrow showing energy being transferred from System 1 to System 2. As each component is added to the energy-transfer model, teachers should ask students what evidence supports it.

“The term ‘high-energy phosphate bond,’ long used by biochemists to describe the P—O bond broken in hydrolysis reactions, is incorrect and misleading as it wrongly suggests that the bond itself contains the energy. In fact, the breaking of all chemical bonds requires an input of energy. The free energy released by hydrolysis of phosphate compounds does not come from the specific bond that is broken; it results from the products of the reaction having a smaller free-energy content than the reactants.”(Source: Nelson, D. L., and M. M. Cox. 2000. *Lehninger: Principles of biochemistry*. 3rd ed. New York: Worth Publishers, p. 504.)

This finding provides evidence that the reaction between glucose + oxygen can transfer energy to the ADP + P_i system to produce ATP + H_2O. We can use this evidence to create an energy-transfer model that includes the cellular respiration system and the ATP system.

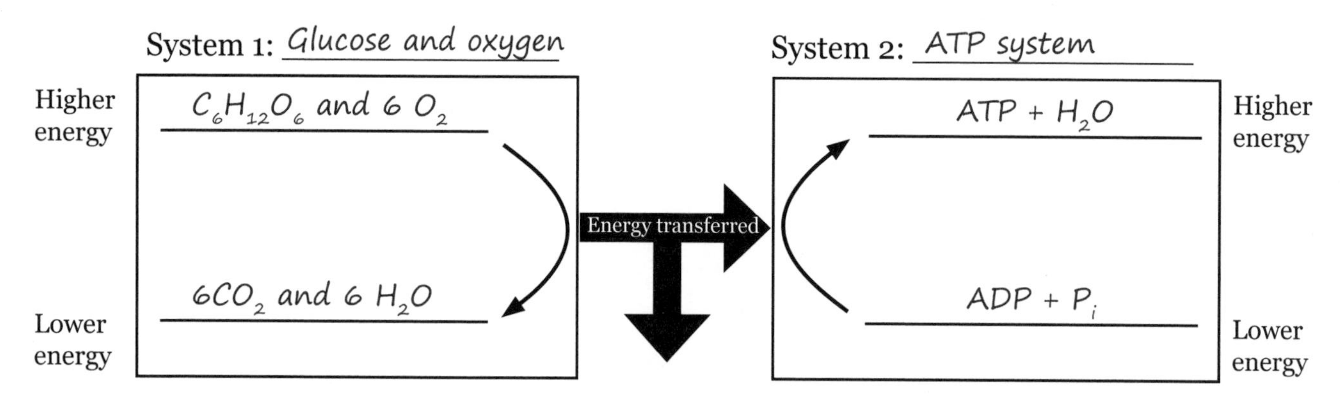

10. Let's see if there is a way to simplify the two energy-transfer models you just created. Compare the ATP system that transfers energy to the muscle system to the ATP system that receives energy from the cellular respiration system.

Propose a way that we can combine the two two-system energy-transfer models into a single three-system energy-transfer model on the diagram below.

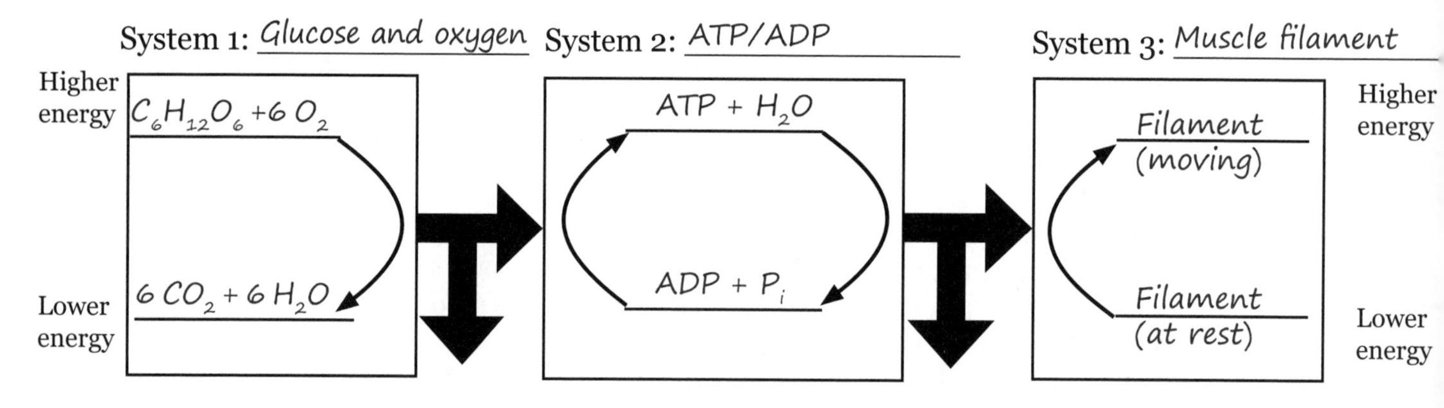

11. In the space below, provide evidence to justify the inclusion of each of the three component systems in your energy-transfer model.

- *Cellular respiration system: The breakdown of glucose requires oxygen and correlates with the production of ATP + H_2O from ADP + P_i*
- *ATP/ADP system: The concentration of ATP decreased and the concentration of ADP increased when muscle was stimulated to contract continuously for 10 minutes. If the muscle rested for an hour, the concentration of ATP returned to normal.*
- *Muscle filament system: Muscle fibers contract when ATP is dripped on them but the muscle fibers do not contract when glucose is dripped on them.*

Teacher Talk and Actions

There are numerous biological reactions and processes for which ATP supplies the energy, and the reaction is usually represented as ATP → ADP + P_i. Although written as a single step, the reaction actually occurs in two steps during which (1) a part of the ATP molecule (phosphoryl, pyrophosphoryl, or adenylate) is transferred and covalently binds to either a reactant or an enzyme, and then (2) the part that was transferred is displaced. This is not a hydrolysis reaction in which ATP reacts directly with water to produce ADP + P_i.

In contrast, some processes do involve direct hydrolysis of ATP. For example, noncovalent binding of ATP, followed by its hydrolysis to ADP and P_i, can provide the energy to cycle some proteins between two conformations, producing mechanical motion. This occurs in muscle contraction (based on *Lehninger: Principles of Biochemistry,* 3rd Edition, p. 504). *MEGA* authors chose to ignore this distinction for high school biology students, which Lehninger rightfully clarifies for university or medical students. The third edition is an outstanding reference and reviews of the fourth edition (which can be purchased for under $20 on eBay) indicate that it is similar. The sixth edition is available free as a pdf at *www.medicosideas.com/lehninger-biochemistry-pdf.*

The actual number of ATP produced is less than 38, even though textbooks list 38 and show where the number comes from (which included 2 ATP for every $FADH_2$ → FAD + 2H+ and 3 ATP for every $NADH_2$ → NADH + H+). The actual measurements weren't 2 and 3 but rather 1.5 and 2.5. When the measurements were made, scientists had an incorrect model of ATP synthesis from the electron transport chain that required whole numbers. They assumed there were measurement errors and rounded up to 2 and 3. However, the updated model of ATP synthesis does not require whole numbers, so the measured 1.5 and 2.5 ratios are now known to be correct. That gives 32 ATP per glucose rather than 38.

12. The next video (Part V) presents the end of the story, an animation of what is happening at the molecular level. Watch the video and describe how the energy-releasing chemical reaction ($ATP + H_2O \rightarrow ADP + P_i$) is coupled to muscle contraction.

The energy released from ATP hydrolysis moves myosin proteins along actin proteins. It is this sliding (or actually, pulling) motion that causes muscles to contract.

13. Nearly 100 years after the initial experiments on muscle were carried out, scientists now have a clear picture of how muscle contraction is coupled to cellular respiration: Energy from cellular respiration is transferred to the $ATP/ADP + P_i$ system in the mitochondria of cells to make ATP (and H_2O). The ATP leaves the mitochondria and travels to the muscle filament system and binds to the head of a single myosin molecule. As the $ATP + H_2O \rightarrow ADP + P_i$ reaction occurs on myosin, energy is transferred to slide the myosin and actin filaments past one another, which results in muscle contraction. The final video (Part VI) puts it all together, showing how chemical reactions provide the energy for the tennis player to win the game (and for all body motion).

Our muscles are amazing machines! Instead of burning glucose in an uncontrolled way, as we observed in the video of the burning donut in Lesson 2.3, the process of cellular respiration involves a multistep sequence of chemical reactions that are carefully controlled so that we can compete in athletic events, bike to school, use cell phones, breathe, and turn over in bed during the night.

Teacher Talk and Actions

Show the video *Muscle Contraction, Part V,* “ATP Hydrolysis ‘Drives’ Muscle Contraction.”

Show the video *Muscle Contraction, Part VI,* “Putting It All Together—From Chemical Reactions to Motion.”

Activity 3: Energy and Muscle Growth

You now know that all people, regardless of whether they are elite athletes, need energy for motion and that the process of cellular respiration provides most of that energy. Does cellular respiration provide energy for muscle growth as well? In this activity, we will try to answer this question.

Procedures and Questions

We know that bodybuilders can greatly increase their strength and the size of their muscles over many years of resistance training. Are their muscle cells just filling with water or are they making more protein polymers, such as actin and myosin? To answer this question, we'll examine one of many studies carried out by sports physiologists.

1. Read about the experiment that scientists conducted, examine the data they collected, and answer the questions that follow with your group.

Experiment 4: Investigating the Effect of Exercise on the Synthesis of Muscle Proteins

Introduction

Scientists recruited 12 healthy males who regularly engaged in resistance training.

Methods

They were advised of the risks associated with the study and provided written informed consent. The volunteers exercised one arm (experimental) but did not exercise the other arm (control). The amount of protein synthesized in both experimental and control arms was measured 4 hours after exercise.

Results

Figure 1 shows the amount of protein synthesized per hour after 4 hours in both experimental and control arms.

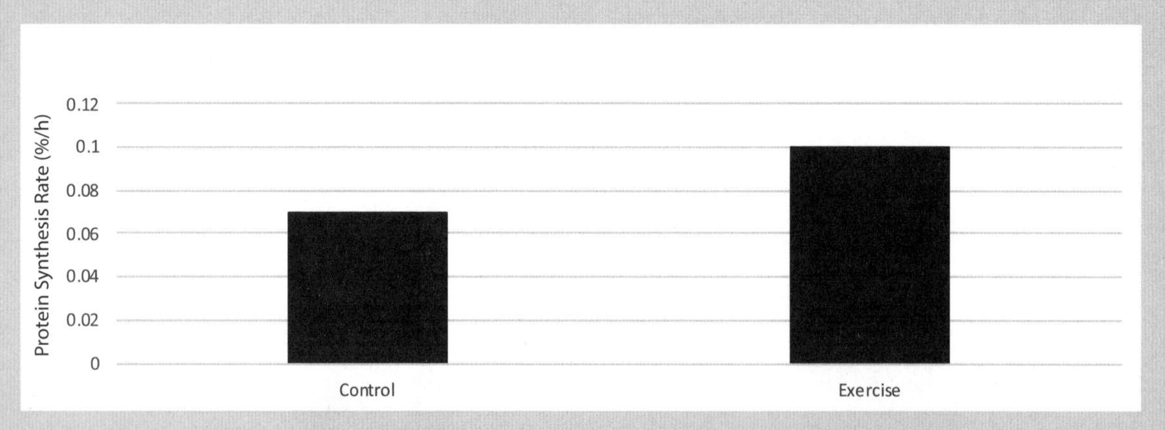

Figure 1. Amount of protein produced per hour in biceps of control and exercised arms measured 4 hours after exercise

Source: Chesley, A., J. D. MacDougall, M. A. Tarnopolsky, S. A., Atkinson, and K. Smith. 1992. Changes in human muscle protein synthesis after resistance exercise. *Journal of Applied Physiology* 73 (4): 1383–1388.

Teacher Talk and Actions

Show the video *Bodybuilder* here (optional).

2. What can you conclude about the effect of resistance exercise on protein synthesis in the biceps muscle?

Resistance exercise increases the rate of protein synthesis.

3. Using what you learned in Chapter 1, write two chemical reactions that you think are involved in muscle growth.

Matter: amino acids → protein + water

Energy: glucose + oxygen → carbon dioxide + water

4. Do these chemical reactions release or require an input of energy? About how much energy? Explain.

Amino acids → protein + water requires energy (Table 2.3 shows that the reverse reaction releases 233 kcal/mole, so forward reaction must require at least 233 kcal/mole)

Glucose + oxygen → carbon dioxide + water releases energy (Table 2.3 shows 670 kcal/mole)

5. Scientists suspected that the same systems involved in providing energy for muscle movement might be involved in providing energy for making proteins needed to build muscle. Various experiments were carried out to investigate the role of ATP in protein synthesis. When ATP was added to a system containing amino acids and other necessary molecules for building proteins (e.g., enzymes), proteins were made. When all the added ATP was converted to ADP + Pi, protein synthesis stopped.

What can be concluded from this experiment?

The chemical reaction ATP + H_2O → ADP + P_i (cellular respiration) can transfer energy to the protein synthesis system.

6. Complete the energy-transfer model below to represent scientists' ideas and your responses to Questions 4 and 5.

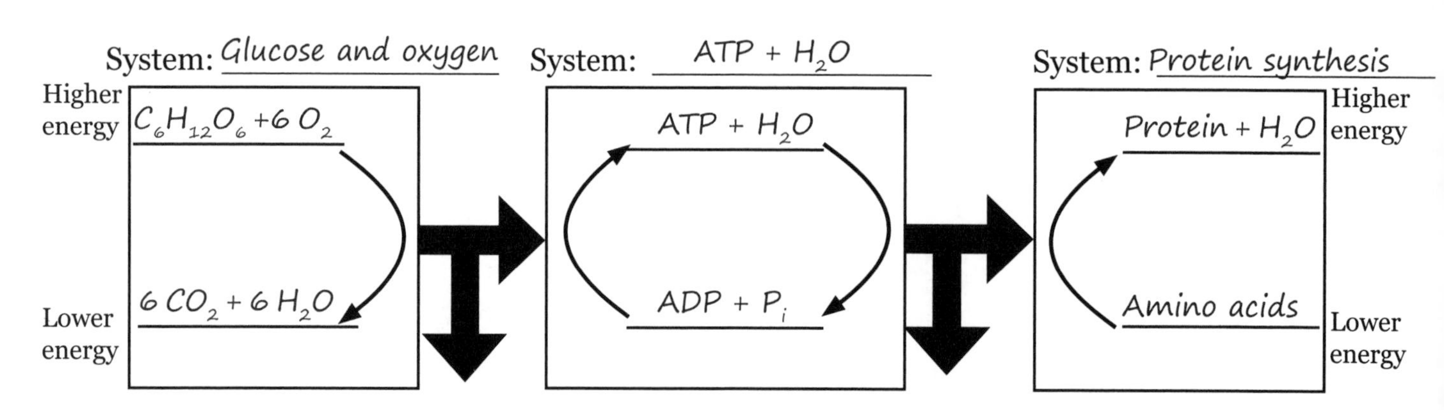

Teacher Talk and Actions

This is a critical question. Be sure that students understand that there are TWO chemical reactions involved—one that builds the proteins and another that provides the energy needed for the building.

If students list just the first reaction, ask, "Does building proteins release or require energy? Where does the energy come from?"

If students list just the second reaction, ask, "What substances must be produced to build muscles? Where does the matter needed to build those substances come from? Does it come from glucose + oxygen?"

Science Ideas

The activities in this lesson were intended to help you understand how muscles get the energy needed to move and grow. Read the ideas below. Look back through the lesson and in the space provided after the science ideas and give evidence that supports each one.

Science Idea #24: Cellular respiration is a chemical process during which bonds are broken between the atoms that make up glucose and oxygen molecules and bonds are formed between the atoms that make up carbon dioxide and water molecules, resulting in a net release of energy.

Evidence:

Evidence that glucose + oxygen are reactants: Experimental data showed that both glucose and oxygen consumption increased during exercise.

Evidence carbon dioxide is produced: During exercise, limewater turned cloudy faster (indicating more carbon dioxide was produced).

Evidence the reaction releases energy: (1) Experimental data showed that the temperature of the leg muscle increased during exercise, AND (2) data in Table 2.3 indicate that energy is released when glucose reacts with O_2 to produce CO_2 + H_2O.

Using ball-and-stick models to represent the reaction, we showed that the only way to form CO_2 + H_2O from $C_6H_{12}O_6$ + O_2 was to break bonds between O atoms of oxygen and between C atoms of glucose and to form new bonds between H and O atoms of water and between C and O atoms of carbon dioxide.

Science Idea #25: Multicellular organisms need energy to move and grow. At the macroscopic level, muscle contraction moves the bones that the muscles are attached to. Cellular respiration provides energy for muscle contraction (motion) and for building muscles during training (growth).

Evidence:

Evidence that multicellular organisms need energy to move: We know from Lesson 2.2 that energy is required for an increase in motion, and the video represented how body motion results from muscle contraction moving bones.

Evidence that growth requires energy: Increased muscle growth is accompanied by increased synthesis of muscle-fiber proteins (actin and myosin) and data from Table 2.5 provides evidence that building protein requires energy.

Evidence that cellular respiration provides energy for muscle contraction: The temperature of the muscle increased during exercise. We know from Activity 1c that the rate of the glucose + oxygen chemical reaction increases with exercise. We have two pieces of evidence that the glucose + oxygen reaction releases energy: we observed that the burning donut gives off light, and Table 2.3 provides data indicating that the reaction of glucose + oxygen releases energy.

Teacher Talk and Actions

Remind students that science ideas are accepted principles that are based on a wide range of observations, not just on the small number of examples they have observed. So Science Idea #25, for example, applies to all multicellular organisms, including dogs and cats and even insects and earthworms—they all obtain energy for motion and growth from cellular respiration.

Science Idea #26: Multicellular organisms need a way to obtain oxygen for cellular respiration and eliminate the carbon dioxide produced. The respiratory system, including the lungs, takes in air rich in oxygen, which passes into blood vessels, and gives off air rich in carbon dioxide that has come from blood vessels.

Evidence:

Evidence oxygen is required for cellular respiration: Experimental data showed that the amount of oxygen consumed increased with exercise.

Evidence carbon dioxide is produced during cellular respiration: The limewater turned cloudy sooner after exercise than after resting for five minutes.

Information on Body System cards (Lesson 1.2) indicates that (a) the respiratory system takes in oxygen-rich air and gives off carbon dioxide-rich air and that (b) red blood cells in the circulatory system transport oxygen to body cells and remove carbon dioxide from them.

Science Idea #27: Some of the energy that is given off during cellular respiration is used to produce ATP and water from ADP and inorganic phosphate. Energy released as heat is used to maintain body temperature.

Evidence:

Evidence that cellular respiration is coupled to ATP synthesis: The text states that scientists had determined that ATP production occurs during cellular respiration, but it didn't provide the data that provided evidence for scientists' conclusions.

Evidence cellular respiration releases energy to the surroundings: There is a correlation between the rate of cellular respiration and muscle temperature—Cellular respiration increased with exercise and so did muscle temperature.

Science Idea #28: The reaction between ATP and water to form ADP and inorganic phosphate provides the energy input for essential energy-requiring processes in living systems, such as muscle contraction and the synthesis of polymers for growth and repair.

Evidence:

When ATP is dripped on isolated muscle fibers, the fibers contract.

When ATP is added to a protein-synthesizing system (containing amino acids and necessary enzymes) proteins are made. Without ATP, the system does not make proteins.

Teacher Talk and Actions

Pulling It Together

Work on your own to answer these questions. Be prepared for a class discussion.

1. Now that you have learned about cellular respiration and muscle contraction and growth, answer the Key Question: **What process in our bodies provides the energy needed for muscle movement and growth, and what is the coupling mechanism the body uses?**

The energy for muscle movement and growth comes from cellular respiration (glucose + oxygen → carbon dioxide + water). Our cells couple cellular respiration to ATP synthesis and then couple ATP hydrolysis directly to sliding myosin along actin proteins. We don't know how cells couple ATP breakdown to biosynthesis.

2. Brown-fat cells are important for helping infants stay warm. Scientists have determined that brown-fat cells carry out cellular respiration but do not produce ATP. What do you think happens to the energy released from cellular respiration in this case?

Because energy is not transferred to the ATP system, it is transferred instead to the surroundings as heat, which keeps the baby warm.

3. In Lesson 2.7, you learned what happens when a bodybuilder adds mass to his body. A similar process happens when living things grow. What do you think happens when a person loses body mass? Use ideas about chemical reactions and systems to explain the decrease in mass.

If someone is losing weight it is because the rate of polymer synthesis from monomers is less than the rate of combustion of monomers. When monomers (glucose, fatty acids, or amino acids) react with oxygen, carbon dioxide is produced. Although the body takes in 2 O atoms for every CO_2 it exhales, there is a net loss of 1 C atom. If the rate of cellular respiration is greater than the rate of polymer (or protein) synthesis, then there will be a net loss of mass due to the C atoms.

Teacher Talk and Actions

Questions 4 and 5 on the following pages expect students to construct energy-transfer models involving four or five interacting systems. Students will need to identify the systems involved and determine the energy transfers between them. If students have trouble getting started, you can guide their thinking with the following questions:

- What systems need to be included in the energy-transfer model? Why do you think so?
- Is energy increasing in any of the systems you identified? How do you know?
- If energy in a system is increasing, where is the energy coming from?
- If energy in a system is decreasing, where is the energy going?

4. At the beginning of Chapter 2, you watched a video of a hamster running on a wheel that caused a light bulb to glow. You are now ready to construct an energy-transfer model that links an energy-releasing chemical reaction happening inside the hamster all the way to the energy transferred to the light bulb. As you watch the video again, do the following:

a. List all the systems you think should be included in your energy-transfer model and provide a reason why you think each system is important to include.

- *Cellular respiration system (in hamster's muscle cells)*
- *ATP/ADP system (in hamster's muscle cells and muscle fibers)*
- *Wheel system (wheel at rest, wheel turning fast)*
- *Light bulb system (light bulb unlit, light bulb lit)*

With each energy transfer from one system to the next, energy should also be transferred to the surroundings as heat.

b. Create an energy-transfer model showing how energy from each system is transferred to the next, starting with the chemical reaction inside the hamster and ending with the light bulb glowing.

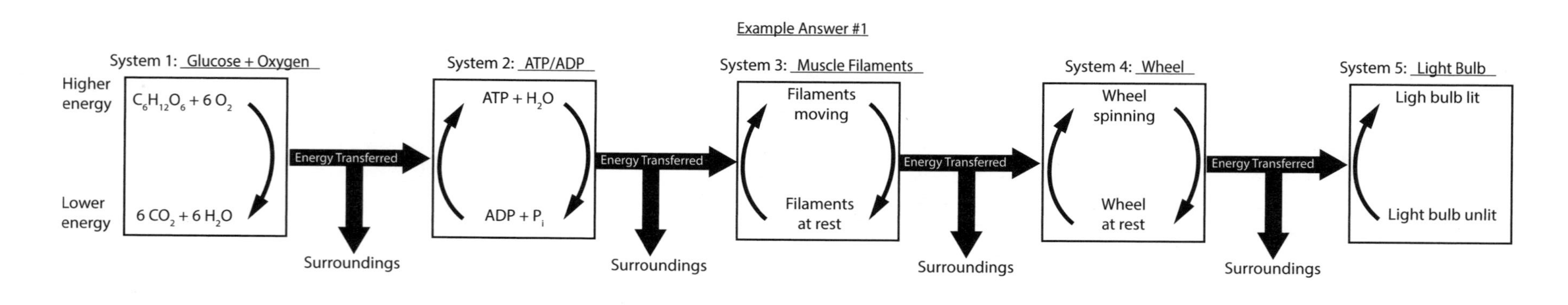

Teacher Talk and Actions

Show the video *Hamster* (optional).

The student response provided for Question 4a describes a quite sophisticated energy-transfer model involving five coupled systems. A four-system model, which leaves out ATP and presents the hamster at the organism level, is also acceptable. Such a model would include cellular respiration, hamster (at rest/running), wheel (at rest/turning), and light bulb (unlit/lit).

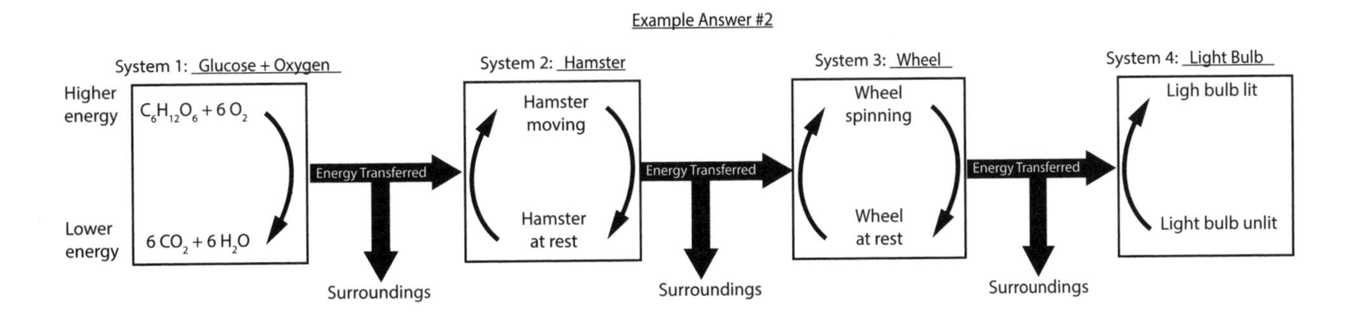

5. As you've just seen, models of matter and energy changes within systems and energy transfer between systems can help us make sense of phenomena involving other animals besides humans, such as hamsters. What about plants?

a. As you watch the videos of plants growing, make notes about the matter changes that are occurring and the energy changes that must take place.

Matter changes: plants producing new leaves, stems, seeds (corn), and roots/tubers (potato). That means plants must make polymers to build these body structures. Plant polymers include starch, a polymer of glucose monomers.

Energy changes: Making starch from glucose requires energy (based on Table 2.5 in Lesson 2.4), which probably comes from cellular respiration (just like other multicellular organisms mentioned in Science Idea #25).

Lesson 2.4 also showed that energy is required to make glucose + O_2 (from CO_2 + H_2O) and that the energy comes from the Sun.

b. List all the systems you think should be included in your model and provide a reason for why you think each system is important to include.

- *Photosynthesis system: Plants need to make glucose, using energy from the Sun.*
- *Cellular respiration system: Plants need to use some of the glucose as fuel to provide energy for making starch and other polymers.*
- *ATP/ADP system: Reasoning from Science Idea #27*
- *Polymer (or starch) system: Plants need to build body structures that are made up of polymers. (Note that plants will use some of the glucose for making polymers. If they use all of it as fuel, they won't have any left to build polymers!)*

c. Create an energy-transfer model showing how energy from each system is transferred to the next.

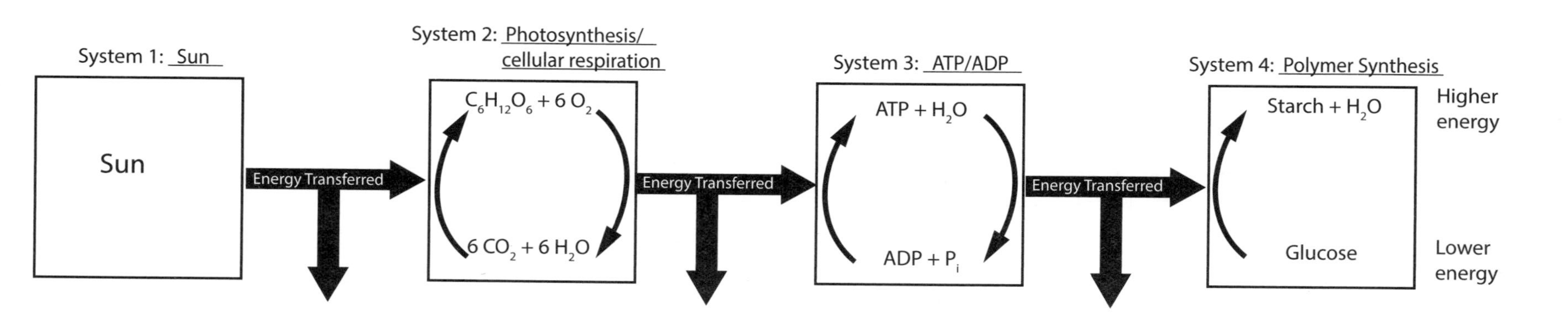

Teacher Talk and Actions

Show the video *Maize Plants Growing*.

We have not tried out this question with students and think it may be quite challenging. Use your judgment about whether to ask students to complete this individually or in groups.

If students/groups struggle, you can guide their reasoning as suggested below and see if they can use it in constructing their models.

If we look carefully at Science Ideas #25 and #26, we see that both refer to multicellular organisms when talking about cellular respiration. From this, we can reason that cellular respiration must occur in all multicellular organisms, including plants. Moreover, Science Idea #27 links cellular respiration with ATP synthesis. From this, we can reason that plants can use the energy from cellular respiration to produce ATP and H_2O from ADP and P_i, just like humans and hamsters.

6. Scientists carried out numerous studies investigating the effects of various poisons on cellular respiration that allowed them to figure out several steps in cellular respiration. Some of those poisons were marketed in diet pills, until it was found that taking these diet pills produced harmful side effects, including death. The federal Food and Drug Administration (FDA) banned the sale of these diet pills in the United States in 1938. These poisons uncouple the cellular respiration system and ATP system. Use the energy-transfer models for muscle contraction and protein synthesis to try to explain why the pills can cause death.

The poison would prevent energy from being transferred from cellular respiration to ATP synthesis. This means that ATP can no longer be produced. If ATP isn't synthesized, muscles can't contract. That means the diaphragm muscle wouldn't contract (so you wouldn't be able to take in oxygen) and the heart would stop contracting (so blood wouldn't be pumped through vessels).

Teacher Talk and Actions

We have not tried out this question with students and think it may be quite challenging. Use your judgment about whether to ask students to complete this individually or in groups.

If students are confused by "uncouple," tell them it is like removing the seesaw in Lesson 2.6. This would be represented by removing the thick energy-transfer arrow from the energy transfer-model.

This is how cyanide poisoning works.

Closure and Link to subsequent lessons:

Congratulations! You have just accomplished an amazing feat—explaining at a molecular level how humans use energy from their food to move and grow. You now know that our bodies

- carry out an energy-releasing chemical reaction called cellular respiration, the reaction between glucose and oxygen to produce carbon dioxide and water.
- use energy from cellular respiration to produce ATP + H_2O from ADP + P_i.
- use energy from ATP hydrolysis to slide muscle proteins actin and myosin past one other, which leads to muscle shortening.
- take advantage of muscles being attached to bones so that muscle shortening causes bones to move toward one another.

Models and science ideas that you had used in earlier lessons to explain simple physical science phenomena like balls colliding and a burning candle warming the air turned out to be equally helpful in explaining observations and data about human motion and growth.

The models and science ideas even helped you explain why plants need energy from the Sun.

Cellular respiration uses oxygen as a reactant and stops if oxygen is not available. Yet we know that some organisms can live without oxygen. Some live deep in the ocean, others can cause food poisoning in humans. How do these organisms obtain the energy they need to stay alive and grow? We'll explore one way to do so in the next lesson.

Lesson Guide
Chapter 2, Lesson 2.8
Growing and Moving Without Oxygen

Focus: Based on data showing that in the absence of oxygen *Lactococcus lactis* bacteria grow more slowly and convert glucose to lactic acid, students conclude that this chemical reaction releases less energy than cellular respiration (and hence can transfer less energy to ATP synthesis). Students then examine data showing that during the initial phase of exercise, when the human body is using very little oxygen, lactic acid production increases with exercise intensity. At the end of the lesson, students are asked to use what they learn about anaerobic and aerobic respiration to make sense of data on what percentage each process contributes to various Olympic running events and to explain why *L. lactis* bacteria grow more rapidly in the presence of oxygen than in its absence.

Key Question: In the absence of oxygen, what chemical reactions provide energy for growth and activity?

Target Idea(s) Addressed

Anaerobic (without oxygen) cellular respiration follows a different and less efficient chemical pathway to provide energy for cells. (Science Idea #29)

Science Practice(s) Addressed

Analyzing and interpreting data
Developing and using models
Constructing explanations

Materials

None

Advance Preparation

None

Phenomena, Data, or Models	Intended Observations	Purpose	Rationale or Notes
Activity 1 *Lactococcus lactis* can grow without oxygen.	The dry cell mass of *L. lactis* increases without oxygen but increases more over 14 hours with oxygen than without oxygen. Students should infer that increased dry cell mass means more cells are produced, which provides evidence for growth.	Provides evidence that (a) bacterial cells can grow, and hence obtain energy, without oxygen but (b) bacterial cells grow more with oxygen. Provide motivation for figuring out how living organisms can obtain energy in the absence of oxygen. Provide example of trying to learn about chemical reactions in humans by studying simpler organisms.	
Activity 2 Without oxygen, *L. lactis* convert glucose to lactic acid.	• When *L. lactis* bacteria were grown on glucose in the absence of oxygen, the glucose concentration decreased linearly from 130 g/l to 60 g/l in 30 minutes and the amount of lactic acid produced increased from 0 60 g/l in 30 minutes. • Models show that two molecules of lactic acid ($C_3H_6O_3$) contain the same number of C, H, and O atoms as one molecule of glucose ($C_6H_{12}O_6$).	Illustrates the role of molecular models in giving us ideas about what chemical reaction might be involved (though models do not provide evidence that this chemical reaction occurs).	
Activity 3 Humans produce lactic acid from glucose.	As the intensity of exercise increases (heart rate increases from 42% to 83% of maximum heart rate), (a) the concentration of lactic acid in the blood increases from 0.5 mmol/liter to over 6 mmol/liter and (b) the concentration of blood glucose decreases from about 5 mmol/liter to 4 mmol/liter.	Provide evidence that bacteria convert glucose to lactic acid during an energy-releasing chemical reaction that doesn't involve oxygen.	Students can use this data in Lesson 2.9 to show that the increase in lactic acid is more than the decrease in glucose. This provides evidence that even though glucose is converted to lactic acid, blood glucose is somehow kept from getting too low.
Pulling It Together Question 4	As the duration of the running event increases, the percentage of energy from anaerobic respiration decreases and the percentage of energy from aerobic respiration increases.	Provide an opportunity for students to use their knowledge of anaerobic and aerobic metabolism to help make sense of world records in running events, raise questions, and incorporate new knowledge and data into their explanations.	

Lesson 2.8—Growing and Moving Without Oxygen

What do we know and what are we trying to find out?

We now know that energy is released when glucose and oxygen react to produce carbon dioxide and water and that the energy released can be used to move muscles or to create polymers (plus water molecules) for building body structures. Humans can also use energy released when fatty acids react with oxygen or when amino acids react with oxygen. All these energy-releasing chemical reactions involve oxygen as a reactant.

You may know that some living things can grow without oxygen. For example, organisms that inhabit deep-sea vents or organisms that cause serious human infections can survive, even thrive, without oxygen. How do these organisms survive? Where do they obtain the energy they need to carry out their energy-requiring life processes like movement and growth?

We know that lots of different chemical reactions release energy, not just those involving oxygen. In this lesson, we will investigate whether organisms, including even humans, use any other chemical reactions that don't involve oxygen to provide energy for life processes.

Answer the Key Question to the best of your knowledge. Be prepared to share your ideas with the class.

Key Question: In the absence of oxygen, what chemical reactions provide energy for growth and activity?

Teacher Talk and Actions

Activity 1: Comparing Growth With and Without Oxygen

We know from earlier lessons that human bodies are complex systems. We saw that it is possible to learn about processes that occur in complex systems by studying those processes in simple systems first. For example, scientists learned about the role of the liver in converting glucose to glycogen (and water) by studying it first in rat livers. It may surprise you that scientists have learned a lot about the chemical reactions involved in human growth by studying the growth of bacteria. In this activity, we will examine experimental data on the growth of bacteria with and without oxygen. Later in the lesson, we will apply what we learn to understand how human muscle cells can function briefly without oxygen.

Procedures and Questions

1. Read the experiment and answer the question that follows.

Experiment 1: The Growth of Bacteria With and Without Oxygen

Introduction

Lactococcus lactis is a type of bacteria that can grow with or without oxygen. Scientists were interested in comparing how fast the bacteria would grow under both conditions. The most reliable way to determine the rate of bacterial growth is to measure the weight of dried bacterial cells at different time points.

Methods

L. lactis were grown in Petri dishes containing water and a glucose polymer called agar. Half of the Petri dishes were grown in the presence of oxygen, and the other half were grown in the absence of oxygen. Every 2 hours for 14 hours, bacteria from each growing condition were separated from the agar, dried, and weighed.

Results

Figure 1 compares the dry cell mass of the bacteria grown with and without oxygen.

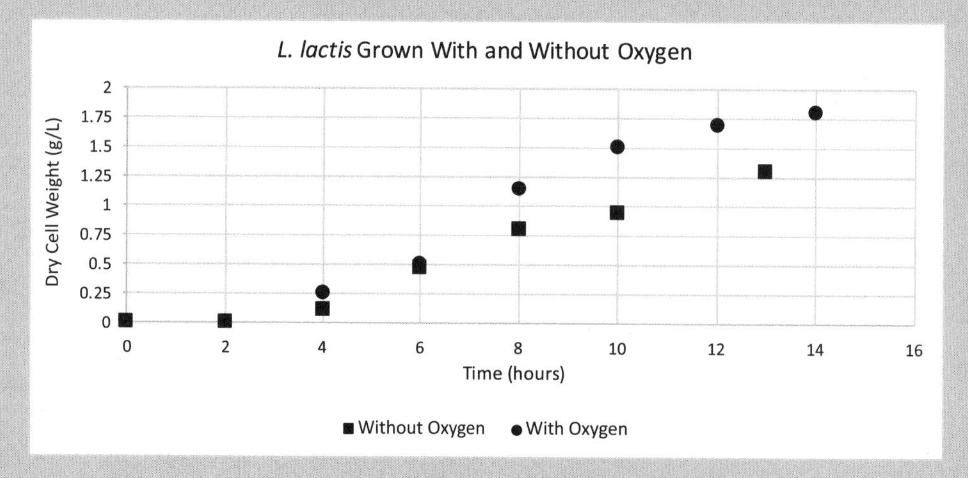

Figure 1. Dry cell weight of *L. lactis* grown with and without oxygen

Source: Lan, C.Q., G. Oddone, D. A. Mills, and D. E. Block. 2006. Kinetics of Lactococcus lactis growth and metabolite formation under aerobic and anaerobic conditions in the presence or absence of hemin. *Biotechnology and Bioengineering* 95 (6): 1070–1080.

Teacher Talk and Actions

2. Compare the growth of *L. lactis* with oxygen to its growth without oxygen.

After eight hours, the dry cell weight of bacteria grown with oxygen is greater than the dry cell weight of bacteria grown without oxygen. This means that the rate of bacterial growth is greater with than without oxygen. However, the bacteria are able to grow and reproduce without oxygen.

Student Edition page 189

Teacher Talk and Actions

Many students have difficulty with the concept of rate as a ratio of amount of mass/amount of time. To simplify, you could ask students to compare amounts at the same time point—for example, 4 hours, 8 hours, and even 12 hours (where students will have to interpolate the amount between 1 and 1.3).

Activity 2: Investigating Growth Without Oxygen

Data from Activity 1 showed that *L. lactis* bacteria grow more when oxygen is present, but the bacteria still grow in the absence of oxygen. We know that growth requires building polymers and that building polymers requires an input of energy. What chemical reaction could be providing the energy? The experiment shown below will help us answer that question.

Procedures and Questions

1. Read about the experiment and answer the questions that follow.

Experiment 2: Inputs and Outputs of Bacterial Growth Without Oxygen

Introduction

Lactococcus lactis can grow in both unsealed and sealed containers of foods such as molasses and beet-sugar. It has been shown previously that *L. lactis* can grow on agar, a polymer of glucose, which means that the bacteria are capable of digesting agar to glucose. In the absence of oxygen, bacteria do not carry out cellular respiration, so they do not produce carbon dioxide and water. Scientists wanted to find out what chemical reaction could be providing the energy for growth.

Methods

L. lactis were grown in Petri dishes containing water and a glucose polymer called agar. As the bacteria reproduced, the scientists measured the concentration of glucose in the Petri dish and the concentration of new substances that were present.

Results

The amount of glucose left in the plate over 90 hours is shown in Figure 1.

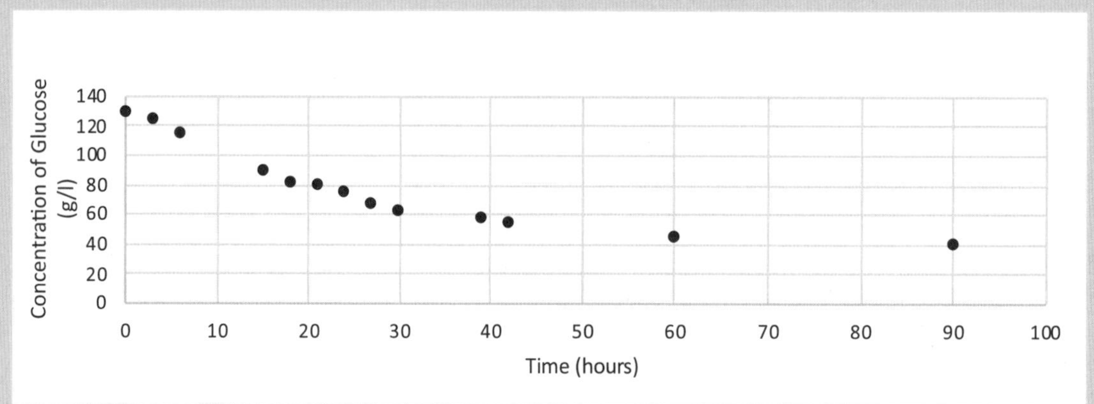

Figure 1. The amount of glucose left in the Petri dish

Teacher Talk and Actions

Ask students how we know that growth requires building polymers. Students should cite Science Idea #11.

Ask students how we know that building polymers requires an input of energy. Students should cite data from Table 2.5, which shows that building protein, carbohydrate, and fat polymers all require an input of energy (note that the values in the table are the energy required for linking two monomers together, so making a huge polymer would require lots of energy).

Results (continued)

Scientists found that as the bacteria grew, a new substance was produced in the Petri dish. The substance is called lactic acid. The amount of lactic acid found in the Petri dish as the bacteria grew is shown in Figure 2.

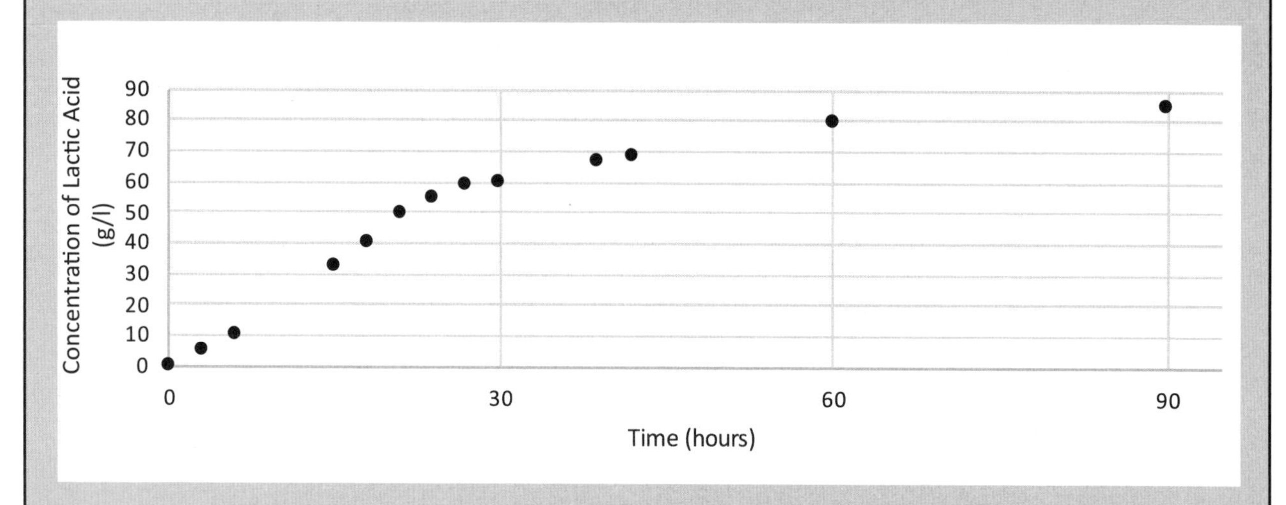

Figure 2. The amount of lactic acid produced by the bateria, and found in the Petri dish

Source: Lan, C.Q., G. Oddone, D. A. Mills, and D. E. Block. 2006. Kinetics of *Lactococcus lactis* growth and metabolite formation under aerobic and anaerobic conditions in the presence or absence of hemin. *Biotechnology and Bioengineering* 95 (6): 1070–1080.

2. Do you have evidence that the bacteria used the glucose to make lactic acid? Why or why not?

Yes. As the bacteria grew, the amount of glucose decreased while the amount of lactic acid increased. The correlation suggests that glucose was converted to lactic acid. The correlation, and the fact that the decrease in glucose and the increase in lactic acid occurred in the same Petri dish, suggests that glucose was converted to lactic acid.

3. Lactic acid has the molecular formula $C_3H_6O_3$. The structure is shown below:

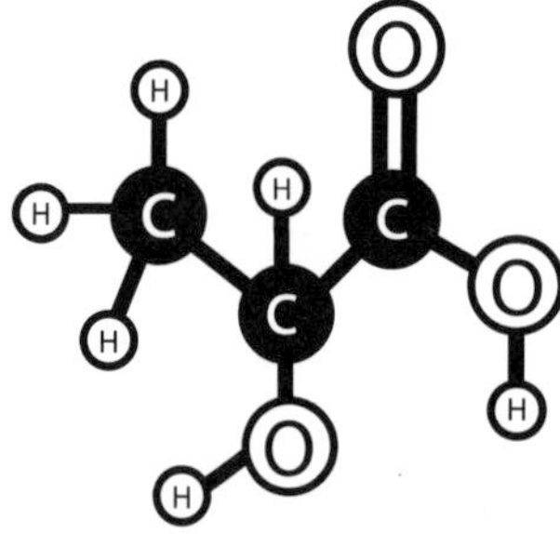

Can the atoms of glucose be arranged to make a molecule of lactic acid? The formula of glucose is $C_6H_{12}O_6$. If yes, how many molecules of lactic acid can be made? Would there be any atoms left that might be a different product?

Yes, the atoms of glucose can be rearranged to form lactic acid. Two molecules of lactic acid can be made from one glucose molecule, and there are no leftover atoms, so there aren't any other products.

Teacher Talk and Actions

4. What could scientists use to prove that the atoms in glucose are rearranged to form lactic acid in bacteria?

Scientists could use data from ^{14}C-labeled glucose and see if the labeled carbon ended up in the lactic acid.

5. In Lesson 2.5 we learned that we could use bond energy calculations to estimate how much energy would be required or released during a chemical reaction. The table below lists the sum of the bond energies for all the bonds in the reactant molecules and for all the bonds in the product molecules when one mole of glucose reacts to form two moles of lactic acid.

Sum of the bond energies in kilocalories (kcal)		
Reactants	$C_6H_{12}O_6$	2,231 kcal
Products	2 $C_3H_6O_3$	2,258 kcal

Using the data in the table above, draw a bar graph to represent the energy changes that occur during this reaction that includes bars for the reactants, separated atoms (6 C, 12 H, and 6 O), and the products.

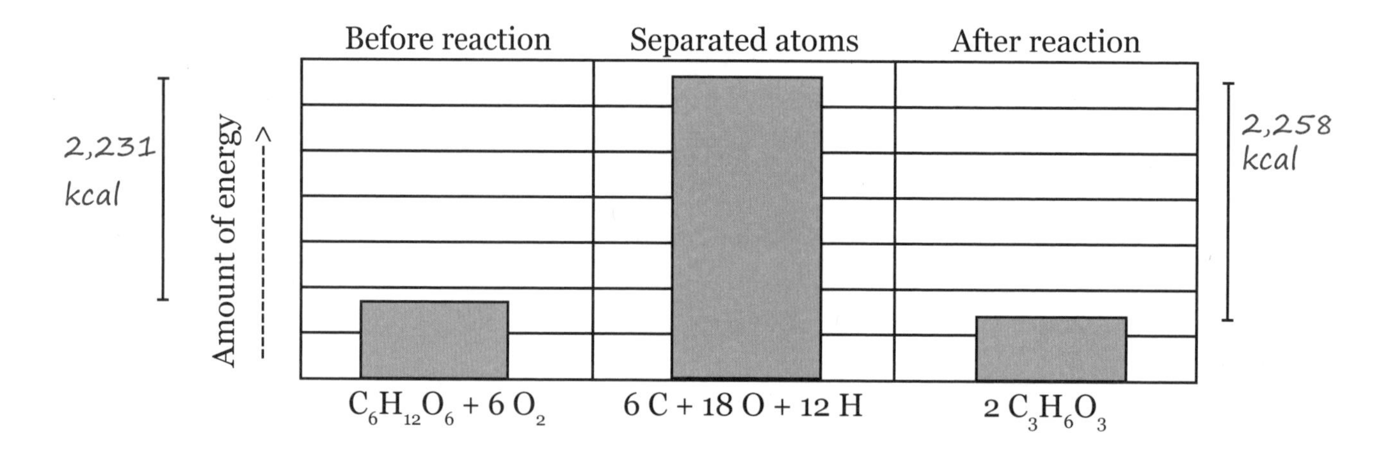

Does this chemical reaction release energy or require an input of energy? Support your answer using the data in the table.

This reaction releases energy because the sum of the bond energies for all the bonds in the products is greater than the sum of the bond energies for the bonds in the reactants. Breaking all bonds of 1 mole of glucose molecules requires an input of 2,231 kcal, and 2,258 kcal will be released when 2 moles of lactic acid form. This means that approximately 27 kcal (2,258 - 2,231) will be released for every mole of glucose reacted.

Teacher Talk and Actions

This is an optional question for students who are more able and interested in mathematics.

If students are struggling, ask,

• What does the sum of the bond energies mean in general? (Answer: The sum of the bond energies is the total amount of energy required to break all the bonds of the molecules listed OR the total amount of energy that would be released as all the bonds form.)

• What do the numbers in the right column mean in this context? (Answer: Because the reaction we are examining is $C_6H_{12}O_6$ [glucose] → 2 $C_3H_6O_3$ [lactic acid], the 2,231 kcal refers to the total amount of energy required to break all bonds of 1 mole of glucose to form separate atoms and the 2,258 kcal refers to the total amount of energy released as 2 moles of lactic acid form.)

• How can the bond energies be represented on the bar graphs? (Answer: The 2,231 kcal can be represented as the difference between the height of the reactant bar and the height of the separated atoms bar, and the 2,258 kcal can be represented as the difference between the height of the separated atoms bar and the height of the products bar.)

6. If the glucose-to-lactic-acid reaction releases 16 kcal/mole, does the glucose-to-lactic-acid reaction release enough energy to convert ADP + P_i to ATP + water?

Yes, the ADP + P_i → ATP + H_2O reaction requires 7.3 kcal/mole.

7. Compare the amount of energy released by the glucose-to-lactic-acid reaction to the amount of energy released by the glucose-to-carbon-dioxide-and-water reaction. Which one releases more energy?

Glucose to carbon dioxide and water releases a lot more energy. 669.8 kcal/mole is much more than 16.4 kcal/mole.

8. How many moles of ATP + water molecules can be produced using energy from reacting one mole of glucose molecules to form lactic acid? How does this compare to the 32 moles of ATP + water molecules that were produced using energy from cellular respiration in Lesson 2.7?

There can only be 2 moles of ATP and water molecules produced using energy from reacting one mole of glucose molecules to form lactic acid. 2 moles is a lot less than 32 moles.

9. Many organisms use the energy-releasing glucose to lactic acid reaction to "drive" the energy-requiring ATP + H_2O from ADP + P_i reaction. The glucose-to-lactic-acid reaction is also called **anaerobic respiration**. Construct an energy-transfer model to show how anaerobic respiration could drive cell division needed for growth.

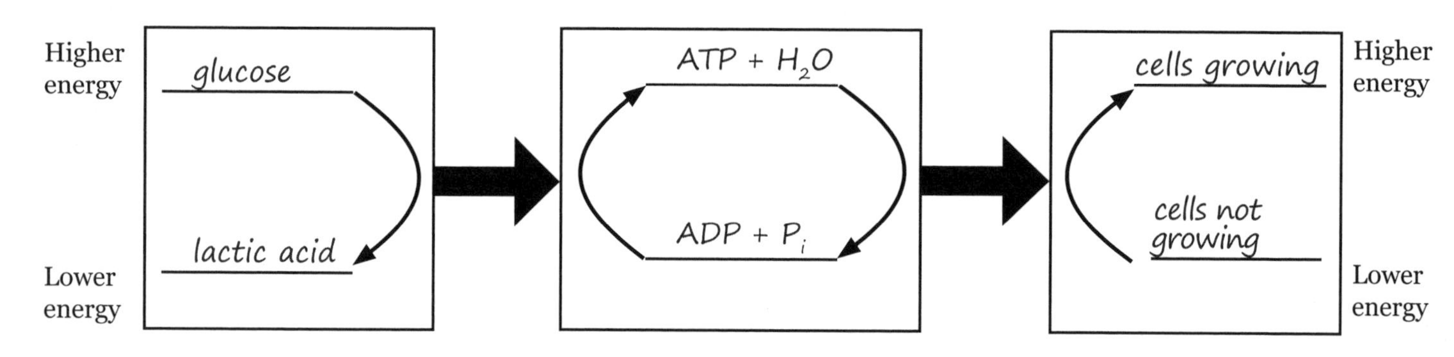

Teacher Talk and Actions

Activity 3: Lactic Acid Production in Human Bodies

Now that we know that some bacteria can grow in the absence of oxygen, we will look at experiments with humans.

Procedures and Questions

1. Read about the experiment and answer the questions that follow.

Experiment 3: Lactic Acid Production During Different Intensities of Exercise in Humans

Introduction

Scientists discovered that during the initial phase of exercise, when the body was using very little oxygen, the amount of lactic acid in the blood increased. They wanted to know whether glucose might be converted to lactic acid, rather than to carbon dioxide and water, when little oxygen was present.

Methods

Scientists asked volunteers to exercise at different exercise intensities (percentage of maximum heart rate) and measured the concentration of both glucose and lactic acid in their blood as they increased their exercise intensity.

Results

The results of their experiment are shown in Figure 1.

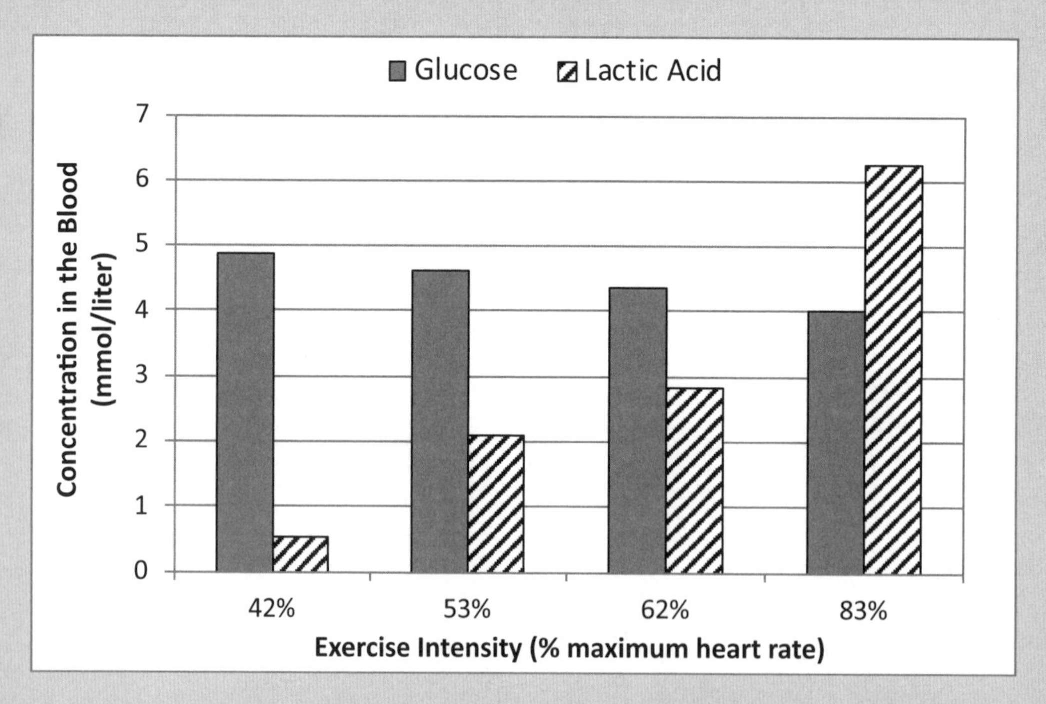

Figure 1. Glucose and lactic acid concentration at different intensities of exercise

Source: Wahren, J., P. Felig, G. Ahlborg, and L. Jorfeldt. 1971. Glucose metabolism during leg exercise in man. *Journal of Clinical Investigation* 50 (12): 2715–2725.

Teacher Talk and Actions

2. Describe what happened to the amount of glucose as exercise intensity was increased.

Glucose concentration decreased as exercise became more intense.

3. Describe what happened to the amount of lactic acid as exercise intensity was increased.

Lactic acid concentration increased as exercise became more intense.

4. The correlation between decreasing glucose and increasing lactic acid provides evidence that humans can carry out an energy-releasing chemical process without oxygen, just as bacteria can. Based on this evidence, create an energy transfer diagram that shows how human bodies use glucose for muscle contraction when there isn't enough oxygen for cellular respiration.

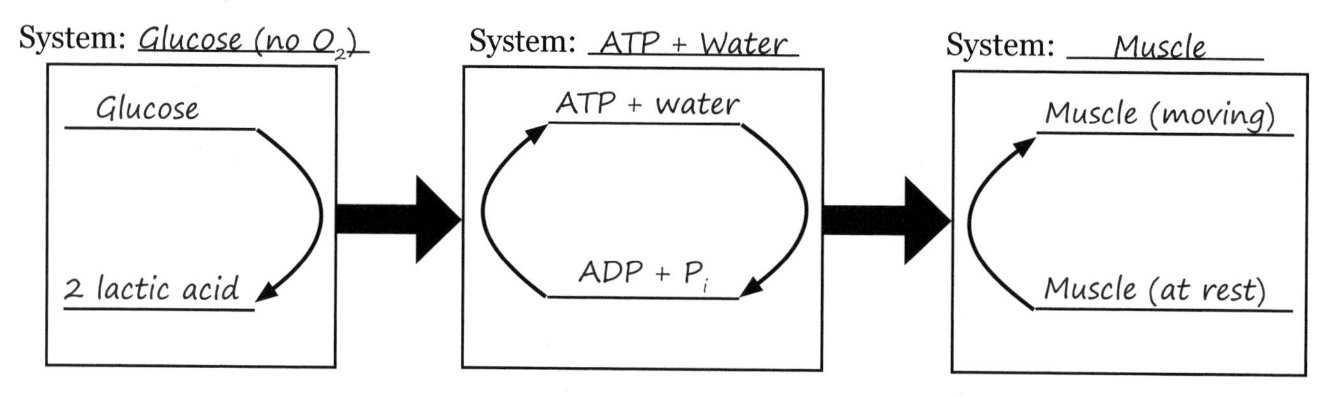

Teacher Talk and Actions

Science Ideas

The activities in this lesson were intended to help you understand how organisms, including humans, get the energy they need when oxygen is not readily available. Read the ideas below. Look back through the lesson and in the space provided after the science idea and give evidence that supports the idea.

Science Idea #29: Anaerobic (without oxygen) cellular respiration follows a different and less efficient chemical pathway to provide energy in cells.

Evidence:

In the absence of oxygen (anaerobic), exercising muscles use glucose to produce lactic acid.

Anaerobic respiration releases less energy than cellular respiration with oxygen. Anaerobic respiration only produces 2 ATP, while cellular respiration produces 32.

Teacher Talk and Actions

While Science Idea #29 is correct and consistent with the *Next Generation Science Standards,* it doesn't tell the whole story. Efficiency is certainly important, but so is speed. While cellular respiration releases enough energy to produce 16 times as much ATP as does anaerobic metabolism, anaerobic metabolism releases energy so much faster than aerobic metabolism (because the latter has many more steps) that it results in more ATP produced per second!

Another factor that must be considered is the fate of the products of the energy-releasing reactions (lactic acid, for anaerobic metabolism of glucose; CO_2 and H_2O, for cellular respiration). For bacteria and other single-celled organisms, waste products diffuse out into the surrounding environment. So, if glucose is abundant, bacteria can take it in, convert it to lactic acid, and excrete the lactic acid, all the while generating ATP faster than if it then converted all the lactic acid to $CO_2 + H_2O$. With an almost endless supply of glucose or other fuels, the low efficiency of anaerobic metabolism is unlikely to reduce bacterial survival or reproductive success.

Humans and other multicellular organisms have a waste-disposal problem because the surrounding environment is far away from the interior. While our body systems efficiently remove CO_2 and H_2O from cells and expel them into the surrounding environment, we don't have an efficient removal system for lactic acid. Instead, our bodies must convert lactic acid to CO_2 and H_2O, a slower process than converting glucose to lactic acid. As a result, humans can rely on anaerobic metabolism for only a few seconds before lactic acid buildup slows or shuts down the process.

Pulling It Together

Work on your own to answer these questions. Be prepared for a class discussion.

1. Now that you have learned about anaerobic respiration and muscle contraction and growth, answer the Key Question: **In the absence of oxygen, what chemical reactions provide energy for growth and activity?**

   Both human muscle cells and *L. lactis* bacteria can use anaerobic respiration involving the reaction of glucose molecules to form lactic acid molecules.

2. Compare and contrast muscle movement and growth when there is oxygen present to when there isn't oxygen present.

   When oxygen is present, muscle cells use cellular respiration. Both cellular respiration and anaerobic respiration are energy-releasing chemical reactions. Both can be coupled to make ATP + water from ADP + P_i.

   When oxygen is present, muscle cells obtain energy from cellular respiration that can be used to make 32 moles of ATP + H_2O per glucose molecule that reacts with O_2.

   When oxygen isn't available, muscle cells can obtain energy from anaerobic respiration. Anaerobic respiration only releases enough energy for the cell to produce two moles of ATP (and water) from each mole of glucose molecules that react with O_2.

Teacher Talk and Actions

3. Using what you have learned so far in this unit, explain why you think more bacteria were produced with oxygen than without oxygen. Your explanation should include ideas about matter and energy.

- Growth requires chemical reactions that produce polymers from monomers (Science Idea #10). Data in Table 2.5, p. 132, show that protein synthesis, carbohydrate synthesis, and fat synthesis all require a net input of energy.
- Both cellular respiration (glucose + oxygen → carbon dioxide + water) and anaerobic respiration (glucose → lactic acid) release energy.
- However, cellular respiration releases much more energy per mole glucose (670 kcal/mole glucose) than anaerobic respiration (16 kcal/mole glucose, which can be calculated by subtracting the energy that can be released from 2 lactic acid [2 x 327 kcal = 654 kcal] from the energy released from glucose 670).
- If more energy is available for building polymers, the bacteria can grow more.

4. Now that you know about aerobic and anaerobic processes that couple glucose metabolism to ATP production, let's see how this knowledge can help us make sense of some amazing athletic feats. The table and figure below provide data on the men's and women's world record holders in four running events that ranged from the shortest event (100-meter sprint) to the longest (42,195-meter marathon).

Use information in the graph or your own calculations to fill in the table.

World Records for Running Events

Running Event (Distance)	Men's Record Holder—Country (Year)	Record Time, Men (seconds)	Average Speed, Men (meters/second)	Women's Record Holder—Country (Year)	Record Time, Women (seconds)	Average Speed, Women (meters/second)
Sprint (100 meters)	BOLT—Jamaica (2009)	9.58	10.44	GRIFFITH-JOYNER—USA (1988)	10.49	9.53
Sprint (200 meters)	BOLT—Jamaica (2009)	19.19	10.42	GRIFFITH-JOYNER—USA (1988)	21.34	9.37
Sprint (1,000 meters)	NGENY—Kenya (1999)	131	7.63	MASTERKOVA—Russia (1996)	148	6.76
Marathon (42,195 meters)	KIMETTO—Kenya (2014)	7,377	5.72	RADCLIFFE—Great Britain (2003)	8,125	5.19

Teacher Talk and Actions

World Record Running Speeds vs. Running Times

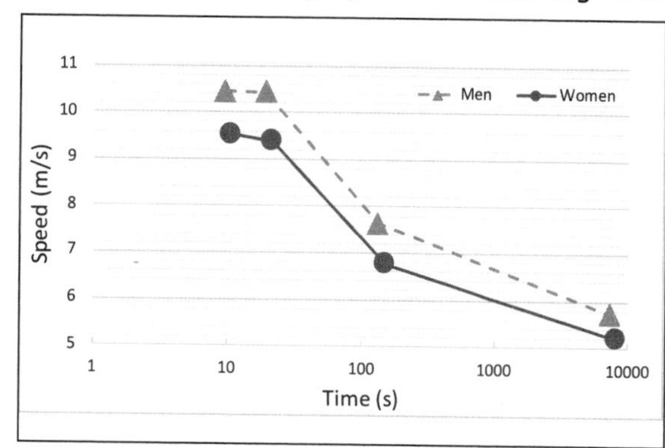

Now use the data from the table and the figure to answer the questions below:

a. In which event did both men and women run at the fastest speed?

100-meter sprint

b. In which event do you think both men and women used the most energy per second, in other words, used energy at the fastest rate? Explain.

The 100-meter sprint, because running faster requires more energy per second than running slower.

c. In which event(s) did both men and women run at the slowest speed?

Marathon

d. In which event(s) do you think both men and women used the least energy per second, in other words, used energy at the slowest rate? Explain.

Marathon, because running slower requires less energy per second than running faster.

e. What trends do you see in the graph?

In general, the longer the race, the slower the speed. However, both male and female record holders run at nearly the same speed for the 100- and 200-meter sprints.

f. We think that running faster uses more energy per second than running slower. Why might that be? Let's consider what we know that might help:

- We know that the chemical reaction between ATP and H_2O provides the energy for muscle contraction and, hence, for running a race.
- We know that two different processes can generate ATP from the energy released from a chemical process involving glucose—one that does not require oxygen (anaerobic) and one that does (aerobic).
- We know that the aerobic process generates more ATP per glucose than does the anaerobic process.

However, the *amount* of ATP produced isn't the only thing to consider. In a race, what also counts is *how fast* body cells can produce ATP. The table on the following page provides data about the rate of ATP production from chemical reactions involving various fuel sources. Examine the table and respond to the questions.

Teacher Talk and Actions

Ask students to think about how running at top speed and jogging at a slow speed are different. For example, how are arm and leg movements different at each speed? How are breathing and heart rate different?

ATP Production by Fuel Source

Fuel Source	Maximal rate of ATP production (mmol per sec)	Total ATP that can be produced* (mmol)
Conversion of muscle glycogen into lactic acid (Anaerobic)	39.1	6,700
Conversion of muscle glycogen to CO_2 + H_2O (Aerobic)	16.7	84,000
Conversion of liver glycogen to CO_2 + H_2O (Aerobic)	6.2	19,000
Conversion of fatty acids from fat to CO_2 + H_2O (Aerobic)	6.7	4,000,000

* Fuels stored are estimated for a 70 kg (154 lb) person having a muscle mass of 29 kg

i. Scientists have determined that the amount of ATP that is immediately available is 669 mmol, which would allow a runner to maintain maximum speed for 5–6 seconds. Will this amount of ATP allow world-record holders to complete a 100-meter sprint? Explain.

No, because even the fastest male runner, Usain Bolt, takes close to 10 seconds to complete the 100-meter sprint. So, he'll need to produce ATP for about 4 seconds after his immediately available ATP is depleted.

ii. What process do you think provides the ATP for the rest of a 100-meter or 200-meter sprint? Explain.

Probably anaerobic metabolism because its rate of ATP production (39.1 mmol per second) is more than double the rate of ATP production from aerobic metabolism (16.7 mmol per second).

iii. What process(es) do you think the marathon runners are relying on? Explain.

Probably combustion of fatty acids (aerobic) and any remaining glucose from liver glycogen because the pace of the marathon runners is the slowest and those processes produce ATP the slowest.

iv. What process(es) do you think the 1,000-meter sprinters are relying on? Explain.

Probably some combination of anaerobic metabolism of glycogen and aerobic metabolism of glycogen, because (a) the pace of the 100-meter sprinters is intermediate and (b) using a mix of anaerobic and aerobic processes would produce ATP at an intermediate rate .

Teacher Talk and Actions

Both muscle and liver store glucose as glycogen. The breakdown of glycogen to glucose occurs rather quickly. The reason the rate of ATP synthesis using energy from muscle glycogen breakdown is more than twice the rate of ATP synthesis using energy from liver glycogen breakdown is because the glucose produced needs to get out of the liver, travel to muscle, and get into muscle. That turns out to be the rate-limiting step.

There are at least two other factors to consider that are not dealt with here—the buildup of lactic acid from anaerobic metabolism and the availability of oxygen. Students may wish to investigate this further after completing Lesson 2.9.

Closure and Link to subsequent lessons:

Lesson 2.8 has introduced you to another energy-releasing chemical reaction that organisms use for motion and growth under some circumstances. Humans can use anaerobic metabolism for short bursts of motion. After a few seconds, their cells shift to cellular respiration to satisfy energy needs and to get rid of the lactic acid produced during the beginning of a race. If you've ever observed athletes at the beginning and end of an event, you know that they are behaving differently at the end of the race than they did before they started. At the end, they are breathing heavily and sweating. In the final lesson we will use what we know about cellular respiration to explain why.

Lesson Guide
Chapter 2, Lesson 2.9
How Body Systems Respond to Exercise

Focus: This is the culminating lesson of the unit. By this point students should know that cellular respiration (a) is a chemical reaction during which glucose and oxygen molecules react to produce carbon dioxide and water molecules, (b) provides energy for energy-requiring chemical reactions involved in muscle contraction and in forming polymers needed to build/repair body structures, (c) releases energy to the cells' surroundings as heat, and (d) occurs faster during intense exercise than at rest. Based only on this knowledge students should predict that during intense exercise, glucose and oxygen levels in muscles would decrease (because they are reactants of cellular respiration), carbon dioxide and water levels would increase (because they are products of cellular respiration), and internal body temperature would increase (because some energy from energy-releasing chemical reactions is always transferred to the muscles' surroundings as heat). Students should also be able to explain their predictions using evidence from experimental data they examined in Lesson 2.7, Activity 1 and Science Ideas #23, #24, #25, and #26, and precursor Science Ideas #7, #8, and #17 learned in earlier lessons. Experimental data they will either collect or examine will provide evidence that, in contrast to their predictions, internal body temperature and the levels of glucose, oxygen, and carbon dioxide stay within a fairly narrow range. In other words, the body must have a way to counteract the changes brought about by intense exercise. A handout on homeostasis provides information about how the body accomplishes that amazing feat.

Key Question: How do various body systems respond to the matter and energy demands of exercise?

Target Idea(s) Addressed

Feedback mechanisms maintain a living system's internal conditions within certain limits and mediate behaviors, allowing it to remain alive and function even as conditions change within some range. (Science Idea #30)

Science Practice(s) Addressed

Analyzing and interpreting data

Obtaining, evaluating, and communicating information

Materials

Activity 1: Videos: *Olympic Athletes* and *Homeostasis*

Activity 3: *Per student or small group:* print or electronic copy of Homeostasis handout based on *Guyton and Hall Textbook of Medical Physiology*, 12th edition; body system hierarchy charts/models (integumentary, muscular, digestive, circulatory, respiratory) prepared in Lesson 1.2; models of glucose and glycogen

Advance Preparation

Organize materials in advance so that each team has materials when class begins.

Download videos and test them in advance.

Phenomena, Data, or Models	Intended Observations	Purpose	Rationale or Notes
Activity 1 Predicting changes in matter and energy in muscle cells during exercise Considering how to test predictions	• Temperature in muscle cells should increase, because cellular respiration provides energy for exercise and releases some energy to the surroundings. The energy released to the surroundings will increase the temperature. (Students could also cite data from Lesson 2.7, which showed that the internal muscle temperature increased by about 0.6°C over three minutes of exercise). • Glucose levels in muscle cells should decrease, because glucose is a reactant of cellular respiration and cellular respiration provides energy for exercise. • Oxygen levels in muscle cells should decrease, because oxygen is a reactant of cellular respiration and cellular respiration provides energy for exercise. • Carbon dioxide levels in muscle cells should increase, because carbon dioxide is a product of cellular respiration and cellular respiration provides energy for exercise. • Water levels in muscle cells should increase, because water is a product of cellular respiration and cellular respiration provides energy for exercise.	Provide opportunity for students to use their knowledge of cellular respiration to predict phenomena.	Data presented in Lesson 2.7 on thigh oxygen consumption and thigh glucose consumption did not directly measure the levels of oxygen and glucose inside muscle cells. The limewater test used in Lesson 2.7 provides evidence that we exhale more carbon dioxide during exercise than at rest but does not provide information about the carbon dioxide levels inside muscle cells.
Activity 2 Data on body temperature of high school cross-country coach before and after a 4.1-mile (31-minute) race Data on body temperature of cross-country team members before and after a 3.1-mile race	• The body temperature of the cross-country coach decreased by about 1.7 °F over a 4.1-mile (31-minute) race. • The body temperatures of both male and female cross-country team members decreased by about 1.8 °F during a 3.1-mile race.	Give students opportunities to consider how to design experiments to test their predictions and to analyze experimental data to see if they are consistent with or challengestheir predictions. Give students the opportunity to explain how the circulatory, respiratory, and integumentary systems maintain body temperature in a normal range.	Do not ask students to collect data without school and parent permission.

Phenomena, Data, or Models	Intended Observations	Purpose	Rationale or Notes
Activity 3 Data on typical oxygen, carbon dioxide, and glucose levels and range of resting muscle cells	• Oxygen, carbon dioxide, and glucose levels within resting muscle cells are maintained within a narrow range.	Give students the opportunity to explain how the circulatory, respiratory, and digestive systems help to maintain body oxygen, carbon dioxide, and glucose levels in muscle cells within a normal range. Give students the opportunity to work together to extract relevant information from a resource and use the information to explain how various body systems help to maintain levels of glucose, oxygen, and carbon dioxide within muscle cells within a normal range during exercise.	

Lesson 2.9—How Body Systems Respond to Exercise

What do we know and what are we trying to find out?

Physical activity provides the greatest demand for energy. In sprint running and swimming, for example, the energy required by active muscles can increase by 120 times or more over resting levels. During less intense but sustained exercise such as marathon running, the body's need for energy increases 20 to 30 times above resting levels. In this final lesson of the unit, you will consider how the body systems you examined in Chapter 1 respond to the body's increased demand for energy.

Answer the Key Question to the best of your knowledge. Be prepared to share your ideas with the class.

> **Key Question: How do various body systems respond to the matter and energy demands of exercise?**

Teacher Talk and Actions

This is the culminating lesson of the unit. By this point students should know that cellular respiration (a) is a chemical reaction during which glucose and oxygen molecules react to produce carbon dioxide and water molecules, (b) provides energy for energy-requiring chemical reactions involved in muscle contraction and in forming polymers needed to build/repair body structures, (c) releases energy to the cells' surroundings as heat, and (d) occurs faster during intense exercise than at rest. Based only on this knowledge, students should predict that during intense exercise, glucose and oxygen levels would decrease (because they are reactants), carbon dioxide and water levels would increase (because they are products), and internal body temperature would increase (because not all the energy released during chemical reactions is used to do "work"—some energy is always transferred to the surroundings). Students should also be able to explain their predictions using evidence from experimental data they examined in Lesson 2.7 and 2.8 and Science Ideas #23, #24, #25, #26, and precursor Science Ideas #7, #8, and #17 in earlier lessons. Experimental data they either collect or examine provide evidence that, in contrast to their predictions, internal body temperature and the levels of glucose, oxygen, and carbon dioxide stay within certain limits. Readings provide information about how the body accomplishes that amazing feat.

Activity 1: Predicting Matter and Energy Changes During Exercise

You learned in Lessons 2.7 and 2.8 that most of the energy for moving muscles comes from cellular respiration, the chemical reaction between glucose ($C_6H_{12}O_6$) and oxygen (O_2) molecules to produce carbon dioxide (CO_2) and water (H_2O) molecules. You also learned that when muscle cells carry out cellular respiration, only some of the energy is available for muscle contraction. Much of the energy is transferred to the cells' surroundings as heat.

Procedures and Questions

1. Think back to the athlete before the race and again after the race. How do you think the athlete's internal body temperature before the race will compare to the internal body temperature right after the race? Justify your prediction.

I predict that internal body temperature will be a lot higher right after the race than before the race because (a) the rate of cellular respiration increases (i.e., fuel is "burned" faster) to provide more energy for more/faster muscle movement, and (b) when the rate of cellular respiration increases, more energy will be released as heat to the cells' surroundings.

2. If the rate of cellular respiration increases, what do you predict will happen to the concentrations of the reactants and products in muscle cells? Complete the table below by indicating what will happen to the concentration of each substance and why you think so.

Substance	What will happen to the concentration of the substance in muscle cells during exercise?	Justification for Prediction
Glucose ($C_6H_{12}O_6$)	*Decrease*	*Glucose is a reactant of cellular respiration, so the faster cellular respiration occurs, the more glucose will be used. If more glucose is used, then the amount of glucose in muscle cells should decrease.*
Oxygen (O_2)	*Decrease*	*Oxygen is a reactant of cellular respiration, so the faster cellular respiration occurs, the more oxygen will be used. If more oxygen is used, then the amount of oxygen in muscle cells should decrease.*
Carbon dioxide (CO_2)	*Increase*	*Carbon dioxide is a product of cellular respiration, so the faster cellular respiration occurs, the more carbon dioxide will be produced. If more carbon dioxide is produced, then the amount of carbon dioxide in muscle cells should increase.*
Water (H_2O)	*Increase*	*Water is a product of cellular respiration, so the faster cellular respiration occurs, the more water will be produced. If more water is produced, then the amount of water in muscle cells should increase.*

Teacher Talk and Actions

Rate is a challenging concept, and *increasing rate* may be even more challenging. Ask students to consider how much gasoline a car will burn in an hour when the car is going 30 mph versus 60 mph. Then ask, if they were to feel the car hood after an hour in both cases, in which case would the hood be warmer.

Show *Homeostasis* video.

You may want to show the *Olympic Athletes* video again to help students focus on what might be happening in the runner's body.

3. How could we test our prediction in Step 1 about what would happen to internal body temperature? What measurements could we make that could provide evidence to support (or contradict) our predictions? Are there any data from previous lessons that we can use?

To test predictions about what would happen to the temperature before and immediately after exercise, we would need to measure the temperature inside muscles. Data in Lesson 2.7 (Figure 1, p. 179), which measured muscle temperature with a probe, shows that leg muscle temperature increased from 37.4 °C to 38 °C after three minutes of exercise.

4. Based on the data in Lesson 2.7, how much should the muscle temperature increase after a 30-minute run?

The leg-muscle temperature increased 0.6°C in three minutes (1.08°F). In 30 minutes (10 times as long), the leg-muscle temperature should increase 6°C (10.8°F).

Teacher Talk and Actions

Optional Question:

How could we test our prediction about the reactants and products of cellular respiration? What measurements could we make that could provide evidence to support (or contradict) our predictions? Are there data from previous lessons that we can use?

Testing predictions about matter: To test predictions about what would happen to the amounts of reactants and products during exercise, we would need to measure the amount of each of the substances inside a muscle cell, which might be difficult to do. We would need to do a muscle biopsy before and right after exercise. We certainly would not be able to get human subjects approval to do it in class!

Activity 2: Testing Our Energy Predictions and Explaining Findings

In the previous activity you predicted that the amount of energy transferred to the muscle as heat could increase the muscle temperature by nearly 11°F during a 30-minute run. Do you think this really happens? You can use data from the experiment described below to help you test your prediction.

Procedures and Questions

1. The following data were collected from a Maryland high school cross-country coach who ran for about 30 minutes. She recorded the weather conditions during her run and she measured oral body temperature before and after the run. She also measured her body weight and heart rate before and after the run. She did not eat or drink anything between measurements, and she did not eliminate any liquid or solid wastes between measurements.

DATE:	24-Oct	
DESCRIPTION OF WORKOUT:	Long Run, Fort McHenry	
OUTDOOR TEMPERATURE:	73°F	
HUMIDITY:	52%	
WEATHER:	Sunny	
LENGTH OF RUN:	4.17 miles	
TIME (PACE):	31:29 min (7 min., 33 sec. pace)	
	BEFORE	AFTER
BODY TEMPERATURE (°F):	98.5	96.8
BODY WEIGHT (lbs)	116.0	115.2
HEART RATE (bpm)	72	114

According to the data, what happened to the coach's body temperature?

It decreased 1.7°F from the start to the end of the run.

2. What additional questions should we be asking, and what data should we collect to help us answer these questions?

- *Is the decrease typical of the coach? (To answer this question, the coach should make similar measurements on another run.)*
- *Is it typical of all female runners? (To answer this question, we would also need to collect similar data from other female runners.)*
- *Is it typical of all runners? (To answer this question, we would also need to collect similar data from male runners.)*
- *Is the data typical of all people, not just trained athletes? (To answer this question, we would also need to collect data from nonathletes, though non-athletes probably couldn't run 4.17 miles!)*

Teacher Talk and Actions

To approach Question 2, lead a discussion using the following questions:

- What variables might affect our measurements besides exercise?

  Weather conditions (e.g., outside temperature, humidity, time of day, whether the sun is shining) and runner variables (e.g., age, gender, whether the runner is a conditioned/trained athlete, body weight).

- How many times should we repeat the experiment?

  Multiple times from the same individual to be sure the data are reproducible

- How many people should we test?

  Students should be led to realize that the range of people tested will inform the claim they will be able to make (e.g., Will we be able to generalize from one person to all people? From females to everyone? From conditioned athletes to everyone?)

3. To see whether similar data would be obtained from members of the girls' and boys' cross-country team, parent and student consent was obtained to collect the temperature data shown below. Temperature was measured with a forehead thermometer.

Athlete (Age) Gender	County Championship Race*	
	Starting Forehead Temperature (°F)	Ending Forehead Temperature (°F)
(15) Male	97.6	95.9
(16) Male	98.0	95.9
(17) Female	97.5	96.8
(14) Female	97.4	94.7

* October 25, 2018, 3.1 miles, Outside Temperature 63°F, Partly Cloudy

Similar data were obtained from a 5.5-mile run on October 25 and a 3.1 mile county championship race on November 2.

4. What can you conclude from the data?

All the runners showed a decrease in body temperature (average decrease was 1.8°F for the above data). We didn't collect data from nonathletes.

5. Based on the data from the cross-country team and the table above, were our predictions correct?

No

6. What do all the data tell us?

The body must have ways to maintain the various substances and physical characteristics within the normal range.

7. To find out how our bodies keep our core temperature from getting too high, use your body system hierarchy maps from Lesson 1.2 and information in the Homeostasis handout to complete the table.

Teacher Talk and Actions

Maintaining Internal Temperature		
Predicted effect of exercise and why	*Internal body temperature will increase, because cellular respiration in muscle cells transfers heat to the muscle cells' surroundings (e.g., blood).*	
What the body must do in response	*Decrease internal temperature by transferring more heat to the body's surface, where it can be transferred to the body's surroundings*	
Whether the response involves positive or negative feedback	*Negative*	
Body systems (organs) involved in response	*Integumentary system (sweat glands in skin)*	*Circulatory system (blood vessels)*
What each organ does in response	*Produce sweat that cools body as it evaporates*	*Dilate vessels supplying skin to bring warmer blood to surface*
Evidence	*Olympic athletes we observed in the videos were sweating after they competed in their event, or I sweat when I exercise.*	*The coach's heart rate was faster after the run than before. This is evidence her blood was circulating through her body faster.*
Effect of training	*Increase capacity to produce sweat*	*Increase cardiac output*

8. Explain in your own words how each body system helps the body to maintain normal temperature.

- *Circulatory system: The amount of blood flowing to body surface (skin) increases, so more blood is in close contact with the air surrounding the body (which allows for more energy to be transferred through the skin to the surroundings and the heart beats faster so blood circulates through the body faster).*
- *Integumentary system: Energy is transferred from the skin to the surrounding air (by conduction and radiation). Energy is also transferred to sweat on the skin, which speeds up evaporation (as sweat evaporates it cools the body). Students should notice that the coach's weight decreases by 0.8 lbs from the start to the end of the race. Some of that is due to sweat that has evaporated from her body.*
- *(Optional) Respiratory system: Increased shallow breathing rate (especially in animals that pant) increases evaporation from air passages.*

Teacher Talk and Actions

In Activity 2, the teacher will lead a whole-group discussion on maintaining temperature. In Activity 3, different groups of students will be assigned to prepare a report on maintaining oxygen, carbon dioxide, or glucose.

Activity 3: Checking Predictions About Matter During Cellular Respiration and Explaining Findings

Materials

For each student or small group
Print or electronic copy of Homeostasis handout
Body system hierarchy charts/models
Models of glucose and glycogen

In Activity 1, you predicted that the amounts of the products of cellular respiration (CO_2 and H_2O) should increase with exercise and the amounts of the reactants of cellular respiration ($C_6H_{12}O_6$ and O_2) should decrease. Does this actually happen? In this activity, you will examine data on what happens to three of the substances—carbon dioxide, oxygen, and glucose—and explain the data.

Procedures and Questions

1. Scientists have collected data about the temperature and resting range of various substances in the fluid surrounding muscle cells and maximum short-term limits without causing death, as shown in the table below.

	Normal Resting Value	Normal Resting Range	Short-Term Nonlethal Limit	Unit
Oxygen	40	35–45	10–1,000	mm Hg
Carbon Dioxide	40	35–45	5–80	mm Hg
Glucose	85	75–85	20–1,500	mmol/L
Body Temperature	98.4 (36.9)	98–98.8 (36.7–37.1)	65–110	°F (°C)

2. What do you notice about the values of the normal resting range of oxygen, carbon dioxide, glucose, and body temperature?

The normal resting range of each substance and body temperature varies only a little bit above and below the normal resting value.

3. The values can go outside the normal range, but only for a short amount of time. For example, the maximum body temperature (110°F) is only 11.2°F higher than the normal resting range. And the maximum nonlethal limit for carbon dioxide (80 mm Hg) is only twice that of the normal resting value (40 mm Hg). Even relatively small changes like these, if sustained for more than a short time, would result in death. What do the data in the table tell us about what our bodies must do?

We just learned that the body can dilate capillaries and produce sweat to help it stay cool during exercise. The body must also have ways to maintain the concentration of oxygen, carbon dioxide, and glucose in the blood within the normal resting range, even when exercising.

Teacher Talk and Actions

Maintaining glucose levels is more complex than maintaining oxygen or carbon dioxide levels. Assign Glucose Homeostasis to the most able groups of students.

	Maintaining Glucose Concentration in Blood		
Predicted effect of exercise and why	Decrease in glucose concentration in blood, because glucose is being used during cellular respiration		
What the body must do in response	Increase glucose concentration in blood		
Whether the response involves positive or negative feedback	Negative		
Body systems involved in response	Muscular	Digestive	Circulatory
What each organ does in response	Muscles break down stored glycogen to glucose	Liver breaks down stored glycogen to glucose, which enters blood	Increased heart rate causes increase in rate glucose is moved from liver to muscles
Evidence	Data from Investigation 3 (Lesson 1.5) indicated that adult male rats stored glucose as glycogen in both the liver and muscles.	Data from Investigation 2 (Lesson 1.5) showed that at the beginning of the experiment (after rats had been fasted for 18-22 hours), their livers had no stored glycogen.	Data from cross-country coach showed increased heart rate at the end of her 30′ run
Effect of training	Carbohydrate loading increases amount of glycogen stored in muscles.		Cardiac output increases with training.

4. Your teacher will assign your group a substance to investigate—oxygen, carbon dioxide, or glucose. Use your body-system hierarchy maps from Lesson 1.2 and information in the Homeostasis handout to complete the table on the next page for your substance. Once you decide which body systems (organs) are involved, list them in the appropriate row in the table. You will need to divide that row and the rows below it so that you have a column for each body system you have identified.

Teacher Talk and Actions

Name of Substance ____________________________

	Maintaining _____Concentration in Blood
Predicted effect of exercise and why	
What the body must do in response	
Whether the response involves positive or negative feedback	
Body systems (organs) involved in response	
What each organ does in response	
Evidence	
Effect of training	

Teacher Talk and Actions

	Maintaining O_2 Concentration in Blood	
Predicted effect of exercise and why	O_2 concentration in blood would decrease, because O_2 is being used during cellular respiration	
What the body must do in response	Increase O_2 concentration in blood	
Whether the response involves positive or negative feedback	Negative	
Body systems (organs) involved in response	Respiratory system	Circulatory system
What each organ does in response	Increase rate and depth of breathing to bring O_2 in faster	Increase heart rate, which pumps blood faster from muscle cells to lungs
Evidence	Runners in video were breathing harder after race.	Heart rate of cross-country coach was higher after 30-minute race.
Effect of training	Minimal	Increase cardiac output and increase amount of hemoglobin (which carries O_2)

	Maintaining CO_2 Concentration in Blood	
Predicted effect of exercise and why	CO_2 concentration in blood would increase, because more CO_2 is being produced during cellular respiration	
What the body must do in response	Decrease CO_2 concentration in blood	
Whether the response involves positive or negative feedback	Negative	
Body systems involved in response	Respiratory system	Circulatory system
What each organ does in response	Increase rate and depth of breathing to remove CO_2 faster	Increase heart rate, which pumps blood faster from muscle cells to lungs
Evidence	Runners in video were breathing harder after race.	Heart rate of cross-country coach was higher after 30-minute race.
Effect of training	Minimal	Increase cardiac output

Science Ideas

The activities in this lesson were intended to help you understand how various body systems maintain constant internal conditions. Read the ideas below. Look back through the lesson and in the space provided after the science idea, give evidence that supports the idea.

Look back through Lesson 2.9. In the space provided after the science idea, give at least one piece of evidence that supports the idea.

Science Idea #30: Feedback mechanisms maintain a living system's internal conditions within certain limits and mediate behaviors, allowing it to remain alive and functional even as conditions change within some range.

Evidence:

During exercise, the rate of cellular respiration increases. As a result, muscles need more glucose and more oxygen and they produce more carbon dioxide that must be eliminated. Data in Lesson 2.7, Activity 1, provide evidence that muscles take up more glucose and more oxygen during exercise than at rest. The limewater test showed that we produced more CO_2 during exercise than at rest. Data in Lesson 2.7, Activity 1, showed that muscles increased in temperature during exercise.

The Homeostasis handout describes how body systems restore and maintain normal levels of each of the following:

Temperature: The skin helps to cool the body down by producing sweat that cools the skin as it evaporates. The circulatory system helps to cool the body down by dilating surface blood vessels, which increases movement of warmer blood to the surface where it can be cooled.

CO_2: The heart beats faster to circulate blood faster, bringing CO_2 produced in muscle cells to the lungs. The depth and rate of breathing increases to increase the amount of CO_2 exhaled.

O_2: The depth and rate of breathing increases to increase the concentration of O_2 taken in, leading to an increase in blood carried from lungs to muscle cells. The heart beats faster, bringing O_2 to muscle cells faster.

Glucose: Both muscles and liver store glycogen, which can be broken down to glucose and used when muscles need increased glucose.

Teacher Talk and Actions

Pulling It Together

Work on your own to answer these questions. Be prepared for a class discussion.

1. Now that you have learned about body systems and maintaining internal environments, answer the Key Question: **How do various body systems respond to the matter and energy demands of exercise?** Include at least one complete example in your answer.

Although exercise shifts body temperature and internal levels of glucose, oxygen, and carbon dioxide away from the normal resting range, body systems respond through negative feedback to restore and maintain levels to normal. For example, as cellular respiration increases internal temperature, blood vessels to skin dilate (to bring more heated blood to the surface where it can be transferred to the environment) and sweat glands produce sweat (which cools the body when it evaporates).

2. Look back to the data in Activity 2 from the cross-country coach and use what you have learned in the *MEGA* unit to explain why her body weight decreased by 0.8 lbs. Your answer should reflect your knowledge of relevant body systems and of the molecules that entered and left her body.

Based on the equation for cellular respiration, 6 moles of CO_2 are given off for every 6 moles of O_2 taken in. Since CO_2 has a carbon atom (in addition to 2 oxygen atoms), CO_2 weighs more than O_2. So, the body loses weight from exhaling CO_2.

According to the equation for cellular respiration, 6 moles of H_2O are also produced for every 6 moles of O_2 taken in. If water evaporates from the skin as part of sweat, then the body also loses weight from the evaporation of H_2O molecules.

Teacher Talk and Actions

3. We have spent several weeks investigating the systems in our bodies that carry out important life functions, the food we eat and what it's made of, and a wide range of phenomena that involve changes in matter and energy. We've also used a set of science ideas to help us explain these phenomena, whether they happen in a nonliving system, such as a solar-powered car, or in a living system, such as the human body. Now we are ready to answer the Key Question for this unit: **How do living things use food as a source of matter for building and repairing their body structures and as a source of energy for carrying out a wide range of body functions?** Write your answer in the space below.

To build body structures for growth and repair, living organisms produce polymers (and water molecules) during energy-requiring chemical reactions between monomers. The energy both plants and animals use to produce polymers comes indirectly from energy-releasing chemical reactions such as the oxidation of glucose, fatty acids, and amino acid monomers and directly from the hydrolysis of ATP to form ADP and inorganic phosphate. The same energy-releasing chemical reactions provide energy for motion. While much of the energy released can be used for growth and motion, a lot of the energy is transferred as heat to the surroundings. Animals obtain monomers from their food, whereas plants make monomers from inorganic substances during energy-requiring chemical reactions that use energy from the Sun.

During both energy-releasing and energy-requiring chemical reactions, atoms are rearranged and conserved; therefore, total mass is conserved. The net amount of energy released or required reflects the difference between the total amount of energy that must be added to break bonds between atoms of reactant molecules and the total amount of energy that is released when bonds form between atoms of product molecules. As with atoms, energy is neither created nor destroyed during chemical reactions.